Photonic Crystals: Physics and Technology

C. Sibilia · T.M. Benson · M. Marciniak · T. Szoplik
Editors

Photonic Crystals: Physics and Technology

 Springer

Concita Sibilia
Università di Roma „La Sapienza"
Dipartimento di Energetica
00161 Roma, Italy

Marian Marciniak
National Institute of Telecommunications
Department of Transmission
and Optical Technologies
Szachowa 1
04-894 Warsaw, Poland

Trevor M. Benson
George Green Institute
for Electromagnetics Research
University of Nottingham
Nottingham NG7 2RD, UK

Tomasz Szoplik
University of Warsaw
Faculty of Physics
Pasteura 7
02-093 Warsaw, Poland

ISBN 978-88-470-0843-4   e-ISBN 978-88-470-0844-1

DOI 10.1007/978-88-470-0844-1

Library of Congress Control Number: 2008932582

Cover concept: Simona Colombo, Milano
Typesetting: le-tex publishing services oHG, Leipzig, Germany
Printing and binding: Grafiche Porpora, Segrate, Milano

Printed on acid-free paper

Springer-Verlag Italia – Via Decembrio 28 – 20137 Milano

springer.com

Contents

Part II Nonlinear Optics in Photonic Crystals

3 Quasi Phase Matching in Two-Dimensional Quadratic Nonlinear Photonic Crystals
Ady Arie, Nili Habshoosh, Alon Bahabad 45

4 Harmonic Generation in Nanostructures: Metal Nanoparticles and Photonic Crystals
Andrea Marco Malvezzi 61

5 Ultra-fast Optical Reconfiguration via Nonlinear Effects in Semiconductor Photonic Crystals
Crina Cojocaru, Jose Trull, Ramon Vilaseca, Fabrice Raineri,
Ariel Levenson, Rama Raj 79

Part IV   Characterisation and Measurements of Nanostructures

9   Near Infrared Optical Characterization Techniques
for Photonic Crystals

10   Characterization Techniques for Planar Optical Microresonators

Part V Simulation Techniques

Introduction

The present book is the result of the activity developed under the COST P11 Action "Physics of Linear, Nonlinear and Active Photonic Crystals". The Action, supported by the European Science Foundation in the framework of the UE-FP6, operated over the period 2003–2007 and successfully coordinated research effort in the field.

In the last decade, considerable advances have been achieved concerning the theoretical understanding, the technical fabrication and the experimental characterization of photonic crystals. Photonic band-gap (PBG) structures, or photonic crystals (PhCs) are composed of a periodic array of macroscopic dielectric and/or metallic unit cells that interact with light in a manner analogous to that in which crystal lattices interact with electrons. The essential property of these structures is the existence of allowed and forbidden frequency bands for light, in analogy to the energy bands and band-gaps of semiconductors. The study of PBG materials is generally considered to have its origin at the beginning of the 1970s in the pioneering work of V. Bykov[1] on spontaneous emission control and was further developed in the late 1980s by Yablonovitch[2] and John[3]. Yablonovitch's and John's work has been the catalyst for more than a decade of intense theoretical and experimental investigation of PBG structures. In fact, during that period a remarkable number of new applications were proposed, including photonic crystal fibres[4], photonic crystal-based integrated optical circuits[5], transparent metal-dielectric stacks[6], and highly efficient micron-sized devices for light emission and nonlinear frequency conver-

---

[1] V.P. Bykov, "Spontaneous emission in a periodic structure", Sov. Phys. JETP **35**, 269 (1972)

[2] E. Yablonovitch, "Inhibited spontaneous emission in solid state physics and electronics", *Phys. Rev. Lett.* **58**, 2059–2062 (1987)

[3] S. John, "Strong localization of photons in certain disordered dielectric superlattices" *Phys. Rev. Lett.* **58**, 2486–2489 (1987)

[4] J.C. Knight, T.A. Birks, P.S.J. Russell, D.M. Atkin, "All-silica single-mode optical fiber with photonic crystal cladding", *Opt. Lett.* **21**, 1547–1549 (1996)

[5] J.D. Joannopoulos, P.R. Villeneuve, S.H. Fan, "Photonic crystals: putting a new twist on light", *Nature* **386**, 143–149 (1997)

[6] M.J. Bloemer and M. Scalora, "Transmissive properties of Ag/MgF2 photonic band gaps", *Appl. Phys. Lett.* **72**, 1676–1678 (1998)

1

sion[7], to name but a few. An up-to-date review of recent advances in the field of PBG structures may be found in reference[8]. Currently the renaissance of plasmon optoelectronics (see, e.g.[9]) brings novel ideas of creating non-linear optical materials by integration of metal nanoparticles with structures possessing periodicity on the wavelength scale.

The study of the physical and optical properties of photonic crystals has generated a burst of new ideas for optical devices and systems. Special mention needs to be made here of photonic crystal silica fibres, which are likely to become in the near future the first application of photonic crystals to the real world of optical communications. For more than four decades, semiconductor physics has played a leading role in almost every field of modern technology. The technologies for tailoring the electronic properties of a number of materials are nowadays well-established. The new frontier is the achievement of the basic knowledge and the technology that allows the properties of artificial photonic structures to be tailored – and to shape their interaction with light. Activity in this field is quite extensive, and the potential that these structures offer is vast. One of the most demanding aspects of research on photonic band-gap structures is the search for suitable criteria and optimum designs for a specific need or application. Thus, the field encompasses basic aspects related to materials research, for example the self-assembling of 3D-photonic crystals (3D-PC), or physical aspects related to the field localizations and enhancing nonlinear interactions in both classical and non-classical regime of interaction and propagation, including novel phenomena in the propagation of short pulses in PhC-fibres.

As mentioned above, this book summarizes the activity developed over four years thanks to the support of European Science Foundation. The main objective of the COST P11 Action was to unify and coordinate national efforts aimed at studying linear and nonlinear optical interactions, including quantum optical features associated with PhCs. Full attention was given to important aspects related to materials research and the ideas and methods for realizing 3D PhCs, together with the development and implementation of measurement techniques for the experimental evaluation of their potential applications in different areas, for example telecommunication with novel optical fibres, lasers, nonlinear multi-functionality, switches, and other potential areas of applications such as display devices, optoelectronics, sensors, etc.

In particular the aims of the Action were:

- Study and develop the basics aspects related to the physics of 2D- and 3D-PC realization, taking into account different methods of realizations (self-assembling, templating, nano-printing all-dielectric and metallo-dielectric PhCs), materials and their characterization.

[7] M. Scalora, M.J. Bloemer, C.M. Bowden, G. D'Aguanno, M. Centini, C. Sibilia, M. Bertolotti, Y. Dumeige, I. Sagnes, P. Vidakovic, A. Levenson, "Choose your color from the photonic band edge: Nonlinear frequency conversion", *Opt. & Photon. News* **12**, 36–40 (2001)

[8] M. Bertolotti, C.M. Bowden, C. Sibilia, "Nanoscale Linear and Nonlinear Optics" – AIP Conference Proceeding, **560**, N.Y. 2001

[9] V.M. Shalaev, "Nonlinear Optics of Random Media", Springer, Berlin, 2000

- Study and design of active (PhC light emitters) and linear PhC structures for light handling and detection.
- Study of the nonlinear optical interactions in 2D- and 3D-photonic crystals, taking into account both quadratic and cubic nonlinear effects, and spatio-temporal responses and effects.
- Study of the pulse propagation in photonic crystals fibres, taking account cubic nonlinear effects, and spatio-temporal effects.
- Experimental study in the microwave regime as a tool for optimizing designs at optical frequencies.
- Study of quantum aspects of propagation and the interaction of fields in PhCs (1D – 2D – 3D) related to the generation of non-classical light states and sources.

The Activity has been focused on the innovative use of technology in PhC realization. It has included basic studies on material properties tied to the method of realization, to the materials selected and their characterization, to the linear and nonlinear optical properties of PhCs and methods to enhance and control them, including quantum aspects of interaction.

In particular the following topics have been addressed:

- Basic studies on material properties allied to the method of realization, to the materials selected and their characterization.
- Study and design of active (PhC light emitters) and linear PhC structures for the generation and control of optical pulses and their detection.
- Development of theoretical models for the study of both quadratic and cubic nonlinear interactions in 1D, 2D and 3D PhCs, taking into account spatio-temporal phenomena and effects.
- Development of theoretical models for the study of pulse propagation in PhC circuits and fibres.
- Study of ultra-fast, all-optical processing possibilities based on cross-gain, cross-phase and four-wave mixing (cubic nonlinear interactions) in multidimensional PhCs – and analysis of the relative merits of these schemes, as well as comparisons with alternative techniques.
- Study of quantum optical aspects of linear and nonlinear propagation in PhCs, to the generation on non-classical light states.

The activities of the COST P11 Action were divided into three working groups:

The activity of Working Group 1, led by C. López and R. Houdré, was concerned with the fabrication and characterization of properties of photonic crystal structures. Working Group 2, led by T. Benson and P. Bienstman, concentrated on the modelling and simulation of proposed structures. Working Group 3, led by K. Panajatov and A. Zheltikov, related to the study of active and non-linear properties of photonic crystals and photonic crystal fibres.

One of the main outcomes of the COST P11 Action has been dissemination and training. The content of the book is, in part, based on lectures presented at the COST P11 Training School jointly hosted in Warsaw in May 2007 by Warsaw University and the National Institute of Telecommunications. The focus of the School was on the work of working groups WG1 and WG3 of the COST P11 Action, concerning the active and non-linear properties of photonic crystals and the fabrication and characterisation of photonic crystal structures. This was supported by a general introduction to the topics covered within the Action and a summary of the challenges faced by and achievements of the modelling and simulation working group, WG2. For convenience the book is divided into five parts covering Basics Properties of Photonic Crystals, Nonlinear Optics in Photonic Crystals, Technology Integration and Active Photonic Crystals, Characterisation Techniques for Photonic Crystals, and Simulation Techniques.

The Editors thank Mariusz Zdanowicz and Olga Bolszo, both attendees of the Training School in Warsaw, for their assistance in the editing of this book.

C. Sibilia
M. Marciniak
T. Benson
T. Szoplik

Part I
Basics

1 Introduction to Photonic Crystals and Photonic Band-Gaps

Richard M. De La Rue[1] and Sarah A. De La Rue[2]

[1] Optoelectronics Research Group, Department of Electronics and Electrical Engineering, University of Glasgow, Glasgow G12 8QQ, Scotland, U.K.
[2] Readable Science, http://www.readablescience.co.uk/

Abstract. Photonic crystals continue to provide a fertile terrain for research effort around the world. Periodic refractive-index variation in two or three spatial dimensions, with a large high-low ratio produces a range of interesting phenomena that offer a variety of possibilities for device functionality. This chapter will introduce some of the basic concepts and the terminology required in order to understand the behaviour of photonic crystals.

Key words: Photonic Crystals, Photonic Band-gaps, Gap-Maps, Photonic crystal channel waveguides

1.1 Introduction

This chapter is concerned with the propagation of light in periodic structures, in particular structures that exhibit *refractive index* periodicity. It has become habitual (though not mandatory) for researchers in the field of optical propagation within and through such periodic structures to refer to them by the moniker '*photonic crystal*'. Indeed, photonic crystals have been a topic of major research interest since the seminal papers of Yablonovitch [1] and John [2] in 1987, although neither paper uses the term *photonic crystal*. Instead, the term was coined shortly after [3] the authors of these key papers had their first meeting, for lunch, in Princeton.

1.2 Brief Historical and Definitional Note

Photonic crystals are optical media with spatially periodic properties. However, this definition is too general to be useful in all contexts, and there has been some debate about the conditions under which it is legitimate to use the term [3]. We shall endeavour to by-pass this controversy simply by saying that if the medium of interest is predominantly periodic in either two or three space dimensions and there is a substantial contrast in the refractive index, it may be called a *photonic crystal*; though a further requirement is that at least one of the characteristic periods in the medium is comparable with the wavelength of the light involved. A characteristic feature of wave propagation in spatially periodic media is the existence of stop-bands, i.e. regions of the frequency spectrum for which, in the ideal case, propagation is completely inhibited. The frequency range over which there is a stop-band along a specific direction depends on both the spatial periodicity along that direction and, where there is a spatially varying refractive index, on the refractive index *contrast*. At a single frequency, there is typically a finite angular range of directions over which the propagation is inhibited. Yablonovitch [1] realised that it might be possible for this angular range to be increased to cover the whole of three-space, given the correct distribution of the refractive index and a sufficiently large range of refractive index contrast. In that situation, it becomes appropriate to talk about a *photonic band-gap*, borrowing partly from the language of solid-state physics, and in particular the physics of *semiconductors*. From a historical perspective, it is interesting to note that the paper by Yablonovitch, which was published in 1987 [1], was anticipated to a substantial extent, and by more than a decade, in a short remark by Bykov [4], near the end of a paper concerned with the principles of distributed feedback (DFB) lasers.

1.3 What Does a Photonic Crystal Look Like?

Although unified by the possession of a periodic structure, which can clearly be seen in micrographs from a scanning electron microscope (SEM), photonic crystals now appear in a diverse range of forms. Even structures that are used in optical regimes, i.e. at wavelengths from the mid-infrared down to shorter wavelengths, extending as far as the ultra-violet, show a high level of diversity. In an SEM micrograph of a cleaved section of synthetic opal (Fig. 1.1), the intrinsic regularity of the square and hexagonal packing of the silica spheres from which opal is characteristically created can clearly be seen, giving the fundamental explanation of its properties. These structures are also present in naturally occurring opals, and therefore pre-date the epoch when mankind learned how to manufacture opal synthetically by millions of years. However, while natural opal often has additional components that

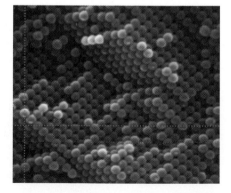

Fig. 1.1 Scanning electron micrograph of bare synthetic opal formed of sub-micrometre (typically ~300 nm dia.) colloidal silica spheres. The cleaved surface shows both *square* packing associated with (100) planes and *hexagonal* packing associated with (111) planes

Fig. 1.2 Scanning Electron Micrograph of high refractive index contrast epitaxial 3D photonic crystal. Note the hexagonal array of air-holes etched into a GaAs/high-x(Al$_x$Ga$_{1-x}$As) DBR mirror stack that has been selectively oxidized

contribute to its cosmetic appearance and appeal, mankind has improved on this technology by controlling the conditions of the natural process. Because of its intrinsically invisible sub-wavelength scale structure, it was only with the advent of electron microscopy that images such as that presented in Fig. 1.1 could be obtained and, at the same time, provide the basis of a scientific explanation for the optical properties exhibited by opal.

Another example of a structure with a large periodic variation of the refractive index in all three spatial dimensions appears in Fig. 1.2. The general process by which the 3D photonic crystal structure shown in Fig. 1.2 has been produced differs considerably from that used to form synthetic opal [5]. The selectively oxidised structure shown is based on a periodically alternating multi-layer of material in the epitaxially grown GaAs/AlGaAs system. Fabrication of this 3D photonic crystal involves electron-beam lithographic (EBL) pattern definition and reactive ion etching (RIE) after the epitaxial growth stage, implying that the structure is formed by a 'top-down' process that is (within the precision limits of the process) deterministic. The subsequent selective oxidation process is also deterministic in that it is almost completely confined to the layers that contain a large aluminium fraction. The chemical composition and thickness of the alternating layers are also deterministic, although the precision is typically no better than one per cent, and the epitaxial growth process used involves an essentially open-loop form of control.

Nonetheless, there is an analogy with synthetic opal, at the atomic scale, in the formation of the above structure. Epitaxial growth of single crystal layers of semiconductors such as gallium arsenide (GaAs) or aluminium gallium arsenide (AlGaAs) involves self-organisation, with the gallium, aluminium and arsenic atoms coming together in a ratio (between the *total* number of group III atoms {Al and Ga} and the total number of group V atoms {As}) that is very close to 1:1. These atoms bind into regular structures that may contain millions of atoms, but have no discernible deviation from perfect spatial periodicity.

1.4 Photonic Band-Structure and the Electron–Photon Analogy

We have already introduced the terms 'stop-band' and 'band-gap' to describe the properties of photonic crystals. We shall now extend the comparison between photonic crystals and semiconductors, and indeed expose its limitations. The propagation of electromagnetic waves in media with a periodic spatial variation of properties, such as the refractive index, leads to the existence of *band-structure*. Single crystal solids are inherently anisotropic, and the corresponding electronic band-structure reflects this anisotropy. The standard way to represent band-structure, in the context of the electronic properties of solids (e.g., single-crystal sections of semi-conducting materials such as silicon, and the III–V compounds gallium arsenide (GaAs) and indium phosphide (InP)), is to generate a graph of energy *vs.* momentum. Because real matter has three space dimensions and we are plotting energy *vs.* momentum as it occurs in 'three-space', graphical representations that can easily be assimilated by the human eye, despite being presented on a 2D surface, are essentially pseudo-3D images that rely on the nature of human perception for their effectiveness. Curves of energy are plotted against momentum along a specified direction, in relation to the axes defining the crystal lattice, then multiple sections are incorporated through rotation about the three principal axes, the orientation of which is determined by the symmetry of the crystal.

Although conductivity in single-crystal semiconductors may exhibit anisotropy, this anisotropy is typically unimportant and ignored in the most widely used applications of semiconductors. In the characteristic situation where a controlled amount of current is passed through part of a semiconductor device, such as a conventional metal oxide semiconductor (MOS) or bi-polar transistor, the electron or hole transport is *incoherent*. The flow of the *drift* current is determined by the acceleration of the electrons or holes that results from the existence of a local electric field, in combination with an almost immediate scattering or collision event that routinely involves lattice vibrations or phonons. The resulting, temperature-dependent, property of the semiconductor is called the *mobility*, and there is a linear relationship between this and the *conductivity*. In many cases, e.g. that of the transport of current through a p-n junction, the existence of a carrier concentration gradient leads to current flow by *diffusion*, but the situation is still typically mediated by phonons, as is implied by the direct, linear relationship between the diffusion constant and the mobility. Therefore, even though it is governed by wave mechanics (i.e., quantum mechanics), the nature of electronic conduction is typically diffusive. However, this is radically different from the situation sought and exploited when propagating light (photons) through photonic crystals.

Photonic crystal structures and their corresponding band-structure are both significantly different from and closely analogous with semiconductors. If the photonic structure that is being investigated or exploited is sufficiently regular (i.e., if the structure is close to perfect), the scattering of light, other than the coherent scattering that leads to the existence of band-structure (i.e., Bragg-diffraction writ large), should be as small as possible. The propagation of optical frequency electromagnetic waves through a photonic crystal typically shows strongly frequency-dependent behaviour, but is nevertheless intrinsically loss-less in the absence of imperfections and absorption. The ability of a monochromatic source of light to launch light into, or from within, a photonic crystal is determined by its detailed characteristics and by the properties of the photonic crystal at the optical frequency of interest. In general, it is not possible to match the source and the photonic crystal structure perfectly and simultaneously over a range of frequencies, even if perfect matching can be obtained at one or more specific frequencies. This unavoidable connection is a characteristic aspect of Maxwell's equations, but it also has quantum consequences. In particular, when considering the amount of light that can be transmitted through a photonic crystal device structure with a light source placed *outside* the structure, there may be a substantial level of reflection, with the level of power reflected being strongly dependent on the optical frequency. In the absence of losses, power conservation implies that light is either transmitted into the photonic crystal, possibly exciting optical wave propagation that is distributed over several modes of the photonic crystal, or that light is 'reflected' by the crystal. In this context, the reflected light includes both light that is specularly reflected by the crystal, and light that is diffracted by the crystal back into the half-space within which it is incident. The direction in which light is diffracted is determined by the photonic crystal lattice periodicities, while the amount of diffracted light is determined by the refractive index distribution in the photonic crystal, as seen through the defined surface of the photonic crystal. Power and energy conservation mean that the sum total of the power in the transmitted, reflected and coherently diffracted optical beams equals that launched into the system.

Therefore, despite the differences between photonic crystals and semiconductors, a direct conceptual transition between the descriptions of 'electronic' and 'photonic' band structure is possible – assuming the immediate conversion, justified by quantum concepts, between energy and frequency – and between momentum and propagation constant.

1.5 Propagation of Light in Periodic Media, Photonic Crystals and Photonic Band-Gaps

Solutions of Maxwell's Equations for propagation in loss-less periodic dielectric media are themselves periodic, and thus provide the Bloch modes or Floquet–Bloch modes of the periodic structure (as explored by Joannopoulos and co-authors [6]). The Bloch modes for a photonic crystal structure at a single frequency are described by the product of a complex exponential propagation term (for each axis along which there is periodicity), multiplied by a term that has exactly the periodicity of the photonic crystal structure, i.e., where $u(x,y)$ is a function that has the periodicities of the photonic crystal, the solution has the form (in 2-space):

$$\exp{(jk_x x)} \times \exp{(jk_y y)} \times u(x,y) \times \exp{(j\omega t)} . \qquad (1.1)$$

The usual convention applies of taking the real part of the whole complex solution to obtain the instantaneous field quantities. This solution has a discrete, two-dimensionally periodic, Fourier-spectrum in reciprocal space (or k-space) for a 2D photonic crystal. The equation immediately below this paragraph is a formal representation of a Bloch mode in a 2D periodic lattice as a summation of space harmonics. The weighting factors $h_{n,m(k)}$ are the Fourier coefficients and $G_{n,m}$ are the reciprocal lattice vectors. The magnitude and phase of the space harmonic components forming the Bloch mode depend on the details of the periodic function specifying the refractive index distribution of the photonic crystal. In our 3D-Universe, the function $u(x,y)$ will also have a z-dependence that is determined by the detailed 'vertical' distribution of the 2D photonic crystal structure, e.g., a waveguide layer with holes in it, typically supported by a substrate that has a refractive index greater than that of air, but less than that of the waveguide layer:

$$H_k(\bar{r}) = \sum_{n,m} h_{n,m(k)} H_0 \exp\left[i\left(\bar{k} + \bar{G}_{n,m}\right)\bar{r}\right] . \tag{1.2}$$

Bloch waves are typically partial standing waves – and can therefore transport optical power. However, they become purely standing waves that transport no power at the Brillouin-zone boundaries, i.e. at the band-edges, where the slope of the ω–k diagram goes to zero and the group velocity is therefore also zero. As the Brillouin-zone boundary is approached, both the *magnitude* of the slope of the dispersion curve for the Bloch mode and the group velocity decrease progressively towards zero. In a typical situation, the shape of the Bloch-mode dispersion curve is parabolic at the zone-boundary and the curve is, as already implied, exactly horizontal. Although time does not permit us to examine the detail, the reader may be interested to read the papers by Lombardet and co-workers [7,8]. In these papers, calculations are described in which the group velocity is calculated via summation, overall all of the Brillouin zones, of the weighted space-harmonic contributions to the total Poynting-vector of the Bloch wave, divided (for normalisation purposes) by the energy density. Although direct estimation of the group velocity as the gradient of the ω–k surface may well be a simpler process, the space-harmonic Poynting-vector summation process could be a useful tool in solving more complex problems that involve the presence of gain, absorption and non-linearity.

For wave propagation in a dispersive situation it is typical to consider both the *phase-velocity*, v_p, defined as the ratio, at any point on the dispersion curve for a particular mode of propagation of the angular frequency (in radians per second) to the propagation constant; and the *group velocity*, v_g, defined as the local gradient of the dispersion curve that relates photon energy to photon momentum. Multiplying Planck's constant by the cyclic frequency or the wave propagation constant, respectively, gives the corresponding energy and momentum. Although it is natural to think of the frequency as being the independent variable and the propagation constant as being the dependent one, the standard convention is nevertheless to display the graph of dispersion with the angular frequency forming the vertical axis and the propagation constant forming the horizontal axis.

In a medium that has a periodically distributed refractive index (in 'real' space), the propagation of the electronic/optical waves also exhibits periodicity in the space of possible propagation constant values ('k-space'). This periodicity in k-space is well understood and leads to the definition of Brillouin zones. For a crystal with lattice constant a, the relevant wave number that defines the Brillouin zone boundaries is $2\pi/a$. In addition, the size and shape of the first Brillouin zone in a two-dimensionally periodic crystal depend on both the lattice constants and on the crystal symmetry. For simple lattices, such as those with square or hexagonal symmetry, there is a single lattice constant that is just the distance between all nearest neighbour lattice points.

Because of the periodic nature of the medium, the allowed solutions of the appropriate wave equation are Bloch waves. Bloch waves are made up of space-harmonics – and the consequence is that there is not a single unique phase velocity that describes the propagation of the Bloch wave. In contrast, the group velocity is defined by the slope of the dispersion surface, which is unique, i.e. the same value is obtained for the group velocity at any point that is equivalent between Brillouin zones. Bloch modes at a single frequency and along a specified propagation direction typically have both forward and reverse travelling space-harmonic components. Situations where the space-harmonic components in corresponding negative and positive parts of the Brillouin zones all have equal magnitude are possible – and there is then no net transport of energy. Such situations are clearly pure standing-wave situations, and the following Bragg condition applies:

$$k = k_B = \pm (\pi/a) \times \text{(non-zero integer)} . \quad (1.3)$$

At the Brilloun-zone boundary, there is a region of frequencies where the dispersion curves are separated by a gap, i.e. there is a band-gap region in which propagation along the specified direction is forbidden. It is usual to talk about the stop-band edges and to label the upper frequency limit of the stop-band as the conduction-band and the lower frequency limit as the valence band. Although these terms only have an immediate physical meaning for the electron probability waves of the solid-state physical theory of semiconductors, they are nonetheless also used for optical frequency electromagnetic waves. On the other hand, for electromagnetic wave propagation in regular arrays of high dielectric constant pillars in air or air-holes in a high dielectric constant background, the bands are more usefully labelled as the air-band and the dielectric band, respectively. The two distinct Bloch modes that can be supported by the photonic crystal may have the same value for their k-vector, but their modal power/energy distributions, and the resulting average energies in each mode, are quite different.

1.6 Quantum Energy–Frequency Relations

In quantum language, the photon energy E (Joules) and the optical frequency f or v are related by:

$$E = hf = hv = (h/2\pi)\omega = \hbar\omega . \quad (1.4)$$

In this equation, f and v are alternative symbolic representations of the cyclic frequency in Hertz, ω is the angular frequency in radian/second, and h is Planck's constant ($\sim 6.6 \times 10^{-34}$ Joules/second). Again using quantum language, there is the relationship between momentum, p, and propagation constant, k, that:

$$p = \hbar k . \qquad (1.5)$$

Plots (in principle) in three-space of the relationship between ω and k describe the properties of a medium from the point of view of wave propagation. For periodic media, there are specific repetitive relationships that are important for an understanding of the properties of the medium. Given an ω–k diagram, it is the ratio of the value of ω along the y-axis to the value of k along the x-axis, in a plot of ω versus k, that gives the *phase* velocity, i.e.:

$$v_p = \omega/k , \qquad (1.6)$$

where ω is the angular frequency (in radians per second) and k is the wave vector magnitude, which has dimensions of inverse length. *Group* velocity is then defined as the *slope* of the dispersion curve, i.e.:

$$v_g = \partial\omega/\partial k . \qquad (1.7)$$

Under conditions of so-called 'normal dispersion' it is usually legitimate to associate the group velocity with the *energy* velocity. But group velocity and phase velocity are both *vectorial* properties of a medium (in three-space). So the group velocity definition should be understood as implying the use of the *gradient operator* from vector calculus.

In considering the propagation of optical waves in media such as photonic crystals, it is often useful to use the free-space ('air') *light-line* as a reference marker – or, alternatively, the *substrate* light-line (e.g. when considering situations such as 2D photonic crystals fabricated in planar waveguides). The light line is a straight line through the origin of an ω–k diagram. The slope of the light line for a given medium is simply given by c/n, where n is the refractive index of the medium. For free-space, the slope of the light line is c, i.e. $3 \times 10^8 \ \mathrm{m\,s^{-1}}$. In typical semiconductor media such as silicon and gallium arsenide, the slope is more than three times smaller, i.e. below $10^8 \ \mathrm{m\,s^{-1}}$. For the layered open waveguide photonic crystal structures that are used in much of photonic crystal device research, the strictly guided Bloch modes must have an effective refractive index that is larger than the external media refractive indices. So the guided Bloch mode dispersion curves must lie *below all* the relevant light-lines. The term '*heavy photon*' has been coined [9] for situations where the group velocity is significantly smaller than the modal phase velocity – and is based on the standard association of particle-like behaviour in wave mechanics with the group-velocity of the waves. However, it is not appropriate to assign mass to such heavy photons. Instead, one may talk about photons that have modified values for their momentum.

1.7 Light Confinement in One Spatial Dimension: The Slab Waveguide

Let us now consider the simple planar slab waveguide, which has a uniformly thick core layer occupying the whole of 2-space and having a characteristic thickness a in the y-direction, i.e. the direction of the normal to the two surfaces that define the slab. The simplest case, but one that is of practical relevance, is a dielectric slab with a refractive index (n_{slab}) that is greater than one, with air (refractive index $n = 1$) above and below the slab (Fig. 1.3). The high refractive index dielectric slab waveguide is a simple example of a structure that provides strong confinement for light. For high refractive index values at optical frequencies (e.g. refractive index $n = 3.5$), the slab has the desirable property of supporting only a single fully confined guided mode of each polarisation type (see below) at thicknesses up to approximately 0.3 μm.

The dispersion equation that describes the propagation of strictly guided electromagnetic waves in an infinite dielectric slab, with lower refractive index media above and below it, is based on solution forms for the field distributions of the discrete mode that have simple trigonometric function dependence in the slab (i.e. the waveguide core) and exponential decay outside. Boundary conditions in the form of continuity of parallel field components across the slab boundaries lead to the dispersion equations for two distinct classes of modes, typically characterised in engineering electromagnetic notation as transverse electric (TE) and transverse magnetic (TM) modes. TE modes have no electric field component parallel to the propagation direction and TM modes have no magnetic field component parallel to the propagation direction. Of greater physical importance, is the fact that the single *electric* field component in the TE modes is characteristically normal to the propagation direction and parallel to the surfaces of the slab, while for the TM modes, the single *magnetic* field component is parallel to the surfaces of the slab. Correspondingly, the main (but not the only) magnetic field component of the TE modes is oriented normally to the

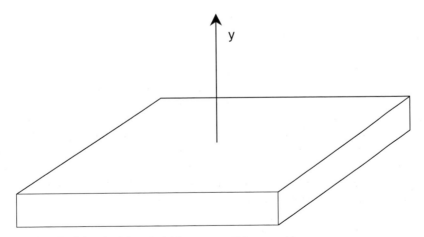

Fig. 1.3 Simple dielectric slab optical (electromagnetic) waveguide

surfaces of the slab, while the main (but not the only) electric field component of the TM modes is normal to the surfaces of the slab. The convention just stated is that used in Joannopoulos et al. [6], as well as the standard electromagnetic engineering textbooks. Unfortunately this is not a universally applied convention, and care must be taken when using popular software packages for modelling of photonic crystal structures, since at least one such package has used exactly the opposite convention!

Although the functions used to describe the modal field distributions for the dielectric slab waveguide are simple trigonometric ones, and the resulting dispersion equation is quite compact, it is a transcendental equation. Therefore the solutions for the propagation constants of the guided modes at a given slab thickness, with given refractive indices and with given wave frequency, have to be found numerically, i.e. by a simple root-solving process. In contrast, it is useful to note that specification of the wave frequency and the propagation constant value can be used, with the same equation, to compute multiple waveguide thicknesses that correspond to the successive modes of the waveguide. (The equation is not a transcendental one as far as the waveguide thickness is concerned.)

An important practical point is that thin slab waveguides are the base for the air-suspended high-index membranes that have been used in many of the 2D photonic crystal structures described in the recent literature, in particular for the realisation of high Q-factor photonic crystal micro-cavities. The weighted average refractive index of a thin dielectric slab with a regular pattern of air holes in it can readily be much larger than that of air, making it possible to support guided and well-confined Bloch modes with a range of effective refractive index values. In particular, the effective guided mode Bloch index can have a value as large as 3, making it possible for full photonic band-gap behaviour to be obtained for in-plane guided propagation.

The mode structure of a uniform dielectric slab that is symmetrically enclosed by half spaces having equal refractive index is shown schematically in Figure 1.4. The vertical axis is a dimensionless normalised frequency, $(\omega a/2\pi c) = a/\lambda_0$, and the horizontal axis is a normalised in-plane wave-vector or propagation constant, $(ka/2\pi) = (a/\lambda_{\text{eff}})$. The fundamental mode of the symmetric slab exists, for a given frequency, down to zero thickness. This property is closely related to that of the fundamental mode of a circular fibre, which also does not exhibit a low-frequency cut-off. In the *asymmetric* situation, e.g. of a thin layer of silicon with air above and supported by a sufficiently thick layer of silica (refractive index $n = 1.45$), the lowest order mode at a given wavelength, e.g. the free-space wavelength of 1.55 μm that is widely used in fibre-optical telecommunications, is cut-off below a finite thickness value that is determined by the refractive index of the slab. The cut-off thickness, at a given wavelength, is slightly different for a TE mode and its corresponding TM mode, implying that there is a narrow range of conditions for which only the lowest order TE mode is supported by the asymmetric slab waveguide. The vertical confinement is characteristically weak for this specific condition.

There are two distinct regions that should be noted in Fig. 1.4. The guided modes (identified with the label n) form a discrete spectrum that always lies between a steeper straight line that corresponds to the velocity of light in air or, to a good approximation, in vacuum, and a less steep straight line that corresponds to

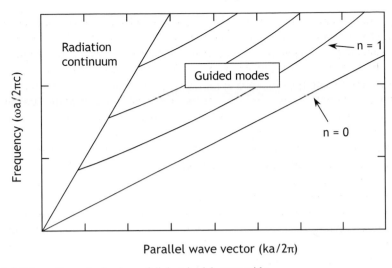

Fig. 1.4 Schematic mode structure of dielectric slab waveguide

the velocity of plane waves in the unconfined (bulk) dielectric medium. (The relatively modest contribution of material dispersion is ignored in this diagram.) The permitted states that are labelled as 'radiation continuum' in Fig. 1.4 have phase velocities that are greater than the free-space velocity of light, and are therefore radiative modes of the slab. Unlike the guided-modes, which have a discrete spectrum, the radiative modes have a continuous spectrum. They may have any phase velocity (along the guide) that is greater than the velocity of light in free-space. If a plane wave is launched in air onto a dielectric slab at an angle that corresponds to an extended state, the wave will be partially transmitted and partially reflected by the slab, emerging at the same angle as the angle of incidence, if the slab has air on both sides. That this behaviour is nevertheless characteristically modal in nature is shown by the fact that the amount of reflected and transmitted light depends on both the angle of incidence and the thickness of the slab, relative to the free-space wavelength and the refractive index of the slab. With the correct choice of parameters, the coupling between a free-space beam and the modes in a photonic crystal structure embedded in a slab waveguide may be sharply resonant in nature, corresponding to large and complementary variations in reflection and transmission.

The partial description that we have given of the dielectric slab modal properties connects back to classic papers on dielectric waveguides, such as that of Marcuse [10], which discussed the propagation losses resulting from a specified level of roughness. In the context of 1D or 2D photonic crystal structures composed of holes etched into slab waveguides, the earlier work on the scattering of light (mainly into radiation modes) due to roughness in a slab waveguide provides a useful starting point, with the roughness treated by Marcuse (and in subsequent work by Payne and Lacey [11]) now being replaced by structural aspects of the photonic crystal. In passing, we may note that the scattering of light due to roughness and irregularity is a continuing concern in the evolution of nano-photonic waveguides such as photonic wires [12] and photonic crystal channel guides [13].

Experiments designed to estimate the characteristic band-structure of planar photonic crystal structures [14, 15] frequently use finite (but large enough) cross-sectional area light beams incident at known angles on the photonic crystal region. Measurements of the reflection and transmission coefficients as a function of frequency then enable identification of the free-space coupled or leaky Bloch modes of the planar photonic crystal. Such measurements are subsequently used to construct the band-structure of the photonic crystal. This technique can be extended to allow identification and characterisation of the guided Bloch modes, via the use of the prism-coupler technique or by illuminating the edge of a finite area photonic crystal region with the probe light beam [16]. The angle of incidence of the light beam immediately identifies the in-plane propagation constant (wave-vector), while rotation of the planar photonic crystal structure around an axis that is normal to the plane of the crystal identifies the direction of propagation with respect to the defining axes of the photonic crystal lattice. The polarisation of the incident light beam must be specified, and the polarisation content of the reflected, transmitted and diffracted light beams must be measured.

Another nomenclature-related discrepancy in the analysis of the guided modes of a dielectric slab is that one standard convention labels the lowest order, strictly-guided, mode as $n = 0$. This is because it associates n with the multiple solutions of the dispersion equation, which are in turn directly associated with the multiple branches of the *tan* and *cotan* functions – and the multiple angles that give the same values of the tan and cotan functions. Our convention here is *not* the one used in Joannopoulos et al. [6].

1.8 Band-Structure in Periodic Media, the 1D Case

Despite the desire of some researchers in the field to restrict the use of the words 'photonic crystal' to media that are periodic in either two or three spatial dimensions, there is much to be gained from understanding the behaviour of optical waves in media that are only periodic along one spatial axis. In this 1D situation, the periodic variation of the electromagnetic properties, e.g. the refractive index, guarantees that there will be band structure, i.e. there will be curves in ω–k space that describe the propagation of electromagnetic waves over a specified range of photon energies, i.e. optical wave frequencies, despite the spacing between the curves implying that propagation is not possible (in the ideal case of a perfect, infinitely repeating lattice).

The representative band structure that can be observed in or calculated for a one-dimensionally periodic dielectric structure with large refractive index contrast between the two constituent layers that form a single spatial period is shown in Fig. 1.5. This pattern is, in principle, merely the first section of a graph that is repeated infinitely-many times along the horizontal (wave-vector) axis in both the positive and the negative directions. The diagram is therefore a representation of the electromagnetic wave dispersion of the periodic system in the first Brillouin zone. Note,

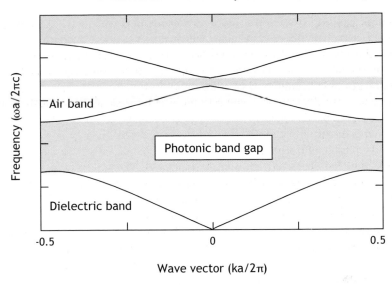

Fig. 1.5 Schematic dispersion (ω–k) diagram of large refractive-index contrast 1D periodic structure

again, that the dispersion curves describing the band structure must intersect the Brillouin-zone edges exactly horizontally.

Brillouin zones are readily defined in ω–k space, even for situations that are only periodic in one space dimension. All Brillouin zones have the same extent along the k-axis, because it is possible to add on a reciprocal lattice vector and obtain a solution that is formally indistinguishable, i.e. unchanged. The Floquet–Bloch modes for a particular lattice are characteristic of that lattice, as their behaviour is determined by the detailed properties of the lattice. These properties include the exact refractive index distribution for optical waves, or the potential distribution for electron waves.

Built-in to the band-structure diagram of a periodic medium are the possibilities of both slow-wave propagation and backward waves. It is important for all researchers currently working in the field of slow light phenomena (and slow electromagnetic wave propagation more generally) to have a proper awareness of key historical aspects of their field. A good introduction to this is found as the web-site:

http://en.wikipedia.org/wiki/Backward_wave_oscillator

It has long been recognised that slow electromagnetic-wave-propagation behaviour on periodic electrode structures could be used as a means of extracting microwave frequency electromagnetic power from a high velocity electron beam in a vacuum, typically with the periodic electrode structure wrapped helically around a cylindrical electron beam. In Amerenglish, devices that exploit this interaction are known as *microwave tubes*. With appropriate construction and choice of parameters, this interaction between electromagnetic waves and electron beam charge oscillations can produce backward wave oscillation, in which the group and phase velocity of the electromagnetic waves have opposite signs. Feedback mechanisms then allow the classic build-up from noise into a characteristic oscillation. A key aspect of travelling-wave type microwave tubes, which may yet find a modern counterpart at optical frequencies, is that the wave propagation is radically modified from that of Bloch waves in a loss-less, gain-free medium. In microwave frequency electron-tubes, the presence of gain enables a useful interaction to take place that involves only a single space harmonic of the periodic structure, i.e. all the important action may be crowded into a single Brillouin zone. Present day microwave tubes are able to operate at frequencies up to 1 THz.

1.9 Band-Structure in Periodic Media, the 2D Case

Moving to photonic crystals with two-dimensionally (2D) periodic lattices radically increases the scope for engineering of the photonic band-structure. A variety of lattice symmetries becomes available, of which the 'triangular' lattice shown schematically in Fig. 1.6 is merely the simplest example. The triangular lattice is so called because the smallest group of holes that defines it has a triangular arrangement. But the lattice has six-fold rotationally symmetry that allows the circular building blocks, in principle, to be close-packed, so it may also be described as 'hexagonally close-packed'.

Included among the possible arrangements of interest for creating photonic crystal structures in two-space are: (i) rectangular lattices with different periodicities along two orthogonal axes; (ii) photonic quasi crystal structures that are strictly not periodic but have rotational symmetries higher than six; and, (iii) 'graphite' lattices that also have six-fold rotational symmetry, but are more open than the triangular lattice and therefore cannot be close-packed.

Practical 2D photonic crystal structures are typically based on planar waveguides, in order to provide confinement along the third spatial dimension. The use

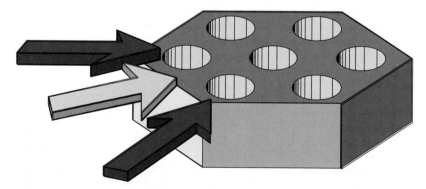

Fig. 1.6 Schematic of section of a 'triangular' photonic crystal lattice of circular air holes in a solid medium. The section has been chosen in order to emphasise the hexagonally close packed symmetry of the lattice. The *two darker arrows* indicate symmetrically equivalent directions, while the *lighter arrow* indicates a direction that exactly is half-way between

of air holes in a continuous high refractive-index medium, rather than high-index pillars in a low index (air) background, provides much better 'vertical' (waveguide) confinement, as well as simple mechanical integrity. The waveguide used may, depending on circumstances, take the form of a suspended membrane in air, e.g. of silicon or an epitaxial III–V semiconductor structure, of a high index core layer on a low index cladding layer, e.g. SOI (silicon-on-insulator) or a low-contrast, but high-index waveguide layer, with holes etched to a sufficient depth, 'from the top', through the waveguide core and into the lower cladding layer.

In contrast to the hexagonal (triangular) lattice, a square photonic crystal lattice has the simple advantage, both for computational modelling purposes and for pattern definition purposes, of being easier to specify in a rectilinear coordinate geometry. For some purposes, the greater degree of anisotropy, i.e. the smaller symmetry level, of the square lattice may even be a useful feature.

Figure 1.7 shows schematically the band-structure of a particular triangular lattice 2D photonic crystal structure. The parameters of the crystal, such as the refractive indices and the hole filling-factor, have been chosen so that there is a full photonic band-gap (PBG), i.e. (as indicated) there is a range of energies where no propagation band is available. As the figure suggests and as will emerge in subsequent chapters of this book, exploiting fully the potential of a band-structure as complex as that which is automatically provided by a 2D periodic lattice of strong light scatterers is a serious challenge – and is an on-going one in research, world-wide. Even if there is no region where a full photonic band-gap or a substantial angular range of photonic band-gap behaviour, many interesting properties become available, e.g. slow-propagation, super-prism, hyper-lens behaviour and auto-collimation, strongly polarisation-dependent effects and largely polarisation-

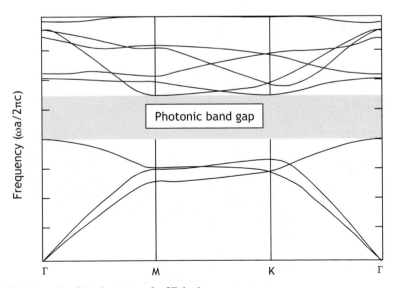

Fig. 1.7 Schematic of band-structure for 2D lattice

independent effects. Changes in the band-structure and the introduction of useful 'states-in-the band-gap' add to the possibilities.

Figure 1.7 follows standard conventions [6] in terms of showing the dispersion along the two high-symmetry directions of a hexagonal photonic crystal lattice, together with the change in the frequency (or, equivalently, photon energy) encountered in moving along the Brillouin-zone boundary between the two high-symmetry directions. Effectively this representation is made up from three chosen cross-sections of the frequency surface in reciprocal lattice space.

1.10 Gap Maps

Gap maps [17] are a representational tool that show graphically, for a given lattice form and a given combination of the high and low refractive-indices of a two material based photonic crystal, regions of the spectrum over which the polarisation-determined phenomenon of 'full' photonic band-gap ('full-PBG') behaviour occurs. These are the regions where propagation in any direction encounters a stop-band. The low index medium involved is usually air, but this is not a mandatory condition. The other variable involved, for circular holes in a high-index solid medium or for high-index pillars in air, is the ratio of the cylinder radius r to the lattice constant a, i.e. the normalised radius. This ratio is directly related to the filling factor, which is defined as the fraction of the total area of the 2D lattice (in plan view) that is occupied by the cylinders. The regions where fully polarisation independent PBG behaviour is obtained simultaneously for both TE and TM polarisations can then

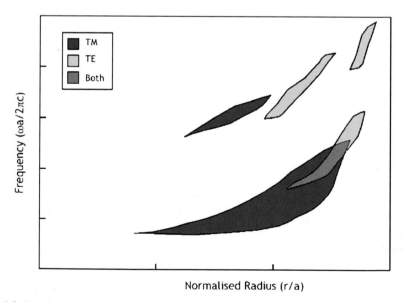

Fig. 1.8 Schematic gap-map

be identified as the characteristically much smaller overlap zones between the PBG regions for the two different polarisations. As Fig. 1.8 shows, such overlap, which may be very desirable for some applications, does not always occur. The other side of the coin, polarisation selectivity, is clearly also potentially useful.

1.11 Channel Waveguides Through Photonic Crystals

The presence of deliberately induced 'defects' in a photonic crystal region is an essential tool for their exploitation in device structures. Associated with such defects, in a manner closely analogous to the effects of doping on the electronic properties of semiconductors, is the creation of 'states in the band-gap'. One specific type of defect of interest is the channel waveguide, which may be obtained, for example, by filling in one row of holes in a photonic crystal lattice – forming a W1 channel, as illustrated in Fig. 1.9. But channel waveguides can also be formed, conceptually, by moving blocks of crystal, making it possible to produce channel waveguides with fractional spacing, e.g. W0.7 or W1.3.

Figure 1.10 shows a possible dispersion curve for a photonic crystal channel wave waveguide, set in the framework of the first Brillouin zone along the main propagation axis and the band structure of the perfectly regular lattice. Although only a single modal dispersion curve is shown in the band-gap region of the band-structure, in general there will be a discrete set of guided modes. The guided-mode dispersion curve is not shown as continuing above the upper band beyond the point where it intersects with it, but this is possible. The guided modal dispersion curve is also shown as continuing for k-vectors down to zero, corresponding to an infinite phase velocity. At the same time, the dispersion curve must be exactly horizontal at its intersection with the frequency axis, corresponding to zero group velocity, with slow propagation behaviour occurring in this neighbourhood.

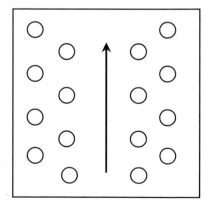

Fig. 1.9 Schematic of W1 channel guide in triangular (hexagonal) lattice of cylinders

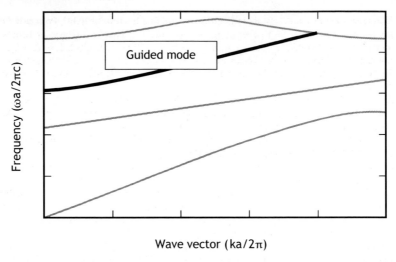

Fig. 1.10 Photonic crystal band-structure for lattice with channel-waveguide type defect state

1.12 Concluding Remarks

This introductory chapter has provided a brief overview of the topic of photonic crystals and the concept of the photonic band-gap; however, there are many more aspects of the topic that have not even been touched on. The chapter [18] immediately following this one provides the reader with an excellent continuation of much of what has been set out in this introductory chapter. In addition, 3D photonic crystals have only been described by implication through the images in Figs. 1.1 and 1.2. The committed reader should look to the book by Joannopoulos and co-workers [6] for a more general understanding of the properties of photonic crystal structures that should suffice to support an odyssey into the research literature, whether the topic is full PBG 3D PhCs, PhC channel waveguides, PhC micro-cavities or photonic quasi-crystal light extraction structures for LEDs. Further recommended reading is exemplified in references [19–21].

References

1. Yablonovitch E., 'Inhibited Spontaneous Emission in Solid-State Physics and Electronics', Phys. Rev. Letts., 15(20), 2059–2062, 18th May (1987).
2. John S., 'Strong Localisation of Photons in Certain Disordered Dielectric Superlattices', Phys. Rev. Letts., 15(23), 2486–2489, 8th June (1987).
3. Yablonovitch E., 'Photonic Crystals: What's in a Name?', Optics and Photonics News, 18(3), 12–13, March (2007).
4. Bykov V., 'Spontaneous emission from a medium with a band spectrum', Kvant. Elektron 1, 1557–1577, (July 1974). [Sov. J. Quant. Electron 4(7), 861–871 (Jan. 1975)]
5. Chong H.M.H. et al., '3D Photonic Crystals BASED ON Epitaxial III–V semiconductor structures for non-linear optical interactions Photonics Europe', Strasbourg, April 2006, paper 6182-28.

6. Joannopoulos J.D. et al., 'Photonic Crystals: mo(u)lding the flow of light', Princeton (1995).
7. Lombardet B., Dunbar L.A., Ferrini R., and Houdré R., 'Fourier analysis of Bloch wave propagation in photonic crystals', J. Opt. Soc. Am. B, 1179–1190, 22(6), June (2005).
8. Lombardet B., Dunbar L.A., Ferrini R., and Houdré R., 'Bloch wave propagation in two-dimensional photonic crystals: influence of the polarization', Optical. Quant. Electron., 37, 293–307, (2005).
9. Astratov V.N., Stevenson R.M., Culshaw I.S., Whittaker D.M., Skolnick M.S., Krauss T.F., and De La Rue R.M., 'Heavy photon dispersions in photonic crystal waveguides', Applied Physics Letters, 77(2), 178–180, 10th July (2000).
10. Marcuse D., 'Mode conversion caused by surface imperfections of a dielectric slab waveguide', Bell Syst. Tech. J., 48, 3187–3215, (1969).
11. Payne F. and Lacey J.P.R., 'A theoretical analysis of scattering loss from planar optical waveguides', Opt. Quantum Electron., 26(10), 977–986, (1994).
12. Sparacin D.K., Spector S.J., and Kimerling L.C., 'Silicon waveguide sidewall smoothing by wet chemical oxidation', IEEE Jour. Lightwave Technol., 23(8), 2455–2461, August (2005).
13. Gerace D. and Andreani L.C., 'Low-loss guided modes in low-loss photonic crystal waveguides', Optics Express, 13(13), 4939–4951, (2005).
14. Coquillat D., Ribayrol A., De La Rue R.M., Le Vassor D'Yerville M., Cassagne D., Jouanin C., 'First Observations of 2D Photonic Crystal Band Structure In GaN-Sapphire Epitaxial Material', Physica Status Solidi A-Applied Research, 183(1), 135–138, Jan (2001).
15. Coquillat D., Ribayrol A., De La Rue R.M., Le Vassor D'Yerville M., Cassagne D., and Albert J.P., 'Observations of band structure and reduced group velocity in epitaxial GaN-sapphire 2D photonic crystals', Appl Phys B-Lasers, 73(5-6), 591–593, Oct (2001).
16. Paraire N. and Benachour Y., 'Investigation of planar photonic crystal band diagrams under the light cone using surface coupling techniques', Applied Physics B, 89, 245–250, (2007).
17. Cassagne D., Jouanin C. and Bertho D., 'Photonic bandgaps in a two-dimensional graphite structure', Phys. Rev. B, 52(4), R2217–R2220, (1995).
18. Viktorovitch P., 'Physics of Slow Bloch Modes and Their Applications', this book, Chapter 2.
19. Krauss T.F., De La Rue R.M. and Brand, S.: 'Two-dimensional photonic bandgap structures at near infra-red wavelength', Nature, 383, 699–702, 24th Oct (1996).
20. Krauss T.F. and De La Rue R.M., 'Photonic crystals in the optical regime – past, present and future', Progress in Quantum Electronics, 23(2), 51–96, March (1999).
21. De La Rue R.M., 'Photonic Crystal Devices: Harnessing the Power of the Photon', Optics and Photonics News, 17(7/8), 30–35, July/August (2006).

2 Physics of Slow Bloch Modes and Their Applications

Pierre Viktorovitch

Nanotechnology Institute of Lyon, CNRS-Ecole Centrale de Lyon, 69134 Ecully, France

Abstract. The unique confinement properties of photonic crystals allow for the storage of photons in confined space (in the wavelength range), for a long time (compared to the period of oscillation). Different confinement schemes can apply for the production of a variety of very compact, spectrally as well as spatially resolved, microphotonic devices, depending upon the regime of operation of the photonic crystal microstructures (photonic band-gap regime, slow Bloch mode regime around an extreme of the dispersion characteristics, or a combination of both). This chapter focuses on two dimensional photonic crystals and on their exploitation along the slow Bloch mode regime.

Key words: Two dimensional photonic crystals, slow Bloch modes, photon confinement

2.1 Introduction

The principal motivations for the emergence of photonic crystals can be summarized in one single word, that is "λ-Photonics", which means the control of photons at the wavelength scale.

Generally speaking, the harnessing of the light consists in *structuring the space* where it is meant to be confined: but there are intrinsic limitations which are related to the undulatory nature of the light and which have been formulated in the famous equations of Maxwell in 1873, providing a consistent picture of the experimental data available then. These limitations are in the heart of the λ-Photonics, whose definition could be *the control of photons within the tiniest possible space during the longest possible time*: this implies to structure space at the wavelength scale, which is the sub-micron range for the optical domain.

The next section (2.2) will present a brief overview of the basic concepts which underlie photonic crystals, with a special emphasis on two-dimensional photonic crystals (2D PhC), which have been the matter, so far, of most of the new applications in terms of device demonstrations. Focus is put on the deep changes of the spatial-temporal characteristics of photons as a result of their "immersion" in the periodic medium formed by the photonic crystal. Section 2.2 will conclude with the presentation of the essential building blocks of the 2D PhC based Integrated Micro-Nano-photonics, which is presently developed along planar technological schemes and is considered as the principal domain of applications of photonic crystals.

It will be shown in Sect. 2.3 that 2D PhC have definitely entered within the realm of practical devices: although 2D PhC have not yet reached the maturity allowing for the mass production and transfer to the market of devices, it must be pointed out the extraordinary flourishing of laboratory demonstrations of Micro-Nano-photonic devices, at a rate which had not been anticipated a few years back. A special attention will be given to surface addressable devices, which have been the matter of very recent developments. In that respect, the concepts of 2.5 Microphotonics based on 2D PhC, which can be considered as a major extension of planar technology through exploitation of the third ('vertical') dimension, will be presented in Sect. 2.4, which will be followed by the concluding section.

2.2 Photonic Crystals: A Brief Overview of Basic Concepts

2.2.1 What Are Photonic Crystals?

A photonic crystal is a medium which the optical index shows a periodical modulation with a lattice constant on the order of the operation wavelength. The specificity of photonic crystals inside the wider family of periodic photonic structures lies in the high contrast of the periodic modulation (generally more than 200%): this specific feature is central for the control of the spatial-temporal trajectory of photons at the scale of their wavelength and of the their periodic oscillation duration.

Figure 2.1 shows schematic views of a variety of photonic crystals with dimensions ranging from 1 to 3.

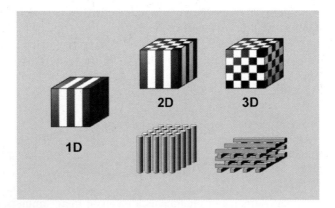

Fig. 2.1 Schematic view of photonic crystals with different dimensionalities

We will restrict the rest of this paper to 2D PhC, which have been the matter of most of the recent developments in the field of Micro-Nano-Photonics and are far more accessible than 3D PhC (initially introduced by E. Yablonovitch in 1987 [1]), from the fabrication point of view.

A real 2D PhC consists in considering a 2D structuring of a planar dielectric waveguide where photons are "index guided", that is to stay vertically confined by the vertical profile of the optical index. Figure 2.2 shows schematic and SEM views of a real typical 2DPC, consisting in a triangular lattice of holes formed in a semiconductor slab.

In the rest of this chapter we will concentrate on the so called membrane approach, where the vertical confinement is strong: guiding of light is achieved in a high index semiconductor membrane surrounded with low index cladding or barrier layers (for example an insulator like silica or simply air: see Fig. 2.3). In single mode operation conditions the thickness of the membrane is very thin, around a fraction of μm; it results that low loss coupling schemes with an optical fibre are not easily achievable, but, the positive counterpart lies in the relaxed technological constraints for the fabrication of the 2D PhC (holes with a shape ratio around unity). Also, the strong vertical confinement, leading to a reduced volume of the optical modes, leads itself to the production of very compact structures, which is essential for active devices to operate at the cost of very low injected power. Another essential asset of the membrane approach will be fully appreciated in Sect. 2.3, where the attention of the reader will be strongly driven to the fact that a brilliant future should be promised to 2D PhC provided that they are not strictly restricted to in-plane waveguided operation and that they may be opened to the third dimension, particularly along the so called 2.5D microphotonics schemes.

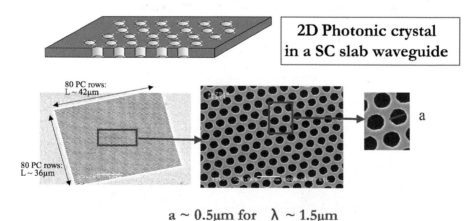

Fig. 2.2 Schematic and SEM views of a real 2D PhC, consisting in a triangular lattice of holes formed in a semiconductor slab

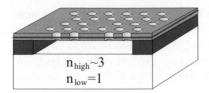
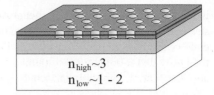

Suspended membrane in air   **Membrane bonded on a low index
                                        layer, silica for example.**

Fig. 2.3 The basic building blocks of the membrane approach

2.2.2 Why Photonic Crystals?

The principal motivation for high index contrast periodical structuring of space, thus resulting in the formation of a photonic crystal, lies in the general objective of Micro-Nano-Photonics, which is the control of optical modes/photons within the tiniest space V during the longest possible period of time τ.

According to the above definition of Microphotonics, it appears natural to grant the optical mode with a merit factor F, which quantifies the properties of the optical mode in terms of the ratio of the time τ during which it remains under control (or its lifetime from the observer/user view-point), over the average real space volume which it fills up during its lifetime.

To put it differently and more precisely, the lifetime τ is the time interval when the user may count on a coherent mode, whose phase remains deterministic, within the volume where he tries to control and confine it. The merit factor can be made dimensionless if normalized to the ratio $\frac{T}{\lambda^3}$, where T is the period of oscillation and λ is the wavelength in vacuum.

$$F = \frac{\tau}{T} \times \frac{\lambda^3}{V} \tag{2.1}$$

with:

$$Q = \frac{2\pi\tau}{T} \tag{2.2}$$

where Q is the traditional quality factor of the mode.

The reader will have noticed that F is proportional to the Purcell factor, which expresses the relative increase of the spontaneous recombination rate of an active medium as a result of its coupling to the optical mode, as compared to the non structured vacuum [2].

It should be pointed out that there exists a variety of ways for the structuring of space, consisting in preventing the propagation of photons along one or several directions, thus resulting in photonic structures with reduced "dimensionality" and optical mode confinement. Refraction phenomena, for example, have been widely used in optoelectronics for the guiding of photons or for their trapping within micro-cavities. The control of photon "trajectory" is based upon the total internal reflection that they experience at the boundary between the external world and the higher index medium where they are meant to be confined. Photonic crystals offer a new strategy for optical mode confinement based on diffraction phenomena. The new avenue opened up by photonic crystals lies in the range of degrees of freedom which they provide for the control of photon kinetics (trapping, slowing down), in terms of angular, spatial, temporal and wavelength resolution.

2.2.3 Photonic Crystal: How Does It Work?

The principal characteristics of the photonic crystal manifest themselves in the so called dispersion characteristics of the periodically structured medium, relating the pulsation ω (eigen-value) to the propagation constants k (eigen-vector) of optical modes, which are the eigen-solutions of Maxwell equations, corresponding to a spatial distribution of the electromagnetic field which is stationary in the time scale. It is appropriate here to speak in terms of dispersion surfaces $\omega(k) = \omega(k_x, k_y)$, real space being two-dimensional.

In a non structured homogeneous dielectric membrane, the dispersion surfaces relate classically to the guided modes of the slab waveguide and show a circular symmetry.

For photonic crystals, which are strongly corrugated periodic structures, strong diffraction coupling between waveguide modes occurs; these diffraction processes affect significantly the dispersion surfaces, or the so called band structure, according to the solid state physics terminology. The essential manifestations of these disturbances consist in:

- The opening of multidirectional and large photonic band-gaps (PBG)
- The presence of flat photonic band edge extremes (PBE), where the group velocity vanishes, with low curvature (second derivative) $\alpha \approx \frac{1}{PBG}$.

These are the essential ingredients which are the basis of the two optical confinement schemes provided by photonic crystals (PBG/PBE confinement schemes) and which make them the most appropriate candidates for the production of a wide variety of compact photonic structures.

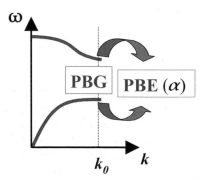

Fig. 2.4 Schematic representation of a photonic band-gap (PBG) and of related photonic band edges (PBE) in the dispersion characteristics of a photonic crystal

PBG Confinement Scheme Using Localized "Defect" or Cavity Modes

In the PBG scheme, the propagation of photons is forbidden at least in certain directions. This is in particular true when they are trapped in a so called localized defect or microcavity and the related optical modes are *localized*: in this case the propagation of photons is fully prohibited. Opening of large PBG (in the spectral range) provided by the photonic crystal, allows for a very efficient trapping of photons, which can be made strongly localized in free space.

PBE Confinement Scheme Using Delocalized Slow Bloch Modes

In the PBE scheme, the photonic crystal operates around an extreme of the dispersion characteristics where the group velocity of photons vanishes. It should be noted however that the dispersion characteristics apply strictly for infinite periodic structure and time and that the concept of zero group velocity is fully true only under these particular extreme conditions. The real common world is actually finite and transitory. It is therefore more appropriate to speak in terms of *slowing down* of optical modes (so called Bloch modes for a periodical structure), which remain however *de-localized*.

It can be shown that the average group velocity of optical *slow Bloch modes* in a photonic structure operating around an extreme of the dispersion characteristics decreases with time t like $v_g \approx \sqrt{\frac{\alpha}{t}}$ (assuming that the dispersion characteristics are isotropic at the extreme whose curvature is α). If we put it in a different way, the lateral extension of the area S of the slowing down Bloch mode during its lifetime τ is proportional to $\alpha\tau$ [3]. It can then be straightforwardly derived that the merit factor of the Bloch mode is simply proportional to $\frac{1}{\alpha}$[1]. As mentioned above, one essential virtue of photonic crystals is to achieve a very low curvature α at the band edge extremes, thus resulting in very efficient PBE confinement of photons and large merit factor of the corresponding slowed down optical modes.

The most efficient confinement of photons can be achieved with the PBG scheme. Record merit factors have been reported in the literature in this way.

The PBE scheme provides weaker confinement efficiency than with the PBG approach, while resulting in an improved control over the directionality or spatial/angular resolution.

It has been explained earlier in this chapter that the vertical confinement of photons is based on refraction phenomena. However, full confinement of photons in the membrane waveguiding slab is achieved only for those optical modes which operate below the light-line (see Fig. 2.5a). This mode of operation is restricted to devices which are meant to work in the sole waveguided regime, where guided modes are not allowed to interact or couple with radiated modes. This is the territory of 2D micro-photonics.

For guided modes whose dispersion characteristics happen to lie above the light line, coupling with the radiated modes is made possible, the guided "state" of the related photons is transitory, and the photonic structure can operate in both waveguide

---

[1] Which is also proportional to the density of photonic modes at the band edge extreme

and free space regimes (see Fig. 2.5b). This is the world of 2D–3D microphotonics, which we will quote later in this paper as 2.5D microphotonics.

The Issue of Vertical Confinement in 2D PhC: Below and Above Light-Line Operation

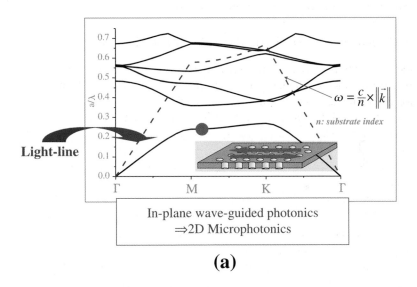

In-plane wave-guided photonics
⇒2D Microphotonics

(a)

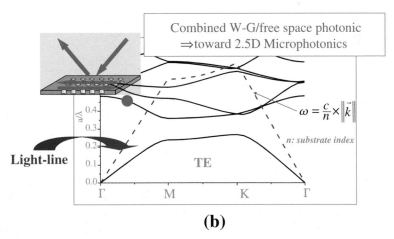

Combined W-G/free space photonic
⇒toward 2.5D Microphotonics

(b)

Fig. 2.5 Below (**a**) and above (**b**) light-line operation of photonic structures based on 2D photonic crystals

2.3 Photonic Crystals: Slow Bloch Modes and Related Devices

Following the pioneering and triggering contributions of S. John (1987) [4] and E. Yablonovitch (1987) [1], it took quite a few years for the modelling and technological tools to reach the degree of maturity requested by the production of the first elementary building block devices, essentially based on 2D PhC. This gradual start has been followed, around 2000, by an ever growing wave of new device demonstrators, so much so that it may be stated, to day, that photonic crystals have entered within the realm of practical devices.

In order to help the reader to find his way within this jungle, we choose to classify the wide range of devices produced so far into four main categories, depending upon whether they operate singly in the waveguide regime or not, and upon whether they make use of the PBG or of the PBE confinement scheme.

This classification is further detailed in the table of Fig. 2.6, which provides a non-exhaustive list per category of the principal device structures demonstrated so far. In the following sections we emphasize devices making use of slow Bloch modes along the PBE confinement scheme and specifically those devices belonging to the fourth category, that is to say devices operating in the waveguide regime while being also opened to the third direction of space: these devices include in their functionality the coupling of guided to radiated modes, which has been the matter of attractive new developments in the recent literature.

It should be mentioned, at this stage, that most of the recent reported achievements, in the literature in terms of device demonstrations, are based on the membrane approach. For passive devices silicon is often used for the membrane material, especially in the silicon on insulator (SOI) configuration, which is fully available in

	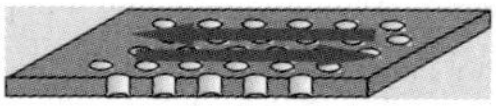	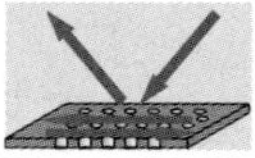
PBG	• Micro-cavities (QED) • Micro-lasers • Guiding / bends • Cavity-guide cascading (add-drop filters) • ...	• Drop filters • ...
PBE	• Directional add-drop filters • Micro-lasers • Super-prism • Pulse compression • ...	• Compact reflectors/filters • Non-linear optics: fully optical micro-switches • Surface emitting Micro-lasers • ... and other devices

Fig. 2.6 Classification of 2D PhC based devices in four main categories

the world of microelectronics. For active devices, III–V semiconductor membranes have been principally used so far: the thin membrane is generally bonded by the molecular bonding procedure on the low index material, such as silica on silicon substrate [5]. This approach presents the definite advantage of lending itself to heterogeneous integration of active III–V optical devices with silicon based passive optical devices and microelectronics.

Photonic devices based on 2D PhC have been principally aimed, so far, at forming the basic building blocks of integrated photonics and are usually designed for in plane waveguided operation. We remind that the operation of photonic integrated circuits based on 2D PhC may be deeply affected by optical losses resulting from unwanted diffractive coupling of guided modes with the radiation continuum.

This problem of optical losses can be approached from a completely different perspective: instead of attempting to confine the light entirely within waveguide structures, the 2D structures can be deliberately opened to the third space dimension by *controlling* the coupling between guided and radiation modes. In this approach, the exploitation of the optical power is achieved by accurately tailoring the optical radiation into free space.

A simple illustration of this approach is the use of a plain Photonic Crystal Membrane as a wavelength selective transmitter/reflector: when light is shined on this photonic structure, in an out-of-plane (normal or oblique) direction, resonances in the reflectivity spectrum can be observed. These resonances, so called Fano resonances [6], arise from the coupling of external radiation to the guided modes in the structures, whenever there is a good matching between the in-plane component of the wave vector of the incident wave and the wave vector of the guided modes (see Fig. 2.7). Accurate tailoring of the spectral characteristics of the Fano resonances (shape, spectral width) is made possible by the design of the 2D PhC membrane (type of 2D PhC, membrane thickness, . . .). If the lateral size of the illuminated membrane is infinite, the spectral width of the resonance is like the inverse of its lifetime τ, that is the lifetime of the guided mode, with $\tau = \tau_c$, where τ_c is simply

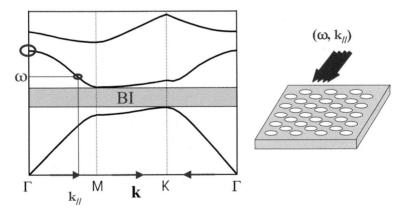

Fig. 2.7 Illustration of the resonant coupling between a guided mode and a radiated mode

the coupling time constant between guided and radiated plane-wave modes[2]. In real devices, the lateral size of the illuminated area is limited, and the lifetime of the resonance is also controlled by the lateral escape rate $\frac{1}{\tau_g}$ of the guided mode out of this area; this escape rate can be considered as a loss mechanism for devices which are designed and meant to operate "vertically". In these real conditions the lifetime of the resonance can be written, under certain conditions as:

$$\frac{1}{\tau} = \frac{1}{\tau_c} + \frac{1}{\tau_g} \approx \delta\omega \qquad (2.3)$$

where $\delta\omega$ is the spectral widening of the resonance. The ability of high index contrast PhC to slow down photons and to confine them laterally, especially at the high symmetry points (or extremes) of the dispersion characteristics allows for a very good control over the lateral escape losses and results in very compact devices.

A variety of passive as well as active devices has been demonstrated in the recent literature. For example, very compact passive reflectors showing a large bandwidth (a few hundreds of nanometres) and consisting in a plain 2D PhC membrane formed in Silicon on silica have been reported [7]. The large bandwidth is obtained for specific designs of the 2D PhC which allow for a very strong coupling rate $\frac{1}{\tau_c}$ of guided modes with the radiation continuum.

The 2DPC membrane can be also designed in such a way as to result in very strong Fano resonances, that is for weak coupling rate $\frac{1}{\tau_c}$. Use of such strong Fano resonances has been made for the demonstration of very low threshold and very compact surface emitting Bloch mode laser [8]. The photonic crystal consists in

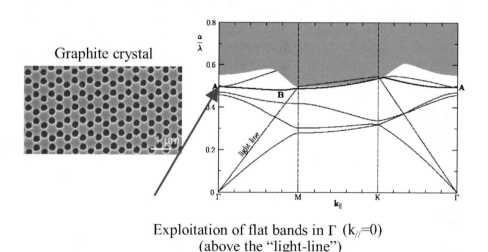

Exploitation of flat bands in Γ ($k_{//}$=0)
(above the "light-line")

Fig. 2.8 Band structure of a surface emitting Bloch mode laser formed in a graphite type 2D PhC

---

[2] Various factors contribute to the control of the coupling time constant, such as the strength of the periodic corrugation, the membrane thickness, the symmetry of the guided mode, …

a graphite lattice (Fig. 2.8), which can be viewed as an array of H_1 coupled cavities, formed in a triangular lattice.

This particular 2D PhC exhibits band edge extremes at the Γ point with very low curvature (Fig. 2.8). One of these extremes is exploited for vertical laser emission. The coupling between wave-guided slow Bloch modes and radiated modes is authorized, but its rate is controlled accurately, allowing for the vertical emission, while retaining the strength of the resonance and, therefore, achieving a weak threshold power.

Emission spectra of the laser are shown in Fig. 2.9 for different hole filling factors f, as well as the spontaneous emission spectrum of the non-structured membrane. The device is optically pumped in a quasi-steady state regime and operates at room temperature. The peak intensity for the optimum filling factor ($f = 19\%$) is larger than the spontaneous emission power by 5 orders of magnitude. For increasing f, as expected, the emission peak is blue shifted as a result of the reduction of the effective optical index of the membrane; in the same time, the emission yield drops rapidly due to the decrease of the modal gain (not shown in the figure). The effective threshold pumping power, for the optimized device, is very weak and does not exceed $40\,\mu W$. The pumped area where the stimulated emission process takes place is very limited and does not exceed 2 to $4\,\mu m$ in diameter: this is a clear demonstration of the outstanding ability of 2D PhC to confine laterally slow Bloch modes, along the PBE scheme.

The graphite lattice 2D PhC active membrane used for surface laser emission is extremely generic and can apply for a large variety of other types of active devices, at the very cheap expense of slight changes in the design of the 2D PhC. Along this line spectacular demonstrations of diverse devices have reported recently, including optical amplifiers and fully optical micro-switches [9–12] for the latter it is made use of electronic Kerr effect, via photo-injection of carriers in quantum wells, to manipulate Fano resonances in the spectral domain (the design of the 2D PhC results

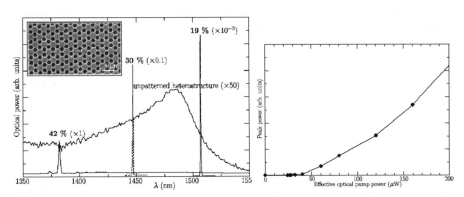

Fig. 2.9 Emission spectra of the surface emitting laser formed in a graphite lattice 2D PhC, for different hole filling factors (f). The plot of the emitted power versus the pumping power indicates a threshold power of $40\,\mu W$ for $f = 19\%$

in Fano resonances lying in the Urbach tail of the quantum wells, instead of the gain maximum as in the case of micro-lasers).

All these devices are convincing illustrations of a planar technological approach resulting in 2D PhC devices freed from the bi-dimensional universe.

2.4 Towards 2.5D Micro-Nano-Photonics

It has been proposed recently a major extension of planar technology, through exploitation of the third ("vertical") dimension by using a so-called multi-layer approach, where the lateral high index contrast patterning of layers would be combined with the vertical 1D high index contrast patterning: it is here more appropriate to think in terms of "2.5 dimensional" photonic structures, in which an interplay between wave-guided-confined photons and radiated photons propagating through the planar multilayer structure occurs [3].

The simplest illustration of this approach is the use of a plain Photonic Crystal Membrane as discussed in the previous Sect. 2.3. If one considers now a multi-layer structure, the strong vertical 1D modulation of the optical index, allows for a fine and efficient "carving" of the density and vertical field distribution of radiated modes, using a limited number of layers. As a result the variety of coupling schemes between optical modes is considerably widened, thus opening large avenues toward new photonic functionality.

In summary, 2.5D Microphotonics, combining lateral 2D PhC and vertical 1D PhC, should provide a very good control over the electromagnetic environment, that is over the distribution of optical modes in 3D real space and time, at a much lower cost than the full 3D approach in terms of technological feasibility: the technology schemes to be adopted are compatible with technological approaches which are normally describable as planar. This multi-layer or multi-level approach is familiar to the world of silicon microelectronics, when it comes, for example, to fabricate the multiple levels of electrical inter-connections; its use for microphotonics extends far beyond, from the viewpoint of widening considerably the range of new accessible functionality and performances.

The 2.5D Microphotonics approach has been successfully applied recently for the production of very low threshold power microlasers [13, 14] and of a new class of optical bistable devices based on the Kerr effect [15]. The basic common building block for these devices is shown is Fig. 2.10. It consists in a graphite lattice 2D PhC active membrane, similar to that presented in the previous section, bonded on to the top of a Bragg reflector formed by high index contrast SiO_2-Si quarter wavelength pairs. It can be shown that the thickness t_G of the top SiO_2 "gap" layer, which supports the bonded 2D PhC membrane, is essential for the performances of both types of devices, in terms of the requested threshold power. This is due to the fact that the resonant coupling rate at the Γ point of the wave-guided slow Bloch mode which is used in these devices, is strongly dependent on t_G: the coupling rate is inhibited for t_G on the order of an odd integer number of quarter wavelength, which results in an increased strength of the slow Bloch mode resonance and, therefore, in

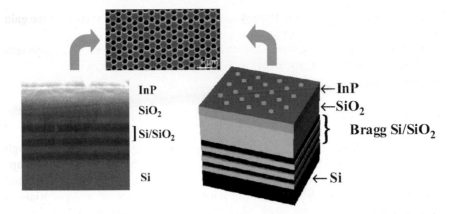

Fig. 2.10 Photonic crystal membrane bonded on top of a Bragg

a significantly reduced threshold power of the device (and *vice versa* for t_G on the order of an integer number of half wavelength).

This is illustrated in a spectacular manner in Fig. 2.11, which shows the gain characteristics of the micro-laser for the two (quarter-wavelength or half wavelength) t_G values. Optical bistabity could be demonstrated for the sole quarter-wavelength t_G case (see Fig. 2.12), corresponding to the strongest mode confinement (inhibition of coupling to the radiation continuum). It should be pointed out that the only difference between these two categories of devices lies in the particular design of the 2D graphite PhC: for the micro-laser, it is managed that the slow Bloch mode resonance wavelength at the Γ point lies close to the gain maximum of the active quantum well medium, whereas, for the bistable device, it is located within the Urbach tail of the quantum well material. Needless to say, therefore, that the building block shown in Fig. 2.10 is very generic.

Other domains of photonics should take advantage of the 2.5D microphotonics approach. For example, the introduction of 2D PhC in MOEMS (Micro Opto Electro

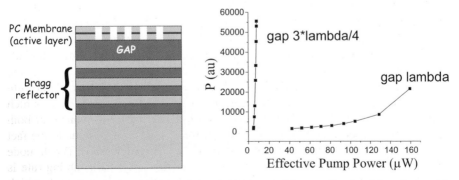

Fig. 2.11 2.5D Photonic crystal micro-laser: the thickness of the top silica "gap" layer has a wide impact on the threshold power of the micro-laser

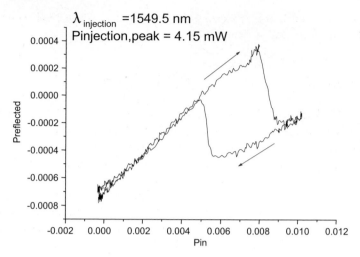

Fig. 2.12 Optical bistability effect in a 2.5D PhC structure as shown in Fig. 2.10

Mechanical) devices shows great promises in terms of widening of the spectrum of (electromechanically actuable) optical functions, achievable with further enhanced compactness structures.

Figure 2.13 shows examples of such 2.5 dimensional MOEMS structures. These new types of photonic structures should be applied in various domains, including optical telecommunications (tuneable or switchable wavelength selective devices, taking advantage of the extra angular resolution provided by the in-plane 1D–2D PhC). Highly selective and widely tuneable 2.5D MOEMS filters have been demonstrated very recently [16]. Also, a new family of hybrid VCSEL, where one of the

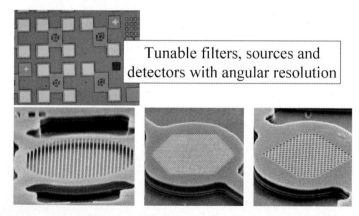

Fig. 2.13 New class of MOEMS devices: structures including several InP membranes suspended in air, with a 1D and 2D PhC formed in the top membrane (SEM view)

traditional Bragg reflector is replaced by PhC membrane reflector has been reported both in the GaAs and InP systems [17].

2.5 Conclusion

The flow of innovations whose threshold has been initiated in the late 1980 by the introduction of the concept of photonic crystals [1,4] is still very close to its source and will inflate in the future to an extent which is certainly beyond our full consciousness: it was simply proposed to extend to the three dimensions of space the field of optics, which was rather confined, yet with very successful outcomes, to the one dimensional world of multilayer optical structures. It is now established that the emergence 3D microphotonics based on full 3D PhC will be significantly delayed, as a result of technological constraints. We hope that the reader will have been convinced that, on the other hand, 2D PhC are fully engaged in the process of innovation and that we are living, in that respect, a true microphotonic revolution. We have shown, in particular for the so called 2.5D microphotonics, where 2D PhC are deliberately opened to the third dimension of space, convincing demonstrations of their ability to generate, in the short run, a wide range of photonic devices ("killer applications"), combining compactness, spatial (angular) and spectral resolution, and whose fabrication meets the standards of the planar technology, familiar to the world of microelectronics.

It appears that the rising trajectory of photonic crystals will not be inhibited in the long range, provided that appropriate tools are made available for their evolution. In that respect, bottlenecks are still to be eliminated and important R&D will have to be deployed for that purpose: this is true for the modelling and design aspects (especially 3D), whose fast and convivial tools are yet to be built; the technological constraints, dictated by the necessity to control the size of the devices at the nanometre scale, are far from being overcome. From the latter point of view, it can be stated that one has really entered within the *Nanophotonic* era.

Acknowledgements This work was partly supported by the EC FP6 network of excellence ePIXnet.

References

1. Yablonovitch E., *Phys. Rev. Lett.*, 63, 2059, (1987).
2. Gérard J.-M. and Gayral B., *IEEE Journal of Lightwave Technology*, 17, 2089, (1999).
3. Letartre X., Mouette J., Seassal C., Rojo-Romeo P., Leclercq J.-L. and Viktorovitch P., *Journal of Lightwave Technology*, 21, 1691, (2003).
4. John S., *Phys. Rev. Lett.*, 58, 2486, (1987).
5. Monat C., Seassal C., Letartre X., Viktorovitch P., Regreny P., Gendry M., Rojo-Romeo P., Hollinger G., Jalaguier E., Pocas S. and Aspar B., *Electron. Lett.*, 37, 764, (2001).
6. Astratov V.N., Whittaker D.M., Culshaw L.S., Stevenson R.M., Skolnick M.S., Krauss T.F. and De La Rue R.M., *Phys. Rev. B*, 60 (24), R16255, (1999).

7. Boutami S., Ben Bakir B., Leclercq J.-L., Letartre X., Rojo-Romeo P., Garrigues M. and Viktorovitch P., Sagnes I., Legratiet L., and Strassner M., *Photonic Technology Letters*, 18, 835, (2006).
8. Mouette J., Seassal C., Letartre X., Rojo-Rome P., Leclercq J.L., Regreny P., Viktorovitch P., Jalaguier E., Perreau P. and Moriceau H., *Electron. Lett.*, 39, 526, (2003).
9. Cojocaru C., Raineri F., Monnier P., Seassal C., Letartre X., Viktorovitch P., Levenson A. and Raj R., *Appl. Phys. Lett.,* 85, 1880, (2004).
10. Raineri F., Vecchi G., Yacomotti A.M., Seassal C., Viktorovitch P., Raj R. and Levenson A., *Appl. Phys. Lett.*, 86, 011116, (2005).
11. Raineri F., Cojocaru C., Raj R., Monnier P., Levenson A., Seassal C., Letartre X. and Viktorovitch P., *Optics Letters*, 30, 64, (2005).
12. Raineri F., Vecchi G., Cojocaru C., Yacomotti A. M., Seassal C., Letartre X., Viktorovitch P., Raj R. and Levenson A., *Appl. Phys. Letters*, 86, 091111, (2005).
13. Ben Bakir B., Seassal C., Letartre X., Regreny P., Gendry M., Viktorovitch P., Zussy M., Di Cioccio L. and Fedeli J.-M., *Optics Express*, 14, 9269, (2006).
14. Ben Bakir B., Seassal C., Letartre X., Viktorovitch P., Zussy M., Di Cioccio L. and Fedeli J.-M., *Appl. Phys. Lett.*, 88, 081113, 2006 and *Virtual Journal on Nanoscale Science & Technology*, 13, no. 10, (2006).
15. Yacomotti A.M., Raineri F., Vecchi G., Raj R., Levenson A., Ben Bakir B., Seassal C., Letartre X., Viktorovitch P., Di Cioccio L. and Fedeli J.-M., *Appl. Phys. Lett.*, 88, 231107, (2006).
16. Boutami S., Ben Bakir B., Leclercq J.-L., Letartre X., Rojo-Romeo P., Garrigues M., Viktorovitch P., Sagnes I., Legratiet L. and Strassner M., *Optics Express*, 14, 3129, (2006).
17. Boutami S., Ben Bakir B., Regreny P., Leclercq J.-L. and Viktorovitch P., *Electronics Letters*, 43, 282, (2007).

Part II
Nonlinear Optics in Photonic Crystals

3 Quasi Phase Matching in Two-Dimensional Quadratic Nonlinear Photonic Crystals

Ady Arie, Nili Habshoosh, and Alon Bahabad

Dept. of Physical Electronics, School of Electrical Engineering, Tel-Aviv University, Tel-Aviv 69978, Israel

Abstract. We analyze quasi-phase-matched conversion efficiency of the five possible types of periodic two-dimensional nonlinear structures: Hexagonal, square, rectangular, centred rectangular and oblique. The frequency conversion efficiency, as a function of the two-dimensional quasi-phase-matching order, is determined for the general case. Furthermore, it is demonstrated for the case of a circular motif. This enables to determine the optimal motif radius for achieving the highest conversion efficiency. As an example for experimental techniques, we discuss the fabrication and nonlinear optical characterization of a rectangularly-poled stoichiometric LiTaO$_3$ crystal.

Key words: Quasi phase matching, nonlinear photonic crystals, nonlinear frequency conversion, second harmonic generation

3.1 Introduction

Quadratic nonlinear photonic crystals are materials in which the second order susceptibility $\chi^{(2)}$ is modulated in an ordered manner, while the linear susceptibility remain constant. They are fundamentally different from the more common linear photonic crystal, in which the linear susceptibility $\chi^{(1)}$ is modulated. One dimensional quadratic nonlinear photonic crystals have been studied for many years, in particular for quasi-phase-matched (QPM) frequency conversion. In recent years, since the introduction of two-dimensional nonlinear photonic crystals (2D NLPhC) by Berger [1], there has been a growing interest in these structures and in their potential applications. They were studied for non-collinear second harmonic generation (SHG) [2], for simultaneous wavelength interchange [3], for third and fourth harmonic generation [4, 5], and proposed for realization of all optical effects, e.g. all optical deflection and splitting [6]. Various methods have been tested for fabrication of 2D NLPhC, including electric field poling [2, 3, 7, 8] and electron beam irradiation [9] of LiNbO$_3$. Recently, two-dimensional nonlinear structures with quasi-periodic modulation have been analyzed [10] and experimentally demonstrated [11–13].

As is well known from solid-state physics [14], 2D periodic structures can be classified by five Bravais lattices: hexagonal, square, rectangular, centred rectangular and oblique, as can be seen in Fig. 3.1. Also shown is the honeycomb lattice, Fig. 3.1f, which is not a Bravais lattice. In order to convert a lattice into a 2D NLPhC, we convolve each one of the lattice points with a nonlinear motif – i.e. some geometrical shape in which the nonlinear coefficient sign is different from the background sign. For example, in ferroelectric crystals, e.g. $LiNbO_3$ and $LiTaO_3$, the sign of the nonlinear coefficient can be locally inverted by a domain reversal. In the first experimental demonstration of 2D NLPhC by Broderick [2], hexagonal lattice was used in $LiNBO_3$, and the motif was also hexagonal. Usually, in other materials, like glasses [15], only the motif is poled thus having a non-zero nonlinear coefficient, whereas the remaining background is un-poled and therefore has a zero nonlinear coefficient.

In order to optimally design and use a 2D NLPhC, one needs to know the conversion efficiency for any given lattice, its dependence on the shape and size of the motif, and its dependence on the QPM orders. Although part of the cases have been discussed in previous works [1, 16], in this work we provide a systematic study on the efficiency dependence of all the structure parameters. We calculate the conversion efficiency of structures made of the five different Bravais lattices and explicitly analyze it for a circular motif. Analysis of additional motifs, e.g. rectangular [17], triangular and hexagonal [18] motifs has been also performed by us recently. Another recent work discussed the effect of elliptical motif, as well as motif orientation,

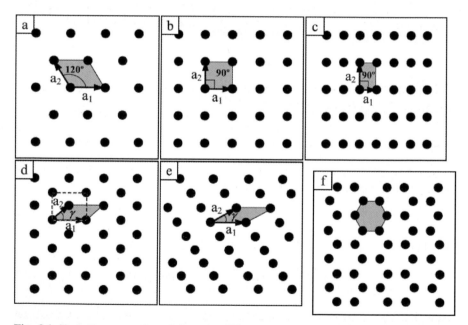

Fig. 3.1 Six lattice types, five of them are different types of Bravais lattices: (**a**) hexagonal, (**b**) square, (**c**) Rectangular, (**d**) centered-rectangular, where the *dashed lines* form a rectangle, (**e**) oblique. Panel (**f**) is a honeycomb lattice. The *gray area* in each lattice refers to the unit cell area

but only in the case of a square lattice [19]. We also determine the optimal radius of the circular motif for some QPM orders in each lattice.

The conversion efficiency for two of the five possible lattices (hexagonal and square), for the case of a circular motif was previously studied by Wang and Gu [16], with identical results for the present analysis of square lattice, but some differences appear for the case of hexagonal lattice. In addition they also analyze a triangular structure. We note that in the photonic crystals community, a "triangular" lattice is what we refer to as hexagonal lattice, while "hexagonal" or "honey-comb" lattice is a hexagonal lattice with a missing point in the middle of each hexagon, as shown in Fig. 3.1f. This "honey-comb" lattice was recently analyzed using the tools presented in this chapter [18], by writing the lattice as the difference between one hexagonal lattice with a base vector of length a, and a second hexagonal lattice with a base vector of length $\sqrt{3}a$. Our results for the hexagonal lattice are different than those of [16] for the "triangular" as well as the "hexagonal" lattices.

A general analysis of quasi-phase-matched interactions in 2D NLPhC is given in Sect. 3.2. This includes a mathematical description of the real and reciprocal lattice, and analysis of the effect of the lattice, motif and interaction area on the generated electric field and intensity. In Sect. 3.3, the normalized conversion efficiency of the five 2D Bravais lattices with a circular motif is described, followed by a specific study of the motif optimum radius for some specific QPM orders. Part of this work was presented in a recent journal publication [17]. An experimental example on production and characterization of two-dimensionally poled stoichiometric LiTaO$_3$ nonlinear photonic crystal is given in Sect. 3.4. The results are discussed and summarized in Sect. 3.5.

3.2 General Analysis of a Periodic Two-Dimensional Nonlinear Photonic Crystal

3.2.1 The Real Lattice

A two-dimensional nonlinear lattice is defined by two primitive, non-parallel vectors a_1, a_2, so that each lattice point is given by

$$r_{mn} = ma_1 + na_2 . \tag{3.1}$$

The lattice is represented by a set of distributed Dirac delta functions:

$$u(r) = \sum_{m,n} \delta(r - r_{mn}) = \sum_{m,n} \delta(r - ma_1 - na_2) . \tag{3.2}$$

Note that this is a generalization of the more common one-dimensional periodic lattice, which is represented by $u(x) = \sum_m \delta(x - m\Lambda\hat{x})$, where Λ is the 1D period. The lattice can be converted into a nonlinear photonic crystal by convolving the lattice points with a suitable nonlinear optical motif. In the 1D case, the motif is a strip having a nonlinear coefficient with a different sign that the background, whereas in the 2D lattice, we can have for example a circular pattern, defining a positive value for the nonlinear coefficient, centred at each one of the lattice points. These motifs are surrounded by a background with a different nonlinear coefficient. The pattern outside the circular motifs, may be linear (zero nonlinearity, as for example is the case for patterns made of poled and un-poled glass), or, may have an opposite sign of the nonlinear coefficient (as is the case in domain inverted ferroelectric crystals). For the case of circular motif, the motif function $s(\boldsymbol{r})$ is given by

$$s(\boldsymbol{r}) = \mathrm{circ}\left(\frac{r}{R}\right) \equiv \begin{cases} 1 & r < R \\ 0 & \text{elsewhere} \end{cases} \tag{3.3}$$

where $r = |\boldsymbol{r}|$. Other shapes that were studied for use in nonlinear photonic crystals include rectangular [17], triangular, hexagonal [18] and elliptical [19] motifs.

If the background is nonlinear (with opposite sign to the motif nonlinear coefficient) the motif function has values of 1 and (-1) instead of 1 and 0. This amount in a DC shift in the Fourier transform of the overall structure function (the result, as implied through Sect. 3.2.3, is doubling of the electric field conversion efficiency for a QPM process). To simplify the following analysis we shall assume first that the background has a zero nonlinear coefficient, and later (in Sect. 3.3) adjust the final results in order to account for non-zero background.

The lattice area is restricted by the crystal physical size and by an effective interaction area. Let us denote the area function as $a(\boldsymbol{r})$. Let us assume that the area function is rectangular with a length L and width W. In that case, the area function is:

$$a(\boldsymbol{r}) = a(x, y) = \mathrm{rect}\left(\frac{x}{L}\right) \cdot \mathrm{rect}\left(\frac{y}{W}\right) . \tag{3.4}$$

The relevant Cartesian component of the nonlinear dielectric tensor as a function of position in a nonlinear photonic crystal can be therefore expressed mathematically as

$$\chi_{ij}^{(2)}(\boldsymbol{r}) = 2d_{ij} \times g(\boldsymbol{r}) = 2d_{ij} \times a(\boldsymbol{r}) \times [u(\boldsymbol{r}) \otimes s(\boldsymbol{r})] \tag{3.5}$$

where d_{ij} is the value of the nonlinear susceptibility tensor for the Cartesian indices i and j, and $\otimes$ is the convolution operator. $g(\boldsymbol{r})$ is a normalized and dimensionless function that represents the space dependence of the nonlinear coefficient function.

3.2.2 The Reciprocal Lattice

Similar to the analysis of crystals in solid-state physics [14], it is useful to define a reciprocal lattice for $u(\boldsymbol{r})$ using two primitive vectors that together with the direct lattice primitive vectors obey the following orthogonality relation:

$$\boldsymbol{a}_i \cdot \boldsymbol{b}_j = 2\pi \delta_{ij} . \tag{3.6}$$

For the 1D case, the lattice points are given by $\boldsymbol{b} = (2\pi/\Lambda)\hat{x}$, whereas in the 2D case, they are given by

$$\boldsymbol{K}_{mn} = m\boldsymbol{b}_1 + n\boldsymbol{b}_2 . \tag{3.7}$$

The reciprocal lattice function is the two-dimensional Fourier transform of the direct (or "real") lattice function:

$$U(f) = \frac{1}{A_{UC}} \sum_{m,n} \delta\left(f - \frac{mb_1}{2\pi} - \frac{nb_2}{2\pi}\right) = \frac{(2\pi)^2}{A_{UC}} \sum_{m,n} \delta(K - mb_1 - nb_2)$$

$$= \frac{(2\pi)^2}{A_{UC}} \sum_{m,n} \delta(K - K_{mn}) . \tag{3.8}$$

Here $A_{UC} = |a_{1x}a_{2y} - a_{1y}a_{2x}|$ is the area of the unit cell [20], f is the spatial frequency in the two-dimensional Fourier space, $K = 2\pi f$ and $a_1 = (a_{1x}, a_{1y})$, $a_2 = (a_{2x}, a_{2y})$.

For the following discussion, it is also useful to calculate the Fourier transform of $g(r)$:

$$G(f) = FT\{g(r)\} = U(f) \otimes A(f) \times S(f) , \tag{3.9}$$

where $A(f)$ and $S(f)$ are the Fourier transform functions of the area function and motif function, respectively. For some specific motif functions, $S(f)$ is known analytically. For example, in the case of circular motif of radius R, the Fourier transform is:

$$S(f) = \frac{R}{f} J_1(2\pi R f), \quad f = |f| = (f_x^2 + f_y^2)^{1/2} . \tag{3.10}$$

The effects of other shapes of motifs were studied in [17–19]. Similarly, the Fourier transform of a rectangular area function of dimensions $L \times W$ is:

$$A(f) = A(f_x, f_y) = LW \sin c(f_x L) \sin c(f_y W) . \tag{3.11}$$

For the case of infinite area, (L, W much larger than the unit cell dimensions), the space-dependent part of the nonlinear susceptibility can be expressed as the following sum:

$$g(r) = \sum_{m,n} G_{mn} \exp(-iK_{mn} \cdot r) . \tag{3.12}$$

This equation shows clearly the relation between the reciprocal lattice vectors and the nonlinear susceptibility as a function of space.

In this case the Fourier coefficient becomes

$$G_{mn} = \frac{1}{A_{UC}} S \frac{K_{mn}}{2\pi} . \tag{3.13}$$

As will be shown in the next section, the conversion efficiency in a 2D NLPhC is proportional to $|G_{mn}|^2$. Equation (3.16) shows the combined effect of the lattice (through the unit cell area), the motif (through its Fourier transform function S) and the QPM orders m, n on the nonlinear process.

3.2.3 Wave Equations in 2D NLPhC

We consider now the case of second harmonic generation in a 2D NLPhC. The results shown here can be easily generalized to other second-order nonlinear processes, e.g. sum frequency generation and difference frequency generation. We assume that a plane wave at the frequency ω propagates in the transverse plane of a 2D NLPhC. This wave generates a second harmonic wave owing to the second order susceptibility of the material. We assume that the fundamental frequency is linearly polarized along one of the 2D NLPhC axes, and we concentrate only in a specific linear polarization of the generated second harmonic wave. The coupling between the two beams is given by the appropriate element of the nonlinear susceptibility tensor d_{ij}.

Let us write the relevant component of the second harmonic wave as

$$\tilde{E}_{2\omega}(r,t) = \tfrac{1}{2} E_{2\omega}(r) \exp\left[i\left(2\omega t - k_{2\omega} \cdot r\right)\right] + \text{c.c.} \tag{3.14}$$

We assume that the nonlinear conversion efficiency is low, hence the pump amplitude can be assumed constant throughout the entire interaction length (non-depletion approximation). We further assume that the slowly varying envelope approximation applies for the second harmonic wave. Under these assumptions it can be shown that:

$$k_{2\omega} \cdot \nabla E_{2\omega}(r) = -i\frac{\omega^2}{c^2} E_\omega^2 \chi_{ij}^{(2)}(r) \exp\left[i\left(k_{2\omega} - 2k_\omega\right) \cdot r\right] . \tag{3.15}$$

If the 2D NLPhC is large enough we can use Eq. (3.12) for the second order susceptibility, and so:

$$k_{2\omega} \cdot \nabla E_{2\omega}(r) = -2i\frac{\omega^2}{c^2} E_\omega^2 d_{ij} \sum_{mn} G_{mn} \exp\left[i\left(k_{2\omega} - 2k_\omega - K_{mn}\right) \cdot r\right] . \tag{3.16}$$

Note that these results can be easily adapted for the 1D case, by simply replacing the 2D Fourier coefficients and reciprocal lattice vectors by their 1D counterpart. Significant build-up of the second harmonic wave requires phase matching, i.e. $k_{2\omega} - 2k_\omega - K_{mn} \approx 0$.

This phase-matching condition is just a crystal-momentum conservation law: the required momentum balance for the interaction is accomplished through a reciprocal lattice vector (RLV). Usually we can assume that if the phase matching condition is achieved by some order (m,n), it would be the only order which contributes to the build-up of the second harmonic while all the other orders contributes negligible oscillating terms.

This process can also be analyzed in Fourier space by integrating Eq. (3.15) above over a rectangular area of width W and length L (see an example [21]). The result is the second harmonic amplitude after an interaction length of L:

$$E_{2\omega}(\Delta k) = \frac{-2i\omega^2 E_\omega^2 d_{ij}}{k_{2\omega} c^2 W} \iint\limits_A g(r) \exp\left(-i\Delta k \cdot r\right) da , \tag{3.17}$$

where $\Delta k = k_{2\omega} - 2k_\omega$ is the phase-mismatch vector. If the integration area is given through the function $a(r)$ defined in Sect. 3.2.1, we can use $g(r) = a(r) \times (u(r) \otimes s(r))$ and set the integration limits to infinity and so:

$$E_{2\omega}(\Delta k) = \frac{\kappa}{W} G(\Delta k) \tag{3.18}$$

where $G(\Delta \mathbf{k})$ is just the two-dimensional Fourier transform of $g(\mathbf{r})$ and κ is a constant defined as:

$$\kappa = \frac{-2i\omega^2 E_\omega^2 d_{ij}}{k_{2\omega} c^2} = \frac{-i\omega E_\omega^2 d_{ij}}{n_{2\omega} c} .$$ (3.19)

From Eq. (3.18) we can see that the field amplitude evaluation for some specific phase mismatch value $\Delta \mathbf{k} = \Delta \mathbf{k}_0$ is proportional to $|G(\Delta \mathbf{k}_0)|$. If phase matching condition is achieved by some order (m,n) then the integral above is dominated by this order and we can write:

$$E_{2\omega}(\Delta \mathbf{k} \approx \mathbf{K}_{mn}) \cong \kappa L G_{mn} \exp\left[-i\left(\frac{\Delta k_{mn,x} L}{2} + \frac{\Delta k_{mn,y} W}{2}\right)\right]$$
$$\times \operatorname{sinc}\left(\frac{\Delta k_{mn,x} L}{2\pi}\right) \operatorname{sinc}\left(\frac{\Delta k_{mn,y} W}{2\pi}\right) ,$$ (3.20)

where $\Delta \mathbf{k}_{mn} = \Delta \mathbf{k} - \mathbf{K}_{mn} = \Delta k_{mn,x}\hat{x} + \Delta k_{mn,y}\hat{y}$.

For perfect quasi-phase-matching $\Delta k_{mn,x} = \Delta k_{mn,y} = 0$ and so:

$$E_{2\omega}(\Delta \mathbf{k} = \mathbf{K}_{mn}) \cong \kappa L G_{mn} .$$ (3.21)

The fundamental and second harmonic amplitudes are related to the corresponding intensities by:

$$I_\omega = \frac{1}{2} n_\omega \sqrt{\frac{\varepsilon_0}{\mu_0}} |E_\omega|^2 , \quad I_{2\omega} = \frac{1}{2} n_{2\omega} \sqrt{\frac{\varepsilon_0}{\mu_0}} |E_{2\omega}|^2 .$$ (3.22)

Hence, the intensity of the second harmonic for the case of perfect quasi-phase-matching after an interaction length L is:

$$I_{2\omega}(L) = \frac{2\omega^2 d_{ij}^2 |G_{mn}|^2}{n_{2\omega} n_\omega^2 c^3 \varepsilon_0} I_\omega^2 L^2$$ (3.23)

and so the interaction efficiency is proportional to the absolute square of the relevant Fourier coefficient $|G_{mn}|^2$. We shall now relate to $|G_{mn}|^2$ as the normalized efficiency.

3.3 Conversion Efficiency for Specific Types of 2D Periodic Structures

3.3.1 General Expressions for Fourier Coefficients

Without loss of generality, we can define the coordinate system so that one of its axes is in the same direction as that of the $\mathbf{a}_1$ primitive vector. Denoting the angle between the two primitive vectors as γ, the primitive vectors of the real and reciprocal lattice are:

$$\boldsymbol{a}_1 = (a_1, 0), \quad \boldsymbol{a}_2 = (a_2 \cos \gamma, a_2 \sin \gamma)$$

$$\boldsymbol{b}_1 = \frac{2\pi}{a_1}\left(1, -\frac{1}{\tan \gamma}\right), \quad \boldsymbol{b}_2 = \frac{2\pi}{a_2}\left(0, \frac{1}{\sin \gamma}\right). \qquad (3.24)$$

We shall now analyze the conversion efficiency of the five possible 2D lattices. For each one of them we present in Table 3.1 the primitive vectors, reciprocal lattice vectors (RLVs) and the unit cell areas.

As was shown in the previous section, the nonlinear conversion efficiency depends on the Fourier coefficient G_{mn} of the normalized nonlinear susceptibility. These Fourier coefficients are presented in Table 3.2 for each one of the lattice types in the case of circular motif. Nearly circular motifs have been demonstrated experimentally in two-dimensionally poled LiNbO$_3$ [7, 9]. A numerical optimization procedure for general shapes of motifs was given by Norton and de-Sterke [22], however its experimental implementation requires to modulate the nonlinear coefficient with high spatial resolution.

Table 3.1 Primitive vectors, RLVs and unit cell area for the five lattice types

Lattice types	Primitive vectors	RLVs	Unit cell area
Hexagonal $\gamma = 120°$	$\boldsymbol{a}_1 = (a, 0)$ $\boldsymbol{a}_2 = a\left(-\frac{1}{2}, \frac{\sqrt{3}}{2}\right)$	$\boldsymbol{b}_1 = \frac{2\pi}{a}\left(1, \frac{1}{\sqrt{3}}\right)$ $\boldsymbol{b}_2 = \frac{4\pi}{a\sqrt{3}}(0, 1)$	$A_{\text{UC}} = \frac{a^2\sqrt{3}}{2}$
Square $\gamma = 90°$	$\boldsymbol{a}_1 = (a, 0)$ $\boldsymbol{a}_2 = (0, a)$	$\boldsymbol{b}_1 = \frac{2\pi}{a}(1, 0)$ $\boldsymbol{b}_2 = \frac{2\pi}{a}(0, 1)$	$A_{\text{UC}} = a^2$
Rectangular $\gamma = 90°$	$\boldsymbol{a}_1 = (a_1, 0)$ $\boldsymbol{a}_2 = (0, a_2)$	$\boldsymbol{b}_1 = \frac{2\pi}{a_1}(1, 0)$ $\boldsymbol{b}_2 = \frac{2\pi}{a_2}(0, 1)$	$A_{\text{UC}} = a_1 a_2$
Centered-rectangular $\gamma \in [0°, 90°]$	$\boldsymbol{a}_1 = (a, 0)$ $\boldsymbol{a}_2 = a\left(\frac{1}{2}, \frac{1}{2}\tan \gamma\right)$	$\boldsymbol{b}_1 = \frac{2\pi}{a}\left(1, \frac{-1}{\tan \gamma}\right)$ $\boldsymbol{b}_2 = \frac{4\pi}{a}\left(0, \frac{1}{\tan \gamma}\right)$	$A_{\text{UC}} = \frac{a^2}{2}\tan \gamma$
Oblique $\gamma \in [0°, 180°]$	$\boldsymbol{a}_1 = (a_1, 0)$ $\boldsymbol{a}_2 = a_2(\cos \gamma, \sin \gamma)$	$\boldsymbol{b}_1 = \frac{2\pi}{a_1}\left(1, \frac{-1}{\tan \gamma}\right)$ $\boldsymbol{b}_2 = \frac{2\pi}{a_2}\left(0, \frac{1}{\sin \gamma}\right)$	$A_{\text{UC}} = a_1 a_2 \sin \gamma$

Table 3.2 Fourier coefficient of a circular motif

Lattice types	Fourier coefficient of a circular motif
Hexagonal	$$G_{mn} = \frac{2R}{a\sqrt{m^2+n^2+mn}} J_1\left(\frac{4\pi R}{a\sqrt{3}}\sqrt{m^2+n^2+mn}\right)$$
Square	$$G_{mn} = \frac{2R}{a\sqrt{m^2+n^2}} J_1\left(\frac{2\pi}{a}R\sqrt{m^2+n^2}\right)$$
Rectangular	$$G_{mn} = \frac{2R}{\sqrt{(ma_2)^2+(na_1)^2}} J_1\left[2\pi R\sqrt{\left(\frac{m}{a_1}\right)^2+\left(\frac{n}{a_2}\right)^2}\right]$$
Centered-rectangular	$$G_{mn} = \frac{2R\cdot 2\cos\gamma}{a\sqrt{m^2+4n^2\cos^2\gamma-4mn\cos^2\gamma}}$$ $$\times J_1\left(\frac{2\pi R}{a\sin\gamma}\sqrt{m^2+4n^2\cos^2\gamma-4mn\cos^2\gamma}\right)$$
Oblique	$$G_{mn} = \frac{2R}{\sqrt{a_1a_2}\sqrt{\frac{m^2a_2}{a_1}+\frac{n^2a_1}{a_2}-2mn\cos\gamma}}$$ $$\times J_1\left(\frac{2\pi R}{\sin\gamma\sqrt{a_1a_2}}\sqrt{\frac{m^2a_2}{a_1}+\frac{n^2a_1}{a_2}-2mn\cos\gamma}\right)$$

According to Eq. (3.10), the Fourier transform of a circular motif with a radius R, depends on a 1st order Bessel function. In Table 3.2 we present the Fourier coefficients for all five Bravais lattice types with a circular motif as function of R, the primitive vectors magnitude and the QPM orders (m,n).

Note that the Fourier coefficients in Table 3.2 are suitable for the case in which the background has an opposite nonlinear coefficient with respect to the motif. This is the case for domain-inverted ferroelectrics. If the background has zero-nonlinearity, the Fourier coefficients in the two tables should be multiplied by $1/2$.

3.3.2 Efficiency for Specific QPM Orders

We shall now examine in more detail the efficiency as a function of the ratio between the motif size and the primitive vectors magnitude, for specific QPM orders. This enables to determine the highest possible efficiency for a given structure and the required dimensions and shape of the motif. Furthermore, it allows determining motif shapes that will completely null the nonlinear conversion efficiency (which, for example, can be useful to nullify unwanted processes). We concentrate on three out of the five lattice types, namely the hexagonal, square and rectangular lattices. For the other two lattices (centred rectangular and oblique), the motif dimension and the efficiency will depend on the angle γ between the two primitive vectors.

We chose two specific QPM orders: $(m,n) = (1,0)$ and $(m,n) = (1,1)$. The first one is usually the most efficient process in a 2D nonlinear structure, however it relies on only one of the two primitive vectors. The second order is usually the most efficient process that relies on both primitive vectors although, as shown below, in some cases one may get higher efficiency with a different choice of (m,n) order.

The normalized efficiency is given as a function of the ratio between the circle radius and a primitive vector length. We limit the motif size, in order to avoid overlap between motifs of adjacent lattice points. For each lattice, this sets the maximum allowed motif size. For the hexagonal and square lattices, the radius of the circular motif has possible values from zero to half the primitive vector magnitude a, whereas for the rectangular lattice, the radius of the circular motif has possible values from zero to half the shortest primitive vector, $\boldsymbol{a_2}$.

In the drawings of 2D NLPhCs in Figs. 3.2 to 3.4, the grey areas represent a certain sign of the nonlinear coefficient, whereas the white areas represent the opposite sign. As can be seen, the order $(m = 1, n = 0)$ is more efficient than the $(m = 1, n = 1)$ in all 3 lattices, as expected. In the case of hexagonal lattice, the maximum efficiency (0.118) is achieved when $R/a = 0.331$, whereas for $(m = 1, n = 1)$, the maximum efficiency is 0.03 when $R/a = 0.439$. In the case of a square lattice, the maximum efficiency of the $(1,0)$ order is 0.158 when $R/a = 0.383$, and for the $(1,1)$ order the maximum efficiency is 0.04 when $R/a = 0.271$. Note that in both cases, the square lattice is more efficient than the hexagonal lattice.

For a rectangular lattice, and for the $(1,0)$ QPM order, the Fourier coefficient is composed of two independent functions:

$$G_{10} = \frac{2R}{a_2} \cdot J_1 \left(\frac{2\pi R}{a_1} \right) . \tag{3.25}$$

The left expression on the right hand side depends only on R/a_2, whereas the right expression is a Bessel function which depends only on R/a_1. The maximum value of G_{10} is therefore achieved when R/a_2 is set to the maximum value of 0.5. In this case, the motifs of adjacent lattice points along the $\boldsymbol{a_2}$ direction are just touching each

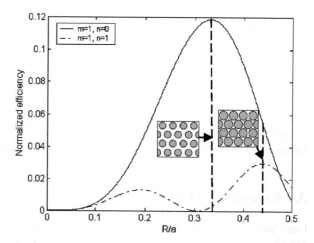

Fig. 3.2 Normalized efficiency of a hexagonal lattice with circular motif as a function of the motif radius to primitive vector magnitude ratio. *Solid* and *dash-dot lines* represent the efficiency curves for QPM orders $(1,0)$ and $(1,1)$ respectively. The *insets* show the shape of the 2D NLPhC for R/a ratios that maximize the $(1,0)$ and $(1,1)$ QPM processes

Fig. 3.3 Normalized efficiency of a square lattice with circular motif as a function of the motif radius to primitive vector magnitude ratio. *Solid* and *dash-dot lines* represent the efficiency curves for QPM orders $(1,0)$ and $(1,1)$ respectively. The *insets* show the shape of the 2D NLPhC for R/a ratios that maximize the $(1,0)$ and $(1,1)$ QPM processes

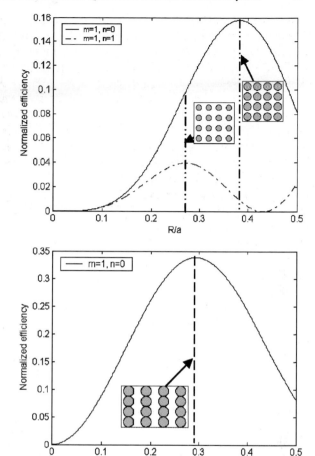

Fig. 3.4 Normalized efficiency of a rectangular lattice with circular motif for $m = 1$, $n = 0$, $R/a_2 = 0.5$ as a function of the motif radius to primitive vector magnitude ratio. The *inset* shows the 2D NLPhC for a specific R/a_1 ratio that maximizes the efficiency

other. Figure 3.4 shows the normalized efficiency $|G_{10}|^2$ as function of R/a_1, when $R/a_2 = 0.5$. Note that in this case, the rectangular lattice is the most efficient structure, and provides significantly higher efficiency with respect to the hexagonal and rectangular lattice. Also note that the shape of the optimal lattice (for phase matching a single process), see inset of Fig. 3.4, is similar to that of a one-dimensional periodic structure. If a rectangular motif is used instead of a circular motif, the optimal lattice shape becomes identical to a one-dimensional structure [17].

For the $(1,1)$ order, the maximum efficiency is similar to the case of a square lattice ($|G_{11}|^2 = 0.04$ when $R/a_1 = R/a_2 = 0.271$).

3.4 Experimental Results

One dimensional nonlinear photonic crystals, as well as two-dimensional nonlinear photonic crystals are usually prepared by electric field poling of ferroelectric

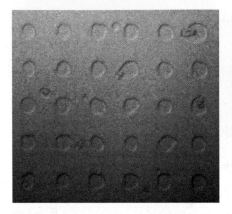 

Fig. 3.5 Microscope picture of the C+ (*left*) and C− (*right*) planes of a two-dimensional nonlinear photonic crystal with a rectangular lattice. The lengths of the two base vectors are 8.5 μm and 9.4 μm, respectively

crystals. Other methods, such as poling by electron-beam [9] or atomic force microscope [23] have also been studied, and have shown to provide very high resolution. However, in these methods, the time needed to pole a reasonable area (say 10 mm^2) is quite long, and in addition it is hard to maintain the exact period. Hence, most demonstrations of nonlinear photonic crystals were made by electric field poling.

In Fig. 3.5 we show a scan of the C+ and C− surfaces of two-dimensionally poled stoichiometric LiTaO$_3$ crystal. The 0.5 mm thick z-cut crystal was grown by the double–crucible Czochralski method at Oxide Co. and doped with 0.5 mol% MgO. A structured pattern of photoresist was contact printed on the C+ face of the sample from a lithographic mask, and uniformly coated by a metallic layer. A rectangular lattice was chosen, with orthogonal base vectors of 8.5 μm and 9.4 μm, respectively. The electrical poling was achieved by applying 1 kV pulses to the crystal surfaces. The total switching charge was 45 μCb/cm^2. The photoresist and the metallic coatings were removed after poling. Although the poling does not change the refractive index of the material, it is possible to distinguish between the positive and negative domains by selectively etching the surface by HF etching.

Nonlinear optical characterization of a nonlinear photonic crystal can be done by measuring the angular dependence of a selected nonlinear process, e.g. second harmonic generation. In Fig. 3.6 we show the dependence of the output angle of the second harmonic of a 780 nm Ti:Al$_2$O$_3$ pump laser, as a function of the pump input angle. The extraordinary polarized pump wave was sent to the crystal that was shown in Fig. 3.5. The crystal was mounted on a rotation stage and held at a temperature of 100 °C. When the crystal wais rotated, at several specific angles in which a reciprocal lattice vector compensated for the phase mismatch of the interactive waves, one observed strong second harmonic signal, as shown for one specific angle in Fig. 3.7. By recording those angles and comparing them to the theoretical

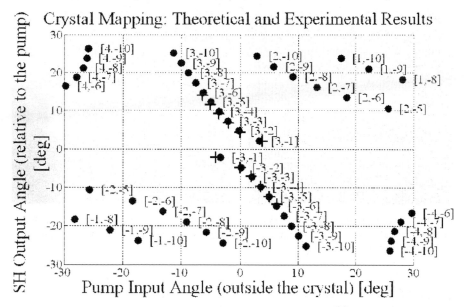

Crystal Mapping: Theoretical and Experimental Results

SH Output Angle (relative to the pump) [deg]

Pump Input Angle (outside the crystal) [deg]

Fig. 3.6 Dependence of the second harmonic output angle as a function of the pump input angle. The theoretical and experimental results are marked by *circles* and *crosses*, respectively. The two values in *parenthesis* are the coefficients of the two reciprocal lattice vectors. This figure is reproduced from the thesis of Nili Habshoosh, Tel-Aviv University (2007)

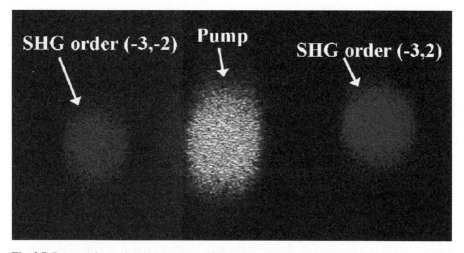

Fig. 3.7 Pump and two simultaneous second harmonic signals from rectangularly poled stoichiometric LiTaO$_3$ crystal. The pump enters nearly parallel to the $(1,0)$ RLV, and two second harmonic beams correspond to the $(-3,2)$ and $(-3,-2)$ QPM orders. This figure is reproduced from the thesis of Nili Habshoosh, Tel-Aviv University (2007)

expectation, one can evaluate the quality of the crystal, and retrieve the reciprocal lattice [9]. The conversion efficiency, as well as spectral and thermal tolerances are obtained by adjusting the input angle of the fundamental to match a specific QPM order, and measuring the second harmonic power as a function of the fundamental power, fundamental wavelength and crystal temperature.

3.5 Conclusion

In this chapter, we have presented a general analysis of quasi-phase-matched conversion efficiency in all possible periodic two-dimensional structures. This analysis enables to design and analyze structures for two-dimensional quasi-phase-matched frequency conversion, in any given periodic lattice and in any chosen QPM order.

We have shown that the conversion efficiency for optical intensities of a QPM process of some (m, n) order scales as $|G_{mn}|^2$. The value of the Fourier coefficient G_{mn} depends both on the choice of the periodic lattice and on the shape and dimensions of the nonlinear motif. A general expression that includes the effects of lattice, motif and QPM order was derived in Eq. (3.13). The conversion efficiency is also determined by the interaction area, defined by the 2D NLPhC area and beam parameters. Specifically, we have analyzed a circular motif, but other motifs have been studied in [17–19]. The values of G_{mn} for all five possible periodic lattice structures are given in Table 3.2.

Some specific examples of conversion efficiency were considered for two QPM orders, namely $(m, n) = (1, 0)$ and $(m, n) = (1, 1)$, for the cases of hexagonal, square and rectangular lattices. The rectangular lattice is the best choice in terms of efficiency for the $(1, 0)$ order, and the optimal condition is such that the circular motifs of adjacent lattice points just touch one another.

An example of experimental fabrication and characterization of nonlinear photonic crystals was given in Sect. 3.4. Currently, the method of electric field poling, which was used to fabricate the stoichiometric LiTaO$_3$ crystal shown in Sect. 3.4, is the most mature technology for making nonlinear photonic crystals. Since it is based on planar techniques, it enables to modulate the nonlinear coefficient in either one or two dimensions. It is desirable to reach three-dimensional modulation as well, which will open new possibilities for applications such as nonlinear beam shaping [24] and generation of vortex beam by nonlinear wave-mixing [25]. A possible solution for constructing such devices is by electric-field poling of ferroelectric materials into thin 2D-nonlinear modulated planar plates and bonding them together [25].

Acknowledgements This work was partly supported by the Israel Science Foundation, grant no. 960/05, by the Israeli Ministry of Science, Culture and Sport and by COST Action P11: Physics of Linear, Nonlinear and Active Photonic Crystals.

References

1. Berger V.: Nonlinear photonic crystals. *Phys. Rev. Lett.* 81, 4136–4139 (1998).
2. Broderick N.G.R., Ross G.W., Offerhaus H.L., Richardson D.J. and Hanna D.C.: Hexagonally poled lithium niobate: A two-dimensional nonlinear photonic crystal. *Phys. Rev. Lett.* 84, 4345–4348 (2000).
3. Chowdhury A., Staus C., Boland B.F., Kuech T.F. and McCaughan L.: Experimental demonstration of 1535–1555 nm simultaneous optical wavelength interchange with a nonlinear photonic crystal. *Opt. Lett.* 26, 1353–1355 (2001).
4. Broderick N.G.R., Bratfalean R.T., Monro T.M., Richardson D.J. and de Sterke C.M.: Temperature and wavelength tuning of second-, third-, and fourth-harmonic generation in a two-dimensional hexagonally poled nonlinear crystal. *J. Opt. Soc. Am. B* 19, 2263–2272 (2002).
5. Saltiel S. and Kivshar Y.S.: Phase matching in nonlinear $\chi^{(2)}$ photonic crystals. *Opt. Lett.* 25, 1204–1206 (2000).
6. Saltiel S.M. and Kivshar Y.S.: All-optical deflection and splitting by second-order cascading. *Opt. Lett.* 27, 921–923 (2002).
7. Ni P., Ma B., Wang X., Cheng B. and Zhang D.: Second-harmonic generation in two-dimensional periodically poled lithium niobate using second-order quasiphase matching. *Appl. Phys. Lett.* 82, 4230–4232 (2003).
8. Peng L.H., Hsu C.C. and Shih Y.C.: Second-harmonic green generation from two-dimensional $\chi^{(2)}$ nonlinear photonic crystal with orthorhombic lattice structure. *Appl. Phys. Lett.* 83, 3447–3449 (2003).
9. Glickman Y., Winebrand E., Arie A. and Rosenman G.: Electron-beam-induced domain poling in LiNbO$_3$ for two-dimensional nonlinear frequency conversion. *Appl. Phys. Lett.* 88, 011103 (2006).
10. Lifshitz R., Arie A. and Bahabad A.: Photonic Quasicrystals for Nonlinear Optical Frequency Conversion. *Phys. Rev. Lett.* 95, 133901 (2005).
11. Bratfalean R.T., Peacock A.C., Broderick N.G.R., Gallo K. and Lewen R.: Harmonic generation in a two-dimensional nonlinear quasi-crystal. *Opt. Lett.* 30, 424–426 (2005).
12. Ma B.Q., Wang T., Sheng Y., Ni P.G., Wang Y.Q., Cheng B.Y. and Zhang D.Z.: Quasiphase matched harmonic generation in a two-dimensional octagonal photonic superlattice. *Appl. Phys. Lett.* 87, 251103 (2005).
13. Bahabad A., Voloch N., Arie A. and Lifshitz R.L.: Experimental Confirmation of the General Solution to the Multiple Phase Matching Problem. *J. Opt. Soc. Am. B* 24, 1916–1921 (2007).
14. Kittel C.: *Introduction to Solid State Physics*, 7th edition, Wiley, New York (1995).
15. Myers R.A., Mukherjee N. and Brueck S.R.J.: *Opt. Lett.* 16, 1732 (1991).
16. Wang X.H. and Gu B.Y.: Nonlinear frequency conversion in 2D (2) photonic crystals and novel nonlinear double-circle construction. *Eur. Phys. J. B.* 24 323–326 (2001).
17. Arie A., Habshoosh N. and Bahabad A.: Quasi phase matching in two-dimensional nonlinear photonic crystals. *Optical and Quantum Electronics* 39, 361–375 (2007).
18. Arie A., Bahabad A. and Habshoosh N.: Nonlinear interactions in periodic and quasi-periodic nonlinear photonic crystals, to appear in: P. Ferraro, S. Grilli and P. de Natale (eds.) Micro/nano engineering and characterization of ferroelectric crystals for applications in photonics. Springer, Germany (2008).
19. Li J., Li Z. and Zhang D.: Effects of shapes and orientations of reversed domains on the conversion efficiency of second harmonic wave in two-dimensional nonlinear photonic crystals. *Journal of Applied Physics* 102, 093101 (2007).
20. Giacovazzo C., Monaco H.L., Artioli G., Viterbo D., Ferraris G., Gilli G., Zanotti G. and Catti M.: *Fundamentals of Crystallography*, 2nd edition, University Press, Oxford, (2002).
21. Russel S.M., Powers P.E., Missey M.J. and Schepler K.L.: Broadband mid-infrared generation with two-dimensional quasi-phase-matched structures. *IEEE J Quant. Electron.* 37, 877–887 (2001).
22. Norton A.H. and de Sterke C.M.: Optimal poling of nonlinear photonic crystals for frequency conversion. *Opt. Lett.* 28, 188–190 (2003).

23. Moscovich S., Arie A., Urenski R., Agronin A., Rosenman G. and Rosenwaks Y.: Non-collinear second harmonic generation in sub-micrometer poled RbTiOPO$_4$. *Optics Express* 12, 2236–2242 (2004).

24. Imeshev G., Proctor M. and Fejer M.M.: Lateral patterning of nonlinear frequency conversion with transversely varying quasi-phase-matching gratings. *Opt. Lett.* 23, 673–675 (1998).

25. Bahabad A. and Arie A.: Generation of Optical Vortex Beams by Nonlinear Wave Mixing. *Optics Express* 15, 17619–17624 (2007).

4 Harmonic Generation in Nanostructures: Metal Nanoparticles and Photonic Crystals

Andrea Marco Malvezzi

Dipartimento di Elettronica and CNISM, Università di Pavia
Via Ferrata 1, 27100 Pavia, Italy

Abstract. The potential of nonlinear optical phenomena in adding functionality to nanostructures like nanoparticles or photonic crystal waveguides are presented. The role of field localization and system resonances is discussed in the light of few examples in which harmonic generation is being drastically enhanced by these effects. In particular the results on single nanoparticle harmonic response are discussed as the basic example of nonlinear interaction of intense radiation with a single nanoparticle. Here the driving resonance mechanism of nonlinear interaction is proven to be due to surface plasmon oscillations that give rise to strong harmonic enhancement when proper resonance conditions are established. This response in further enriched when assemblies of nanoparticles are considered next, illustrating the role of the surrounding medium, particle shape and orientation and temperature. Switching from metals to semiconductors leads to exploitation of photonic resonances as opposed to plasmonic ones of previous examples. In photonic planar structures the nonlinear response gains also from the notion of quasi-phase matching introduced by the periodicity of the system and from exploiting leaky (or quasi-guided) optical modes that considerably expand the possibilities for an optimised output. Examples based on second and third harmonic generation from such systems are discussed in details. Finally, potential new routes for exploiting higher harmonic efficiencies or very broad-band resonance conditions are briefly mentioned.

Key words: metal nanoparticles; photonic crystals; harmonic generation; plasmonic resonances; photonic resonances

4.1 Introduction

The research towards compact and efficient nonlinear (NL) photonic devices has greatly improved in recent years in view of potential applications in integrated photonics. The technological requirements are for optimisation of NL processes together with miniaturization and device integration. Given the general structure of the NL interaction between an electrical field E and the material response, in terms of polarization $P = \chi^{(1)}E + \chi^{(2)}: EE + \chi^{(3)}: EEE$, one may act on susceptibilities χ, local fields and propagation in order to enhance a particular NL process. This translates into a research oriented towards new materials and new geometries. In particular, novel effects can be obtained when material properties are modulated at the sub-wavelength scale, that is by exploiting nanostructures. Two different approaches in this respect are represented by metal nanoparticles (NP) and patterned

optical waveguides. They have been extensively studied [1, 2] and their NL properties will be briefly presented and discussed here. Metal NP aggregates show optical
properties that are substantially different from individual isolated particles or bulk
materials. By controlling the aggregation and the geometry (size, shape, distance)
in which NP are organized, a full range of new materials can be assembled. Indeed,
NP linear and nonlinear [3–8] optical properties have been found to depend strongly
on these parameters. A second class of novel NL materials is obtained by modifying optical waveguides by patterning a photonic structure thus changing linear and
nonlinear optical characteristics. In both cases system specific mechanisms for enhancing the optical response enter into play that can be exploited and optimised in
several applications.

In this chapter the accent is on harmonic generation in metal NP systems and
in photonic waveguides. The purpose is here to illustrate the complementary features of the microscopic mechanisms involved, i.e. plasmonic versus photonic resonances. The nanometre size of the relevant elements makes them particularly attractive for integration in compact devices providing novel functionalities. In the
following section the origin of the NL response in the simplest nanostructure, an
isolated nanosphere, is described and the main features of the corresponding second harmonic scattering intensities are briefly reviewed. Random aggregates of NPs
in monolayers are then described in Sect. 4.3. Here, the NL interaction gives rise
to a macroscopic polarization vector and therefore the coherence properties of the
pump are replicated in the harmonic field in reflection and transmission. Section 4.4
is devoted to harmonic generation in patterned waveguides giving rise to a photonic crystal (PhC) structure. Optimisation of harmonic generation can be achieved
in different geometries by the use of quasi-guided modes and multiple resonances
with the photonic structure. Moreover NL diffraction can be exploited and provides
extra harmonic conversion. Finally, recent developments of harmonic generation in
multiple microcavities and studies on broadband phase matching will be briefly described.

4.2 Metallic Nanospheres

Let us consider first a collection of metallic nanospheres dispersed into a liquid,
so that their mutual distance is of the order of several wavelengths. The expected
NL harmonic signal originating from such a system is made by the superposition
of the contributions from the isolated particles, since no correlation between them
occur [9]. For this reason the effect is described by a Hyper Rayleigh scattering coefficient rather than by NL susceptibilities. The latter are instead used to describe
effects dealing with a macroscopic polarization in condensed systems. In isolated
NP the nonlinearity of the response is described by a Rayleigh scattering dipole
moment $\mu_i = \alpha_{ij}E_j + \beta_{ijk}E_jE_k$ that is quadratic in the pump field through a first
hyper-polarisability β_{ijk} coefficient responsible for second harmonic [SH] generation [10]. For N_{NP} NPs per unit volume in solution this leads to an intensity of the

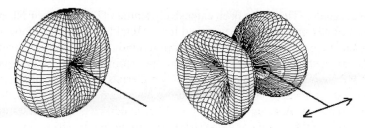

Fig. 4.1 Polar diagrams of the dipole (*left*) and quadrupole (*right*) second harmonic emission from a nanosphere. *Arrows* indicate direction and polarization of the exciting radiation

second harmonic signal given by

$$I_{2\omega} = GI_\omega^2 \left(\beta_s^2 N_s + \beta_{NP}^2 N_{NP} \right) \exp\left(-\sigma N_{NP} l \right) \tag{4.1}$$

where the subscript s refers to solvent, σ is the absorption cross section, l the path length and G a constant. The signal therefore exhibits the usual quadratic dependence with pump I_ω and is linear with NP number density. Experiments on metallic NP indicate for β_{NP} values largely exceeding those of superfluorescing molecules [11–14].

To proceed into the details of the NL mechanisms a model has been recently put out [15, 16] by using the multipole expansion of the NL macroscopic polarization of the scattered fields leading to a comprehensive picture of the second harmonic generation from a metallic sphere embedded into a uniform dielectric. In particular, it turns out that the emission from the sphere arises from both induced electric dipole and electric quadrupole moments at the second-harmonic frequency. The former requires a nonlocal excitation mechanism in which the phase variation of the pump beam across the sphere is considered, while the latter is present for a local-excitation mechanism. In contrast to the ω^4 dependence in the linear scattering case the second-harmonic intensity is found to scale as ω^6 and is determined by a combination of surface and bulk nonlinearities. The two contributions are shown [17], for the pure dipole and quadrupole cases in Fig. 4.1 where angular distributions of the irradiated SH power are drawn. The direction of both incident propagation and polarization vectors are shown. While the non-local dipole term results in a polarization independent pattern with maxima in the plane perpendicular to the incident direction, for the quadrupole case zeros are present both in the polarization direction and perpendicularly to it. These characteristic features exhibit different intensity dependence with NP size and can be used for diagnostic purpose [18, 19].

4.3 Nanoparticles in Monolayers

A second system to be discussed is a single monolayer of NP deposited on a substrate. These monolayers can be readily obtained by evaporation–condensation in the Vollmer–Weber mode and subsequently embedded in transparent substrates [20].

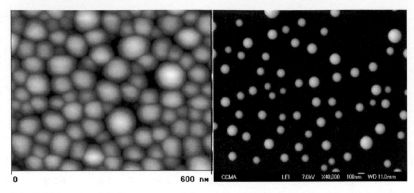

0   600 nm

Fig. 4.2 AFM images of Gallium nanoparticles on Sapphire (*left*) and Graphite (*right*) substrates

The particle density distribution is self-determined by the wetting between the liquid metal and the substrate. Fairly monodispersed size distributions, determined by the amount of evaporated material, can be obtained, as Fig. 4.2 shows.

Laser pumping of these systems gives rise to coherent SH signals in the transmitted direction as well as in reflection [21]. By changing the NP size one obtains a resonant behaviour for the NL transmittance/reflectance. Similar evidence of an increase in the optical NL response is obtained when these metal NP's are subject to temperature cycling across the liquid (at room temperature) to solid phase transition at ≈ 150 K [22, 23]. Here the effects of hysteresis are amplified by significant factors with respect to the linear case [24]. All these observations concur to the notion of an enhanced optical response in the NL domain for materials structured on a nanometre scale. The main mechanism for this behaviour turns out to occur when the incident radiative field resonates with surface plasmons on the metal NP. The electric field E_i on a NP can be written in terms of the impinging field E_0 and dielectric constants of the material, $\varepsilon(\omega)$, and of the medium ε_m:

$$E_i = E_0 \frac{3\varepsilon_m}{\varepsilon(\omega) + 2\varepsilon_m} . \tag{4.2}$$

E_i is maximized when $\varepsilon(\omega) + 2\varepsilon_m = 0$. The resulting resonance frequency, occurring when

$$\omega_{res}^2 = \frac{\omega_p^2}{1 + 2\varepsilon_m} \tag{4.3}$$

is derived from the plasma frequency ω_p of the metal through which the real part of the dielectric constant can be expressed as $\varepsilon_1 \approx 1 - \omega_p^2/\omega^2$. Since plasma frequencies occur in general in the ultraviolet, the resonance condition falls more naturally in the range of the generated second harmonic when near IR laser pulses are used. Figure 4.3 illustrates this point by showing the linear optical absorbance of a Ga NP monolayer embedded in TiO_2 for different average NP sizes. The peak wavelength position shifts from 360 nm for 3 nm radii to 420 nm for 40 nm radii.

Fig. 4.3 Absorbance of
a monolayer of Ga NP em-
bedded in TiO$_2$ matrix for
different NP sizes. From the
bottom curve average NP radii
are 3, 5, 10, 15, 20, 30 and
40 nm

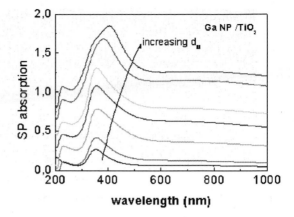

The position of the resonances can be reached with the second harmonic signal generated by a Ti sapphire laser tuned between 750 and 850 nm. These resonance peaks are observed in the second harmonic signal. Furthermore, the harmonic signal exhibits a strong dependence on the angle of incidence and on the orientation of the sample, e.g. the angle of rotation of the sample around its normal [25]. These facts indicate a complex and yet not completely explored interplay of shape, orientation and patterning of nanoparticles on observed nonlinearities.

Figure 4.4 shows the observed NL transmittance $T_{nl} = I(2\omega)/I(\omega)^2$ as a function of the NP radius and with two different matrices. The curves are strongly affected by the cover material confirming the plasmonic origin of the effect [26].

Several models for the NL interaction in nanostructured materials have appeared in the literature [27] using different levels of approximation. In particular, explicit expressions for the SH amplitudes in reflection from semi-infinite centro-symmetric bulk materials [28], from continuous thin films, and from spherical NPs embedded in homogeneous dielectrics [29] have been derived. In the present case, for a detailed description of the interaction one should take into account the effects of

Fig. 4.4 Size dependence of
SH nonlinear transmittance as
measured on Ga monolayers
embedded in SiO$_x$ (*squares,
red*) and TiO$_2$ (*circles, green*).
The *lines* represent the results
of model calculations discussed in the text. Reproduced
from [26] with permission of
The Electrochemical Society

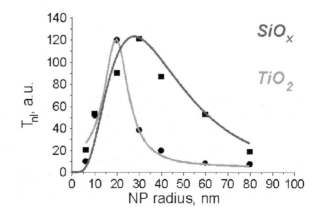

non-symmetric shape of the NPs and evaluate explicit expressions for the dipolar contribution at the NP surface. An alternative approach [30] consists instead of introducing a field enhancement factor $F(\omega, 2\omega)$ accounting for the increased field at the surface in the expression of the SH intensity $I(2\omega)$

$$I(2\omega) \propto \left|\chi^{(2)}\right|^2 F(\omega, 2\omega) \cdot I^2(\omega) \qquad (4.4)$$

where $I(\omega)$ and $I(2\omega)$ are the fundamental and SH intensities, $\chi^{(2)}$ is the NL susceptibility of the system. $F(\omega, 2\omega)$ can be expressed as:

$$F(\omega, 2\omega) = Q_{NF}^2(\omega)Q_{NF}(2\omega)R^2 \qquad (4.5)$$

where $Q_{NF}(\omega)$ and $Q_{NF}(2\omega)$ are averaged local fields at NPs boundary, at fundamental and SH frequency respectively and R is the (average) NP radius. The two factors Q can be calculated in the frame of the Mie theory of scattering by using an effective dielectric function for the whole medium obtained in a Maxwell–Garnett approach. The model is in this way independent on the particular modelling of the NL interaction, the material properties being determined by the effective dielectric function used. Satisfactory agreements have been obtained for Ga NP monolayers embedded in SiO_x using dielectric constant data form the literature [26].

4.4 Planar Photonic Structures

A third example of the ability of guiding and localizing radiation in materials is provided by structures evolving from e.g. planar waveguides in which a layer of high refraction index is sandwiched between two low index materials, one of them possibly air. In general, these structures are made by high refraction index materials that provide the necessary index contrast for the guiding effect. In contrast to previous cases, semiconductors are being used in place of metals. Therefore the main resonances one expects are not of plasmonic origin. These in fact would rely on the generation of electron oscillations in e.g. an electron–hole plasma, a condition requiring very strong and transient photoexcitation of the material. The appropriate approach for semiconductor-based structures is then exploiting resonances originating from spatial modulation of the medium through the notion of photonic crystal.

One- and two-dimensional photonic crystals can be readily obtained from planar waveguides. Starting from such structures a spatial modulation of the in-plane index of refraction can be introduced by engraving into the waveguide patterns of holes, valleys, pillars or fences [31, 32]. The flexibility of this design allows one to exploit NL effects in the resulting PhC and further enhance them by tailoring the photonic mode dispersion. A considerable effort throughout the years has been devoted to the optimisation of the design of such structures for harmonic generation [33]. More generally, enhancement of the harmonic/nonlinear response has been reported in experiments on photonic systems in one [34, 35], two [36, 37] and three dimen-

Fig. 4.5 Schematic diagram of a photonic band structure in 1D. The *shaded area* represents the light cone that can be accessed by the exterior via quasi-guided modes

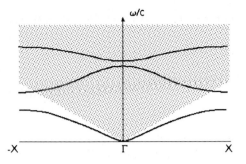

sion [38]. Several media have been explored, from crystalline semiconductors (Si, GaAs, GaN, InP, ZnO) to amorphous and organic materials.

As far as photonic waveguides are concerned, an important feature of this experimental approach is that the photonic band structure can be studied and resolved by employing waves that do not propagate along the waveguide itself. These radiation modes were initially described in terms of vertical cavity resonances of an optically active patterned photonic slab [39]. By impinging on the surface of the waveguide with radiation of angular frequency ω above the light line (or light cone) defined as $\omega > ck_{\parallel}$, with $k_{\parallel}$ the component of the wavevector parallel to the waveguide, quasi-guided modes [40] may be excited inside the waveguide (see Fig. 4.5). Conversely, the enhanced signal emitted after the interaction with the optically active material can be collected outside the structure. Thus reflectance from the leaky modes above the light line carries information on the photonic band structure of the patterned waveguide.

As shown in Fig. 4.6 below, the frequency and momentum matching between impinging field and photonic modes marks the reflectance spectrum with characteristic modulations. This fact allows the experimental determination of the photonic band structures across most of the Brillouin zone by rotating the waveguide around the normal and repeating the measurements [36]. This technique offers substantial advantages in flexibility over the direct evaluation of the transmitted spectrum across the waveguide that requires a careful polishing of the sides and becomes critically dependent on surface finish and optical coupling of the radiation into the system. Figure 4.6 illustrates an example of variable angle reflectance measurements taken in a 1D SOI (Silicon On Insulator) PhC and the derived band system accessible with this technique [42]. These measurements have also been extended below the light line by using attenuated total reflection techniques that allow direct coupling from above of the probing radiation into the modes of the waveguide [43].

The shape of the photonic bands of a particular patterned waveguide can also be obtained via NL interaction by detecting i.e. the SHG or THG signals in reflection from the waveguide. For materials with a nonzero NL susceptibility the reflected harmonic signal appears modulated at the same angular positions of incidence as the linear reflectance due to the interaction with the quasi-guided modes.

In Fig. 4.7 the case of a GaAs/AlGaAs 2D photonic waveguide patterned with a square lattice of square air holes is reported [44] at fixed angle of incidence

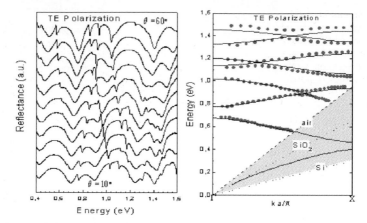

Fig. 4.6 a) Variable angle reflectance spectra for a 1D SOI PhC. The dispersion of spectral features with angle of incidence is clearly visible. **b)** Calculated (*continuous lines*) photonic bands of a SOI 1D PhC and experimental values (*points*) as derived from variable angle reflectance measurements above the (air) *light line*. *Lines* corresponding to fixed values of the angle of incidence are also shown. Figures taken from [41] © 2005 IEEE

Fig. 4.7 Linear and nonlinear reflectivities at 45 deg. incidence of a GaAs/AlGaAs waveguide patterned with a square array of air holes as a function of the azimuth angle. Panels **a)** and **c)** refer to the PhC while panel **b)** to the unpatterned waveguide. Figure reprinted with permission from [44]. Copyright (2003) by the American Physical Society

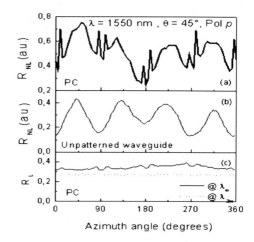

(45 degrees) by varying the azimuth, i.e. by rotating the sample around the normal to the face.

The NL reflectance R_{NL} defined as $I^{(2\omega)}/(I^{(\omega)})^2$ contains the superposition of two components. The first one is originated by the structure of the GaAs second order susceptibility $\chi^{(2)}$ and the other is due to the interaction with the photonic modes of the structure. The latter appears substantially more pronounced than the corresponding one measured in *linear* reflectivity. The deeper modulation in the nonlinear case may be of use in the derivation of photonic band structures. In fact one can derive the band information by exploring the NL reflectance above the light cone at different photon energies. The peaks observed reveal the occurrence of a resonance between the pump and/or harmonic radiation with a particular photonic band

Fig. 4.8 Second harmonic
signal in reflection from the
GaAs/AlGaAs PhC discussed
in the text. The SH signal
is obtained by excitation
with fs laser pulses at the
wavelengths indicated. Figure
reprinted with permission
from [44]. Copyright (2003)
by the American Physical
Society

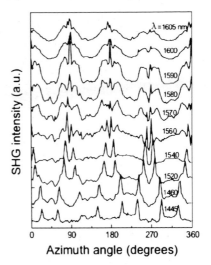

of the waveguide. An example of this is shown in Fig. 4.8 that depicts the SHG intensity versus azimuth angle at different laser wavelengths. The diagram is fully equivalent to that of Fig. 4.6a in the variation in position of the resonant peaks with photon energy. The position of the peaks uniquely determines the photonic band in the Brillouin zone. This procedure can fully explore the band structure when also the dependence on angle of incidence is exploited.

Another effect worth mentioning in the NL interaction between radiation and the photonic structures is the appearance of NL diffraction effects due to the periodicity of the medium. In the NL case, however, a special form of Bragg condition for diffracted beams occurs for the nth harmonic component [45]:

$$k'_\parallel(n\omega) = nk_\parallel(\omega) + G . \tag{4.6}$$

Here $k'_\parallel(n\omega)$ is the diffracted in-plane component at the nth harmonic, $k_\parallel(\omega)$ is the parallel component of the incident wavevector and G is a 1D reciprocal lattice vector of the photonic structure. When the PhC is of 1D type the reciprocal vector G is made to coincide with the plane of incidence. In this case the above equation reduces to the familiar, though NL, grating equation:

$$\sin\theta' = \sin\theta + m\frac{\lambda}{na} \tag{4.7}$$

connecting the mth order angle of diffraction θ' to the angle of incidence θ, pump wavelength λ, harmonic order n and grating constant a. $m = 0$ corresponds to reflection while $n = 1$ to linear diffraction. This kinematic law has been well verified in the case of 1D SOI PhCs made by Silicon parallel stripes on a SiO_2 substrate with a 1 μm period and 0.24 μm thick (air filling factor $f = 0.7$) using 100 fs, 810 nm laser pulses. Figure 4.9 illustrates the experimental points obtained from spectral

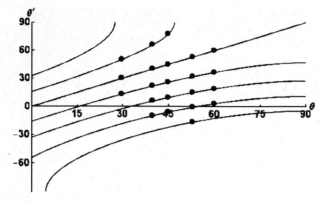

Fig. 4.9 Diffraction orders for the third harmonic signal at 270 nm in the (θ, θ') plane generated by the 1D SOI photonic crystal discussed in the text. Picture taken from [42] with kind permission from Springer Science and Business Media

orders $m = -4$ to $m = +1$ and well matching the theoretical behaviour. It is to be noted that the number of diffracted orders for the harmonic signal outnumbers that of fundamental radiation by a factor n.

The presence of higher order harmonic components diffracted at high diffraction orders becomes more complicated when dealing with a 2D patterning of a plane waveguide. Images of the diffraction patterns observed are quite suggestive and illustrative of all diffraction occurrences since a direct and complete visualization of the diffraction pattern is revealed. Figure 4.10 for instance refers to a triangular pattern of air holes drilled in a SOI waveguide in which a line defect (of W1 type) made by a missing hole line every ten in the ΓX direction is present. The panel at right illustrates the NL diffraction pattern generated by the third harmonic of 810 nm fs laser pulses impinging at normal incidence on the sample. The diffracted beams are intercepted in reflection by a paper screen in front of the sample that emits fluorescence radiation in the blue upon excitation by the 270 nm diffracted radiation. Two sets of diffraction patterns are present. The bright, well-separated spots are generated by the triangular patterning while the W1 defect gives rise to the more finely spaced lines. The image appears distorted by the very slanted position of the camera.

A more immediate application of optical harmonics in PhCs is efficient frequency conversion, a property that preludes to significant extension of functionalities in PhC devices. As already mentioned, a natural way to improve conversion efficiency is to exploit resonances between pump radiation and photonic bands. In this case an enhanced pump field provides extra excitation to the NL process. This scheme is sometimes referred to as resonant–nonresonant (RNR) process [46]. A similar situation may also occur when the harmonic signal generated inside the PhC becomes resonant with a photonic mode, giving rise to a nonresonant–resonant (NRR) process. The simultaneous occurrences of both resonances gives rise to the optimised resonant–resonant (RR) process. In this case the search of the optical conditions for double resonances can be eased by the so-called quasi-phase matching condi-

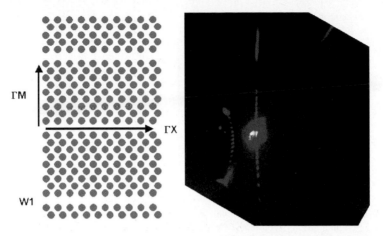

Fig. 4.10 Image of third harmonic signal generated by a W1 waveguide with the line defect along ΓM in the vertical direction and with the pump impinging at normal incidence. The pattern is created on a screen parallel to the sample surface. The *bright spot at the centre* is the aperture for the pump beam. Besides the line defect diffraction along ΓM, different diffraction spots due to the triangular PhC lattice are visible. On the *left* and on the *right* a *curved line* of weak spots between the prominent first order diffraction represents the mixed diffraction orders between triangular pattern and defects

tion [47] in which the phase matching in the propagation of pump and harmonic signal in the periodic NL medium may be satisfied through a reciprocal lattice vector, i.e. $k'_{\parallel}(n\omega) = nk_{\parallel}(\omega) + G$, as Fig. 4.11 shows. The RR process has been predicted and calculated [48] for a 2D PhC made in a 130 nm thick GaAs waveguide patterned with a square array of air holes. Extremely high ($>10^6$) enhancements in specular reflection in the quasi-guided mode regime have been predicted.

In fact, experiments performed on GaN 1D PhCs pumped at 795 nm have demonstrated [46] an enhancement factor close to 10^4 for SHG with respect to the nonpatterned waveguide. Further experiments on the same kind of samples produced with a different growing technique have produced enhancement factors as high as 10^5 [49].

When material symmetry does not allow a nonzero second order susceptibility, then third order effects become the leading ones in the NL interaction. This is the case with centre-symmetric structures like Si based systems. As Fig. 4.10 shows, considerable amounts of third harmonic signal in the UV range starting from $\approx$810 nm can be obtained. It is worth noting that at this third harmonic wavelength (270 nm) the absorption coefficient of Si amounts to $\approx 3 \times 10^6$ cm^{-1} leaving a very small fraction of the 26 nm thickness of the Si waveguide to emit radiation escaping outside. The improvement in NL reflectivity from the simple SOI waveguide is measured to be $\approx$100 times. Such an increase is attributed to the effect of field enhancement at the edges of the walls of the pattering that increase the amplification of the NL process. Using IR radiation at 1.55 µm, instead, only a fraction of the energy of the pump pulse is stored into the material. Conversely, the entire thickness

Fig. 4.11 Scheme of the quasi-phase matching in a photonic band system for SHG

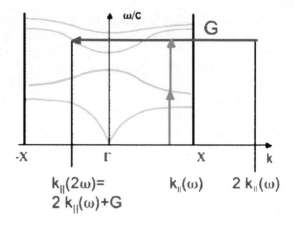

$$k_{||}(2\omega)=$$
$$2\,k_{||}(\omega)+G \qquad\qquad k_{||}(\omega) \qquad 2\,k_{||}(\omega)$$

of the waveguide participates to the emission process. The enhancements observed in the IR with respect to the plane waveguide are of the order of 200 as compared to factor of 100 found for the visible-UV case, a result probably due to the reduced effect of surface scattering that scales as λ^4. The dispersion of the resonances in the TH reflectance as a function of the angle of incidence are shown in Fig. 4.12 below and appear qualitatively well described by model calculations using the scattering matrix method of Whittaker and Culshaw [50] implemented with a NL extension for the evaluation of electric field and NL polarization in space. Needless to say that the dispersion of the observed TH peaks closely follows the calculated photonic band structure.

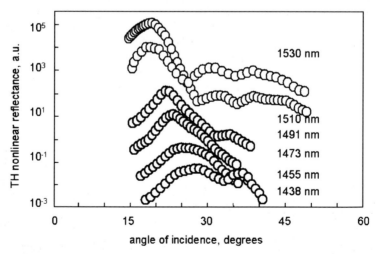

Fig. 4.12 Experimental dispersion of TH nonlinear reflectance as a function of angle of incidence. Note the logarithmic vertical scale. Picture taken from [42] with kind permission from Springer Science and Business Media

Wide band-gap semiconductors bypass the problem of transparency at both fundamental and harmonic wavelengths from the IR into the UV. Therefore, they certainly are materials of choice in this photon energy range, especially since they exhibit substantial NL effects. As an example, GaN exhibits a NL second order susceptibility $\chi^{(2)}$ a factor of 10 smaller than that of GaAs, however it is known to have it enhanced by the presence of piezo-electric fields generated at the walls of the photonic patterning [51]. Moreover GaN is known to have a very high optical damage threshold and a good mechanical stability. Resonant conditions for SHG in GaN waveguides made of a triangular lattice of air holes have been found both theoretically and experimentally for 900 nm excitation [52, 53]. The occurrence of the resonance is marked by a distinctive peak that exceed by a factor $\approx$250 the SH signal of the bare waveguide. Due to the rotation symmetry of the triangular lattice, the matching occurs at 16 degree incidence every 60 degrees in azimuth rotation along the ΓM direction. At the very same position an extremely high UV peak at 300 nm is observed due to third harmonic resonant generation. This occurrence shows that resonant conditions for both second and third harmonic generation hold simultaneously in this configuration. In fact, the resonance conditions reached refer to the single photon pump absorption, a process that is common to both kinds of NL processes. $\chi^{(3)}$ has been measured in the near IR in connection with GaN/AlN quantum well transitions. Values around $2 \times 10^{-16}\,\mathrm{m^2\,V^{-2}}$ have been obtained [54, 55]. The measured conversion in THG with respect to the unpatterned waveguide amounts to a factor $\approx$3400.

The pumping scheme adopted here, with a photon energy smaller than half the gap energy of GaN naturally induces in the material three-photon luminescence from band-gap recombination of the carriers, accompanied by the characteristic yellow band attributed to inter-gap structural defects in GaN. Due to the strong resonant excitation of the third harmonic signal the fluorescence signal closely follows the angular pattern of the pump and exhibits a clear cubic intensity dependence. With the simultaneous presence of $\chi^{(2)}$ and $\chi^{(3)}$ in GaN the question whether the origin

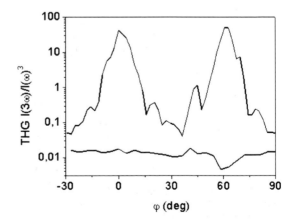

Fig. 4.13 Resonant third harmonic reflectivity at 300 nm from GaN waveguide patterned with a triangular lattice of air holes. For comparison, the *lower curve* refers to the unpatterned waveguide

of the cubic resonance is due to a direct third order or to a cascaded of two second order effects requires a more detailed study of the resonances occurring in the PhC.

A further strategy for increasing the strength of the interaction between radiation and photonic structure can arise from a reduced group velocity of light within the NL crystal. This consideration naturally leads to devise schemes in which the resonance occurs at photonic band positions that exhibit little dispersion, i.e. an horizontal dependence versus momentum. Starting from the various patterning geometries applied to 2D waveguides, a method to introduce in the photonic Brillouin zone similar bands is by adding periodic defects into the structure. The resulting superlattice can be tailored to interact with the existing band system to produce low group velocity additional bands. In particular the so-called W1 waveguide, based on a periodic missing row every n in the ΓM direction, has been modelled theoretically [56] in the case of a triangular pattern of air holes in a SOI waveguide. In agreement with theoretical predictions, optical linear measurements [43] indicate the presence of such bands, suggesting this possibility. Indeed, as illustrated in Fig. 4.10 strong NL effects in the third harmonic signals are observed originating from diffraction from such defects. Similar results have been observed also by using 1.55 μm pump radiation. Taking into account the experimental condition of a limited laser beam size that illuminates only few periods of the superlattice structure, added to the consideration of a (strongly absorbing) vertical cavity of one-wavelength size, the yield in harmonic crystal appears to be remarkably high.

4.5 Conclusions

NL optics represents one of the key elements of novelty among the properties of nanostructures. It is instrumental in the research aiming at the optical control and conditioning of optical signals. This challenging goal can be now pursued at the nanoscale level thanks to the development in design and fabrication of photonic and plasmonic structures. The knowledge of the microscopic mechanisms governing the interaction of radiation fields with matter in various geometries and periodicities has guided the research and the technology towards integrated NL systems. Restricting now our views to the harmonic generation in nanosystems, we see that both photonic and plasmonic devices introduce resonant NL response in the nanoscale. The intensity of the harmonic signal and its specificity in terms of width of the resonance, sensitivity to surface conditions, geometry of the generation and spatial characteristics are all interesting features in a number of potential applications. In particular, harmonic generation can be used as a diagnostic tool for the determination of the photonic band structure of PhC. The sensitivity of this diagnostic method relies on the high discrimination in intensity of the signal and in the intrinsic dependence on perturbing effects such as structural and geometrical imperfections and on the photonic surface conditions. This latter aspect can be used for devising schemes for the detection of e.g. bio-molecules through selective functionalisation of nanostructure surfaces. By exploiting the superior sensitivity of the NL response towards surface conditions efficient detection techniques can be at hand. Several other aspects have however still to be fully explored.

Two developments that are outside the topics described here are nevertheless worth mentioning, based on the knowledge acquired in PhC waveguides. The first refers to the extrapolation of the concept of photonic resonance to NL propagation. The idea is based on coupled microcavities made up by stacks of Bragg resonators. They can be also considered as one-dimensional PhCs. This system can benefit from two resonance conditions to be simultaneously fulfilled, i.e. photonic resonances and phase matching. The former can be made to occur when pump and/or harmonic fields are resonant with band-edge states being formed by the periodic structure. The latter may occur when the pump is tuned to the nth order stop band while the harmonic field is in resonance with the $2n$th order stop band. This configuration has been demonstrated theoretically and experimentally to scale as N^6 with the number of periods of the structure [35, 57]. Very recently, however evidence of even faster scaling have been numerically obtained for non-phase matched structures [58, 59]. A second recent development is in the direction of extending phase matching condition in planar photonic waveguides for all directions of propagation for a suitably designed patterning of the system [60]. The scaling of the lattice constant provides a broad tuning of the harmonic emission. These findings can open the way to compact frequency converter devices without phase matching limitations.

In summary, harmonic studies conducted on NPs and photonic structures have shown interesting potential for nanophotonic devices. Developments in theory and in device technology will bring fully guided NL optics into reality with true integration of different functionalities on the same device. Further topics of interest include coupling together plasmonic and photonic resonance mechanisms and extend the investigations in the field of parametric NL effects. This latter aspect is of extreme interest in the development of compact and efficient sources of entangled photon pairs.

Acknowledgements I gratefully acknowledge here stimulating and illuminating discussions with several colleagues on many occasions. In particular I am indebted with M. Galli, D. Bajoni, M. Belotti, M. Patrini, G. Guizzetti, A. Stella, G. Vecchi, L.C. Andreani, M. Agio, D. Gerace, M. Liscidini, A. Balestreri, N. Sinha (Università di Pavia), R. Kofman, P. Cheyssac, M. Allione (Université de Nice – Sophie Antipolis, France), Y. Chen, D. Peyrade (CNRS–LPN, Marcoussis, France), D. Coquillat, M. Le Vassor d'Yerville, J. Torres, C. Gergely (Université Montpellier II, France), S. Lurdoudoss, C. Olson (KTH, Sweden) and A. Postigo (CSIC, Madrid).

The material presented in this chapter has been partially developed under COST Action P11: 'Physics of Linear, Nonlinear and Active Photonic Crystals'. Partial support from the European Network of Excellence 'Phoremost' is also gratefully acknowledged.

References

1. Kreibig U., Vollmer M.: Optical Properties of Metal Clusters, Springer Series in Materials Science Vol. 25, Springer, Berlin (1995).
2. Zheludev N.J., Stockman M. and Zayats A. (eds): Special issue on fundamental aspects of nanophotonics. J. Opt. A: Pure Appl. Opt. **8**, S1–S295 (2006).

3. Boyd G.T., Raising T., Leite J.R., and Shen Y.R.: Local-field enhancement on rough surfaces of metals, semimetals, and semiconductors with the use of optical second-harmonic generation. Phys. Rev. B **30**, 519–526 (1984).

4. Zhang X., Stroud D.: Numerical studies of the nonlinear properties of composites. Phys. Rev. B **49**, 944–955 (1994).

5. Aussenegg F.R., Leitner A., Gold H.: Optical second-harmonic generation of metal-island films. Appl. Phys. A **60**, 97–101 (1995).

6. Faccio D., di Trapani P., Borsella E., Gonella F., Mazzoldi P. and Malvezzi A.M.: Measurement of the third-order nonlinear susceptibility of Ag nanoparticles in glass in a wide spectral range. Europhysics Lett. **43**, 213–218 (1998).

7. Lamprecht B., Krenn J.R., Leitner A. and Aussenegg F.R.: Resonant and Off-Resonant Light-Driven Plasmons in Metal Nanoparticles Studied by Femtosecond-Resolution Third-Harmonic Generation. Phys. Rev. Lett. **83**, 4421–4424 (1999).

8. Malvezzi A.M., Allione M., Patrini M., Stella A., Cheyssac P., and Kofman R.: Melting-Induced Enhancement of the Second-Harmonic Generation from Metal Nanoparticles. Phys. Rev. Lett. **89**, 087401 (2002).

9. Williams D.J.: Introduction to Nonlinear Optical Effects in Molecules and Polymers. Wiley, New York (1991)

10. Clays K. and Persoons A.: Hyper-Rayleigh scattering in solution. Phys. Rev. Lett. **66**, 2980–2983 (1991).

11. Antoine R., Pellarin M., Palpant B., Broyer M., Prevel B., Galletto P., Brevet P.F., Girault H.H.: Surface plasmon enhanced second harmonic response from gold clusters embedded in an alumina matrix. J. Appl. Phys. **84**, 4532–4536 (1998).

12. Vance F.W., Lemon B.I. and Hupp J.T.: Enormous Hyper-Rayleigh Scattering from Nanocrystalline Gold Particle Suspensions. J. Phys. Chem. B **102**, 10091–10093 (1998)

13. Galletto P., Brevet P.F., Girault H.H., Antoine R. and Broyer M.: Size dependence of the surface plasmon enhanced second harmonic response of gold colloids: towards a new calibration method. Chem. Commun. 581–582 (1999).

14. Johnson R.C., Li J., Hupp J.T. and Schatz. G.C.: Hyper-Rayleigh scattering studies of silver, copper, and platinum nanoparticle suspensions. Chem. Phys. Lett. **356**, 534–540 (2002).

15. Dadap J.I., Shan J., Eisenthal K.B. and Heinz T.F.: Second-Harmonic Rayleigh Scattering from a Sphere of Centrosymmetric Material. Phys. Rev. Lett. **83**, 4045–4048 (1999).

16. Dadap J.I., Shan J., Heinz T.F.: Theory of optical second-harmonic generation from a sphere of centrosymmetric material: small-particle limit. J. Opt. Soc. Am. B **21**, 1328–1347 (2004).

17. Hao E.C., Shatz G.C., Johnson R.C. and Hupp J.T.: Hyper Rayleigh scattering from silver nanoparticles. J. Chem. Phys. **117**, 5963–5966 (2002).

18. Wang H., Yan E.C.Y., Bourguet E. and Eisenthal K.B.: Second harmonic generation from the surface of centrosymmetric particles in bulk solution. Chem. Phys. Lett. **259**, 15–20 (1996).

19. Nappa J., Revillod G., Russier-Antoine I., Benichou E., Jonin C. and Brevet P.F.: Electric dipole origin of the second harmonic generation of small metallic particles. Phys. Rev. B **71**, 165407 (2005).

20. Søndergård E., Kofman R., Cheyssac P. and Stella A.: Production of nanostructures by self-organization of liquid Volmer–Weber films. Surf. Sci. **364**, 467–476 (1996).

21. Tognini P., Stella A., Cheyssac P. and Kofman R: Surface plasma resonance in solid and liquid Ga nanoparticles. J. Non-Cryst. Solids **249**, 117–122 (1999).

22. Kofman R., Cheyssac P. and Garrigos R.: From the bulk to clusters: Solid–liquid phase transitions and precursor effects. Phase Transit. **24**, 283–342 (1990).

23. Lereah Y., Kofman R., Penisson J.M., Deutscher G., Cheyssac P., Ben David T. and Bourret A.: Time-resolved electron microscopy studies of the structure of nanoparticles and their melting. Phil. Mag. B **81**, 1801–1819 (2001).

24. Malvezzi A.M., Allione M., Patrini M., Stella A., Cheyssac P. and Kofman R.: Melting-Induced Enhancement of the Second Harmonic Generation from Metal Nanoparticles. Phys. Rev. Letters **89**, 087401 (2002).

25. Malvezzi A.M., Patrini M., Stella A., Tognini P., Cheyssac P. and Kofman R: Second harmonic generation in Ga nanoparticle monolayers. Eur. Phys. J. D**16**, 321–324 (2001).

26. Achilli S., Allione M., Patrini M., Stella A., Malvezzi A.M. and Kofman R.: Nonlinear Optical Response of Nanoparticle Single Layers: Polarization and Structural Effects. In Cahay M., Bandyopadhyay S., Hasegawa H., Leburton J.P., Seal S., Urquidi-Macdonald M., Guo P., Koshida N., Lockwood D.J., Stella A. (eds) Nanoscale Devices, Materials, and Biological Systems: Fundamentals and Applications, ECS Proceedings Volume 2004-13, pp. 325–333. The Electrochemical Society, Inc., Pennington, NJ, USA (2004).

27. Boyd R.W., Gehr R.J., Fischer G.L. and Sipe J.E.: Nonlinear optical properties of nanocomposite materials. Pure Appl. Opt. **5**, 505–512 (1996).

28. Bloembergen N., Chang R.K., Jha S.S. and Lee C.H.: Optical Second-Harmonic Generation in Reflection from Media with Inversion Symmetry. Phys. Rev. **174**, 813–822 (1968).

29. Götz T., Buck M., Dressler C., Eisert F., and Träger F.: Optical second harmonic generation of supported metal clusters: size and shape effects. Appl. Phys. **60**, 607–612 (1995).

30. Muller Th., Vaccaro P.H., Balzer F., Rubahn H.-G.: Size dependent optical second harmonic generation from surface bound Na clusters: comparison between experiment and theory. Opt. Comm. **135**, 103–108 (1997).

31. Soukoulis C.M. (Ed): Photonic Band Gaps and Localization. Plenum, New York (1993).

32. Brunne R., Geissler E., Messerschmidt B., Martin D., Soergel E., Inoue K., Ohtaka K., Ghatak A. and Thyagarajan K.: Advanced Optical Components. In F. Trager (Ed) Handbook of Lasers and Optics, pp. 419–502. Springer, New York (2007).

33. Slusher R.E. and Eggleton B.J.: Nonlinear Photonic Crystals. Springer, Berlin (2003).

34. Balakin A.V., Bushuev V.A., Mantsyzov B.I., Ozheredov I.A., Petrov E.V., Shkurinov A.P., Masselin P. and Mouret G.: Enhancement of sum frequency generation near the photonic band gap edge under the quasiphase matching conditions. Phys. Rev. E **63**, 046609 (2001). [11 pages]

35. Dumeige Y., Sagnes I., Monnier P., Vidakovic P., Abram I., Mériadec C., and Levenson A.: Phase-Matched Frequency Doubling at Photonic Band Edges: Efficiency Scaling as the Fifth Power of the Length. Phys. Rev. Lett. **89**, 043901 (2002). [4 pages]

36. Malvezzi A.M., Cattaneo F., Vecchi G., Falasconi M., Guizzetti G., Andreani L.C., Romanato F., Businaro L., Di Fabrizio E., Passaseo A. and De Vittorio M.: Second harmonic generation in reflection and diffraction by a GaAs photonic-crystal waveguide. J. Opt. Soc. Am. B **19**, 2122 (2002).

37. Mondia J.P., van Driel H.M., Jiang W., Cowan A.R. and Young J.F.: Enhanced second-harmonic generation from planar photonic crystals. Opt. Lett. **28**, 2500–2502 (2003).

38. Martorell J., Vilaseca R. and Corbalan R.: Second harmonic generation in a photonic crystal. Appl. Phys. Lett. **70**, 702–704 (1997).

39. Pottage J.M., Silvestre E. and Russell P.S.J.: Vertical-cavity surface-emitting resonances in photonic crystal films. J. Opt. Soc. Am **A18**, 442–447 (2001).

40. Pacradouni V., Mandeville W., Cowan A., Paddon P., Young J.F. and Johnson S.R.: Photonic band structure of dielectric membranes periodically textured in two dimensions. Phys. Rev. B **62**, 4204–4207 (2000).

41. Galli M., Bajoni D., Belotti M., Paleari F., Patrini M., Guizzetti G., Gerace D., Agio M., Andreani L.C., Peyrade D. and Chen Y.: Measurement of Photonic Mode Dispersion and Linewidths in Silicon-on-Insulator Photonic Crystal Slabs. IEEE J. On Selected Areas In Communications (J-SAC) 23, 1402–1410 (2005)

42. Comaschi C., Vecchi G., Malvezzi A.M., Patrini M., Guizzetti G., Liscidini M., Andreani L.C., Peyrade D. and Chen Y.: Enhanced third harmonic reflection and diffraction in Silicon on insulator photonic waveguides. Appl. Phys. B **81**, 305–311 (2005).

43. Galli M., Belotti M., Bajoni D., Patrini M., Guizzetti G., Gerace D., Agio M. and Andreani L.C.: Excitation of radiative and evanescent defect modes in linear photonic crystal waveguides. Phys. Rev. B **70**, 081307 (R) (2004).

44. Malvezzi A.M., Vecchi G., Patrini M., Guizzetti G., Andreani L.C., Romanato F., Businaro L., Di Fabrizio E., Passaseo A. and De Vittorio M.: Resonant second-harmonic generation in a GaAs photonic crystal waveguide. Phys. Rev. B **68**, 161306 R (2003). [4 pages]

45. Freund I.: Nonlinear Diffraction. Phys. Rev. Lett. **21**, 1404–1406 (1968).

46. Torres J., Le Vassor d'Yerville M., Coquillat D., Centeno E. and Albert J.P.: Ultraviolet surface-emitted second-harmpnic generation in GaN one-dimensional photonic crystal slabs. Phys. Rev. B **71**, 195326 (2005).
47. Bloembergen N. and Sievers A.: Nonlinear optical properties of periodic laminar structures. Appl. Phys. Lett. **17**, 483–486 (1970).
48. Cowan A.R. and Young J.F.: Mode matching for second harmonic generation in photonic crystal waveguides. Phys. Rev. B **65**, 085106 (2002).
49. Bouchoule S., Boubanga-Tombet S., Le Gratiet L., Le Vassor d'Yerville M., Torres J., Chen Y. and Coquillat D.: Reactive ion etching of high optical quality GaN/sapphire photonic crystal slab using CH_4–H_2 chemistry. J. Appl. Phys. **101**, 043103 (2007).
50. Whittaker D.M. and Culshaw I.S.: Scattering-matrix treatment of patterned multilayer photonic structures. Phys. Rev B **60**, 2610 (1999).
51. Schmidt H., Abare A.C., Bowers J.E., Denbaars S.P. and Imamoglu A.: Large interband second-order susceptibilities in InxGa1–xN/GaN quantum wells. Appl. Phys. Lett. **75**, 3611–3613 (1999).
52. Coquillat D., Torres J., Peyrade D., Chen Y., Legros R., Lascaray J.P., Le Vassor d'Yerville M., De La Rue R.M., Centeno E., Cassagne D. and Albert J.P.: Equifrequency surfaces in a two-dimensional GaN-based photonic crystal. Opt. Express **12**, 1097–1108 (2004).
53. Coquillat D., Vecchi G., Comaschi C., Malvezzi A.M., Torres J. and Le Vassor d'Yerville M.: Enhanced second- and third-harmonic generation and induced photoluminescence in a two-dimensional GaN photonic crystal. Appl. Phys. Lett. **87**, 101106 (2005).
54. Rapaport R., Chen G., Mitrofanov O., Gmachl C., Ng N.M. and Chu S.N.G.: Resonant optical nonlinearities from intersubband transitions in GaN/AlN quantum wells. Appl. Phys. Lett. **83**, 263–265 (2003).
55. Hamazaki J., Matsui S., Kunugita H., Ema K., Kanazawa H., Tachibana T., Kikuchi A. and Kishino K.: Ultrafast intersubband relaxation and nonlinear susceptibility at 1.55 μm in GaN/AlN multiple-quantum wells. Appl. Phys. Lett. **84**, 1102–1104 (2004).
56. Andreani L.C. and Agio M.: Intrinsic diffraction losses in photonic crystal waveguides with line defects. Appl. Phys. Lett. **82**, 2011–2013 (2003).
57. De Angelis C., Gringoli F., Midrio M., Modotto D., Aitchison J.S. and Nalesso G.F.: Conversion efficiency for second-harmonic generation in photonic crystals. J. Opt. Soc. Am. B **18**, 348–351 (2001).
58. Centini M., D'Aguanno G., Sciscione L., Sibilia C., Bertolotti M., Scalora M. and Bloemer M.J.: Non-phase-matched enhancement of second-harmonic generation in multilayer nonlinear structures with internal reflections. Opt. Lett. **29**, 1924–1926 (2004).
59. Liscidini M., Locatelli A., Andreani L.C. and De Angelis C.: Maximum-Exponent Scaling Behavior of Optical Second-Harmonic Generation in Finite Multilayer Photonic Crystals. Phys. Rev. Lett. **99**, 053907 (2007). [4 pages]
60. Centeno E., Felbacq D. and Cassagne D.: All-Angle Phase Matching Condition and Backward Second-Harmonic Localization in Nonlinear Photonic Crystals. Phys. Rev. Lett. **98**, 263903 (2007). [4 pages]

5 Ultra-fast Optical Reconfiguration via Nonlinear Effects in Semiconductor Photonic Crystals

Crina Cojocaru[1], Jose Trull[1], Ramon Vilaseca[1], Fabrice Raineri[2], Ariel Levenson[2], and Rama Raj[2]

[1] Universitat Politècnica de Catalunya, Departament de Física i Eng. Nuclear
Colom 11, 08222 Terrassa, Spain
[2] Laboratoire de Photonique et de Nanostructures (CNRS UPR 20)
Route de Nozay, 91460 Marcoussis, France

Abstract. We analyze different possibilities to control and modify the optical response of a photonic crystal via nonlinear effects induced by an external optical control beam. We consider two different nonlinear schemes in which such active optical control may be achieved. The first one is based on the second order nonlinear interaction within a one-dimensional photonic structure with a defect. We show that using a control beam at the second harmonic frequency we can induce important changes in the reflection and transmission of a signal at the fundamental frequency. In the second scheme, the nonlinearity is achieved via electronically resonant optical Kerr effect, through optical injection of carriers. We report on wide wavelength tuning of a non defective two-dimensional photonic crystal resonance observed in reflectivity. Moreover, we prove that the same structure shows different optical functions such as frequency shifting, switching, amplification and lasing, if it is used under appropriate stimulus. These very fast effects may be used for the design of an ultra-fast all-optical modulator or switching device to be integrated in a photonic circuit.

Key words: photonic crystals; nonlinear optics; integrated photonic circuits

5.1 Introduction

It is believed that future developments in optical networks will rely crucially on all-optical processing and on its inherent potential to provide improved high speed data transmission and processing with lower power consumption. Photonic integrated circuits have the potential to be as important in the 21st century as electronic integrated circuits have been in the 20th century. The interaction-free nature of light enables completely new switching architectures that are not possible in electrical communication. In the search for all-optical devices, photonic crystals (PhC) [1] are increasingly evoked. These special optical materials, with their periodic lattice dimensions of the order of the wavelength of light provide a very elegant way of controlling the light propagation. They offer new possibilities of engineering the dispersion properties of materials, which yield photonic band-gaps of forbidden transmission frequencies and determine both the phase and group velocities. Near the high symmetry points of the photonic band structure, or within the resonances introduced by the presence of defects, the group velocity approaches zero and the photons experience an increased lifetime within very small distances [2]. This leads to a strong localization of the electromagnetic field that may be used to enhance the nonlinear interactions between light and matter or between light and light. This immediately suggests the means of fulfilling many of the promises of nonlinear optics through nonlinear interactions in PhCs.

Since the initial proposition of the PhC concept, an increasing amount of research has been dedicated to its design and fabrication, leading to innovations both in optics and in material science. After pioneering work in the microwave domain, it's transposition to the optical frequencies was slowed down by the technological difficulties in the fabrication of three-dimensional (3D) optical PhCs. This led researchers to divert their efforts essentially towards one-dimensional (1D) and two-dimensional (2D) PhC where one could greatly profit from the tremendous advances made in nano-technology. Besides, 2D PhCs form the most practical platform for a compact integration of all-optical circuits comprising elements such as micrometer size lasers, filters, modulators, etc.

By combining the capability of 2D PhCs to manipulate light with the nonlinear material response, versatile reconfigurable multifunctional devices may be envisaged. By "reconfigurable" we mean, for example, that we may modify the optical response of a PhC, or that a single element may be attributed different functions such as slowing down of pulses, pulse shaping, active filtering or frequency shifting by changing the optical characteristics of the external stimulus. The final aim is to arrive at all-optical functions and achieve efficient all-optical signal processing. For this, two principal ingredients are needed: firstly, well defined photonic states, guided modes or cavity resonances, capable of carrying or stocking the optical information are required; secondly, a strong nonlinear response, able to optically manipulate the information with reasonable optical consumption is needed. 2D semiconductor PhCs give an independent handle on both these aspects and allow a fine tuning of both the photonic modes and the electronic or intrinsic nonlinear response. Usually nonlinear operation is hampered by high-power requirements.

Several passive operations have already been achieved exploiting sharp bends and guide-microcavity coupling and are opening the way to passive micro-photonic applications. Low-threshold laser operation has also been demonstrated in active 2D PhC dielectric membranes with embedded quantum wells (or boxes), exploiting either confinement in high-Q microcavities [3], or field localization at band-edge singularities both for in-plane emission [4], and for vertical surface emission [5].

Although nonlinear effects like gain and absorption saturation are the main ingredients for laser operation, these nonlinearities or their refractive counterpart have been used only very recently in PhC's for modifying or controlling the response of these structures to an external stimulus. A fair number of theoretical and a few experimental studies have dealt with optical Kerr effect in 2D PhCs [6] They predict a dependence of the photonic band-gap position on the stimulus intensity [7], self trapping and solitonic wave generation [8] for sufficiently high incident intensities. As an alternative to the devices that rely on intensity dependent changes induced in cubic materials, it has been demonstrated recently that the phase-shift induced via intrinsically fast quadratic cascaded nonlinear interaction is capable of performing all-optical processing. In both cases, a new effective dispersion relation is induced in the medium resulting in a shift of the effective index of refraction.

5.2 Ultra-fast Optical Tuning of a Photonic Crystal Response via Quadratic Nonlinearities

In the last decade, the search for optical phenomena capable to perform efficient all-optical processing lead to the proposal of several all-optical switching devices based on the phase shift induced via intrinsically fast and low loss second order cascading nonlinearity [9–12]. In such quadratic nonlinear interaction a wave at the second harmonic (SH) frequency induces a coherent change in the wave at the fundamental frequency detuned from phase matching. This nonlinear effect is capable to perform all-optical processing that could replace the one obtained in devices that rely on intensity dependent changes induced in cubic nonlinear materials.

In 1999 we proposed theoretically an active nonlinear mirror with a quadratic nonlinearity [13]. The device was designed as a 1D microcavity implemented with two parallel mirrors highly reflecting at the fundamental wavelength, and filled with a quadratic nonlinear material. The performance of such a device relies on the optically induced changes in the reflection and transmission of a weak beam at a given frequency ω_F (considered as signal in an optical network), via quadratic nonlinear interaction with a control beam at its second harmonic (SH) frequency ω_{SH}. Reference [13] shows numerically that this nonlinear mirror can operate from almost full transmission when the SH is turned off, to a high reflectivity when the control SH signal is turned on. If materials with high enough second order nonlinear coefficients ($\chi^{(2)}$) and strong control beam intensities are implemented, about 100% changes in the reflection of the signal would be achievable. However, the development of a device based on such phenomenon is limited by the coherent nature of the cascaded quadratic nonlinear interaction because in a real device this input phase difference would be very difficult to control. Reference [13] shows that we may reduce, or even eliminate this phase dependence of the induced nonlinear reflectivity either by increasing the linear reflectivity of the cavity's mirrors, or the SH beam intensity.

Using realistic values for the $\chi^{(2)}$ coefficients corresponding to the available materials that could be implemented in such a microcavity, we would need extremely high control beam intensities in order to induce more than 50% reflectivity changes. In 2001 we have done an experimental study using a microcavity filled with a very thin KTP crystal. Using reasonable control beam intensities changes of the order of 10% have been induced in the PhC response [14]. A further increase in the magnitude of this effect would suppose either higher control intensities (which is not acceptable from the practical point of view), the use of materials with higher nonlinear coefficients, or design of better geometries that would allow the enhancement of the nonlinear effect.

We propose here a new configuration for the device described above that, combined with a high nonlinear semiconductor material, increases considerably the nonlinear effect. The structure is a 1D periodic photonic crystal etched onto a semi-conductor waveguide that consists of a bottom layer of $Al_{50\%}GaAs$ (low refractive index) epitaxially grown on a GaAs substrate and a top layer of $Al_{30\%}GaAs$ (high refractive index). A transverse periodic 1D lattice is patterned perpendicular to this waveguide, as represented schematically in Fig. 5.1. This structure may be seen as a 1.5D PhC because the light is confined in z direction by the periodic structure and in the vertical direction by the waveguide. There are three reasons that make us to chose $Al_{30\%}GaAs$: (i) it is a highly nonlinear material; the maximum value of its nonlinear tensor ($\chi^{(2)} = 120\,pm/V$) corresponding to [110] crystallographic direction, which is exactly the direction of the beam propagation in our geometry; (ii) it is transparent around the wavelengths of 1550 nm and its second harmonic, which we choose for the nonlinear interaction; (iii) there is an advanced fabrication PhC technology based on AlGaAs devices and our device could be easily integrated as a block in a photonic circuit.

Our structure consists of six periods of two alternating layers of $Al_{30\%}GaAs$ (118 nm thick) and air (399 nm thick). The optical length of each layer is taken to be equal to a quarter of the fundamental wavelength ($\lambda_{FF} = 1550\,nm$). Such structure is known as a quarter-wave Bragg mirror and it has a bang-gap centred at 1550 nm, but no gap appears at its SH frequency. The periodicity of the structure is broken by a central defect (1704 nm thick) to excite a transmission mode within its first-order Bragg reflection band. The geometry of the structure (number and thickness of the vertical layers and of the defect) has been chosen in order to provide a narrow photonic resonance around 1.55 μm and a low reflectivity at the SH wavelength. The length of the defect has been selected, as will be stated later, to optimize the nonlinear effect.

We consider a low-intensity fundamental wave ("signal") at λ_{FF} and a high intensity SH ("control beam") simultaneously propagating through the structure described above. To study their propagation and cascaded nonlinear interaction we first study the nonlinear optical response of the structure using a simplified 1D analytical model based on the *transfer matrix method*. We start with the coupled complex amplitude nonlinear wave equations for both fundamental and SH electric

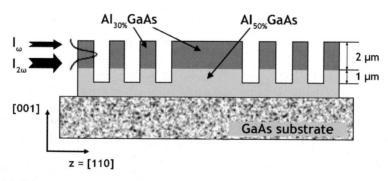

Fig. 5.1 Schematic representation of the one-dimensional periodic structure

fields:

$$-\frac{\partial^2}{\partial z^2}E_\omega(z) - k_\omega^2 E_\omega(z) = \frac{k_\omega^2}{n_\omega^2}\chi^{(2)}E_{2\omega}(z)E_\omega^*(z) \tag{5.1a}$$

$$-\frac{\partial^2}{\partial z^2}E_{2\omega}(z) - k_{2\omega}^2 E_{2\omega}(z) = \frac{k_{2\omega}^2}{2n_{2\omega}^2}\chi^{(2)}E_\omega^2(z) \tag{5.1b}$$

where: $E_\omega(z)$ and $E_{2\omega}(z)$ are the complex amplitudes of the fundamental and of the SH waves propagating in z-direction, respectively, k_ω and $k_{2\omega}$ are the wave numbers of the fundamental and of the SH waves, respectively, while n_ω and $n_{2\omega}$ are the refractive indices of the nonlinear material corresponding to the fundamental and SH frequencies, respectively. $\chi^{(2)}$ is the quadratic nonlinear susceptibility of the nonlinear material. We assume that both fundamental and SH are plane waves propagating in z-direction, normal to the layers of the structure. To properly account for all reflections of the fundamental and SH fields at all linear and nonlinear interfaces, we keep all second order spatial derivatives in Eqs. (5.1a) and (5.1b). In our numerical solution, we consider the energy transfer from the SH to the fundamental wave or vice-versa in both forward and backward directions. We also assume that the absorption of the material is negligible. The two coupled equations for the fundamental and SH fields are numerically integrated in steps much shorter than a wavelength, using the method of variation of constants described in detail in Ref. [13]. This solution is propagated through rest of the periodic structure using the transfer matrix method and setting the usual boundary conditions at either end of the structure, that consider the fields incident from the right to be zero and the fields incident from the left having fixed amplitudes and phases. The amplitude and phase of both electric fields can be calculated at any point within the structure. We may derive then the reflectivity and transmission coefficients, the effective refractive index of the structure and the group velocity for both fundamental and SH waves, accounting for the cascading nonlinear interaction between the two beams. The output of this model gives the changes induced by the nonlinearity in all these parameters and consequently in the PhC optical response.

We apply this method to the structure described above (see Fig. 5.1). When the signal is incident only into the structure, the linear reflectivity in the neighbourhood of 1550 nm and 775 nm is shown in Fig. 5.2a and 5.2b, respectively. One can see that a very narrow symmetric resonance appears at 1557.8 nm which we will consider as fundamental wavelength. The full width at half maximum (FWHM) of this resonance is equal to 0.1 nm, indicating a high quality factor of the resonator, $Q = 15{,}500$. At this wavelength, multiple reflections within the cavity lead to a strong localization of the field and to an important increase of the optical density of the fundamental beam. At the same time, the control beam at λ_{SH} is almost completely transmitted, since the structure has no gap and it is not resonant at the SH wavelength (Fig. 5.2b).

When both control and signal beams are simultaneously incident onto the structure the quadratic nonlinear interaction between the fundamental and the SH beams

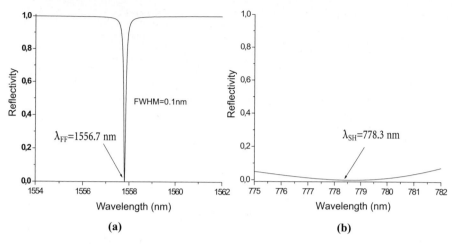

Fig. 5.2 Linear reflectivity of the structure as a function of wavelength in the neighbourhood of λ_{FF} (**a**) and of λ_{SH} (**b**)

within the PhC induces changes both in the effective index of refraction and in the effective group velocity in the neighbourhood of the resonance [15]. The strength of such effect is governed by the product $(1/n_{2\omega})\chi^{(2)}|E_{2\omega}|$ between the quadratic nonlinear susceptibility $\chi^{(2)}$ and the value of the SH field amplitude $|E_{2\omega}|$ and it will not depend on the intensity of the fundamental field. The combination of these two nonlinear effects leads to a shift of the resonance and at the same time to a change of its amplitude, as can be seen in Fig. 5.3, where the intensity of the control beam is 2.3 GW/cm^2. The shift of the resonance can occur to either side of the linear resonance because, as expected from a cascaded nonlinear interaction, the induced changes in the optical parameters of the structure show a strong dependence on the input phase difference ($\Delta\phi = 2\phi(\omega) - \phi(2\omega)$, where $\phi(\omega)$ and $\phi(2\omega)$ are the phases of the fundamental and SH waves, respectively).

However, even if the shift depends on the relative phase difference, if we look what happens at the wavelength of the signal, we observe that the reflectivity changes from almost 0% when the control beam is not present, to a value that ranges between a minimum and a maximum limits, depending on the value of $\Delta\phi$ when the control beam acts onto the structure. This effect can be seen in Fig. 5.4a where the nonlinear reflectivity changes at the signal frequency are shown as a function of the relative phase difference for different intensities of the control beam. For a control beam intensity of 0.5 GW/cm^2 (at the focal point) the reflectivity changes from nearly 0 in the linear case to a value between 53% and 260%. Note that there is a range of $\Delta\phi$ where the reflectivity is higher that 100%. For these input phase differences there is a transfer of photons from the SH to the fundamental field.

Normally, in a real device based on this effect it would be very difficult to control and to keep constant the input phase difference between the signal and the control beams. This is why we would like to decrease as much as possible this input phase dependence of the nonlinear response of our structure, and trying to obtain similar

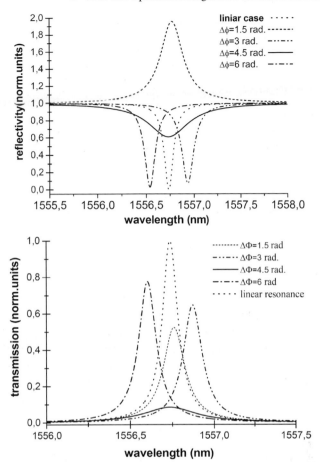

Fig. 5.3 Reflectivity and transmission as a function of wavelength when the input phase difference is equal to: 1.5 rad (*dashed line*), 3 rad (*dashed dot dot line*), 4.5 rad (*continuous line*), 6 rad (*dotted dashed line*). The *dotted line* corresponds to the linear reflectivity

responses for all possible $\Delta\phi$ values. This input phase dependence can be decreased and even eliminated it we decrease the resonance width, and/or if we increase the value of the product $\chi^{(2)}|E_{2\omega}|$. If we think in a future application of this effect in a switching device, make the resonance narrower means to work with longer pulses which is not interesting for such applications.

In order to find the optimum value of the defect length we have made a study of the difference between the minimum and maximum values of the nonlinear reflectivity obtained when the relative phase difference is changed (maxima and minima in Fig. 5.4a) as we modify the thickness of the defect while keeping fixed the control beam intensity. The result, shown in Fig. 5.5 for the particular case when the intensity of the control beam was set to $0.5\,\text{GW/cm}^2$, shows that there are a certain values for the defect thickness where the maximum and minimum values of

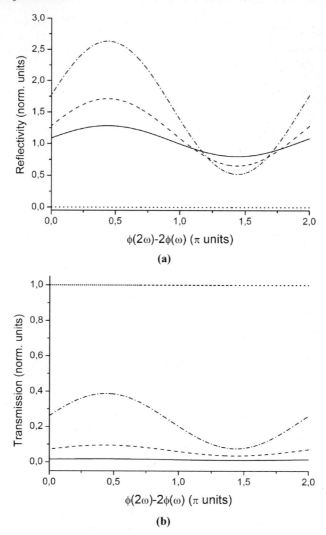

Fig. 5.4 (**a**) Reflectivity and (**b**) transmission of the signal ($\lambda_{FF} = 1557.8$ nm) as a function of the relative phase difference between fundamental and SH waves. The *dashed dot line* corresponds to a SH intensity of 0.5 GW/cm^2, the *dashed line* to 1.3 GW/cm^2 and the *continuous* one to 5 GW/cm^2. *Dotted line* corresponds to the linear case

the reflectivity change are very close each other. The minimum phase dependence is obtained for $L = 1703$ nm (for higher values of the defect length the change in reflectivity is similar but the resonance width decreases so as commented before it is not suitable for short pulse propagation). Setting the defect thickness to this value we show in Fig. 5.4a and b the changes induced in the reflectivity and in the transmission, respectively, for three different values of the SH intensity. One may see that using a control beam of 1.3 GW/cm^2 the PhC response in reflection and in transmis-

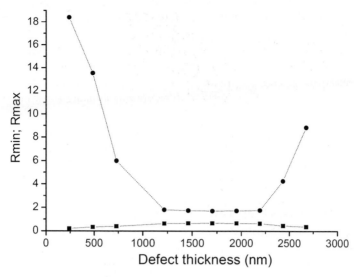

Fig. 5.5 Maximum and minimum changes induced in the reflectivity of our structure at 1557 nm by a control beam of 0.5 GW/cm^2 (dashed dot line in Fig. 5.4) as a function of the defect thickness

sion becomes much less sensitive to the input phase difference. If we could further increase the control beam intensity up to 5 GW/cm^2 the reflectivity would change from 0% to a value between 80% and 120%, which means that we would have drastically reduced the input phase dependence of the reflectivity changes, obtaining at the same time a very good contrast between the linear and nonlinear cases. If we consider the modification of the transmission of the PhC we see that for the same value of the control beam we practically switch the transmission between 100% in the linear case to less than 5% for any value of the input phase difference.

The time duration of such optical modification of the PhC response is only limited by the resonance width or by the pulse duration, since the nonlinear interaction that produces the effect is intrinsically fast.

Once we have found the best configuration of the structure and we have improved the nonlinear induced response of the PhC using the 1D model we have to take into account the whole structure as it should be constructed in a real device. This means that we have to consider the 1D PhC but also the waveguide in which it is etched. This guided PhC structure may be seen as a 1.5D PhC. Waveguiding in 1D or 2D PhC, offers a twofold advantage: it allows a supplementary field enhancement due to transverse field confinement and it reduces problems related with the spatial walk-off between the interacting fields. Nevertheless, guiding operation introduces new drawbacks associated, in particular, with the need to achieve guiding at two very distant wavelengths such as those of the fundamental and the SH. Also, the use of periodic media with high index contrast of indices between the linear and nonlinear layers induce an additional source of losses. These intuitive arguments show that special care should be taken to avoid or at least minimize radiation losses. Recently, a study concerning the optimization of the losses at 1.5 μm and at it's SH has been done in a similar structur [16]. This study shows that losses sensibly depend on the thickness of the $Al_{30\%}GaAs$ layer and on the etching deepness. This study applied to our structure shows that the configuration that provides the minimum losses consists in a 1 µm thick $Al_{30\%}GaAs$ layer and a 3 µm deep etching. We have used a commercial FDTD code to perform the simulation of the linear propagation of these two beams trough the structure. After an optimization of the parameters of the structures that reduces as much as possible the losses, we have obtained a quality factor for the photonic resonance at $\lambda_{FF} = 1557$ nm of 15,000, which is equivalent to the one considered in the 1D simulation results obtained previously with the simplified model. The transmission of the SH non-resonant field was also calculated.

5.3 Multifunction Operation in Two-Dimensional Semiconductor Photonic Crystal Slabs

Here we present experimental results on lasing operation, dramatic frequency shifts (up to 19 times the line-width of a PhC resonance) via Kerr nonlinear optical effect and the potential to obtain ultra-fast switching speeds using a 2D PhC slab with just kW level pump powers. The theoretical model we developed, based on a nonlinear coupled mode theory accurately explains the experimentally observed nonlinear phenomena.

The 2D PhC slab is made of high index contrast InP-based material with pattern lattice of cylindrical holes. The maximum optical confinement obtained in PHC structures is in the suspended membrane configuration where the structure is flanked by air layers offering maximum index contrast. However, in order to efficiently manage the thermal budgeting, we opted for an alternative configuration designed for a maximum evacuation of heat, where the membrane is in contact with a thermally draining medium. The active membrane is a hetero-structures incorporating quantum wells. In 2D PhC slabs without defect in periodicity, the coupling between the quasi-continuum of states of radiation modes of the Fabry–Perot resonator formed by the air/PhC slab/substrate and the low group velocity photonic modes sharp resonances occur in the reflectivity spectrum for light incident normal to the plane of periodicity. Such resonances, signature of the coupling between a discrete mode and a continuum of states, are of a general nature and known as Fano resonances (FR) [17]. In PhCs these resonances are used in the linear optical regime as a means of investigating optical modes above the light line [18]. In our work, we use one of these FRs to explore the nonlinear optical response in a 2D PhC slab. Even though the PhCs are designed for in-plane guided operation it is more practical to address them normal to the patterning not just for experimental comfort but more importantly for 2D matrix array signal processing. The narrow spectrum of the FRs attests to the enhancement of the optical mode density. Hence the use of FRs is particularly interesting in that it can be used for obtaining efficient laser effect and increased nonlinear interactions because of the high Q factor. The narrow spectrum is also well adapted for switching operation as a shift of one full width at half maximum (FWHM) of 0.5nm with high contrast may be activated rapidly.

The 2D PhC slab used in our experiments is a graphite-lattice structure patterned on an InP membrane incorporating an active medium composed of four InAs$_{0.65}$P$_{0.35}$/InP (50 Å/200 Å) quantum wells with its photoluminescence spectrum centered at 1450 nm. The $\lambda/2n$ optical thickness of the membrane is chosen to ensure a high photon lifetime and hence an increased in-plane nonlinear interaction time due to a strong field confinement in the vertical direction. The heterostructure is transferred onto a heat sinking silicon wafer by SiO$_2$-SiO$_2$ wafer bonding. The $20 \times 20 \mu m^2$ 2D graphite lattice of cylindrical air holes was patterned through a silica mask into the semiconductor by electron beam lithography followed by reactive ion etching. The period of the etched PhC was 791 nm with an air filling factor of 20%. A scanning electron micrograph (SEM) picture of the fabricated structure is shown in the left side of Fig. 5.6. The photonic band structure corresponding to the sample described above is calculated using a three-dimensional plane-wave expansion method and is shown in Fig. 5.6 (right). Around the Γ point (point A in the Fig. 5.6) where the flatness of the band indicates a very small average group velocity a distinct FR occurs at the desired wavelength. The structure was designed to have a FR far into the electronic Urbach tail, at 1540 nm that is about 90 nm from the maximum of the luminescence peak.

The sample was excited by 120 fs-long pump pulses at 810 nm provided by a Ti:Sapphire laser. The nonlinear response was elucidated using 150 fs-long probe pulses around 1540 nm obtained from an Optical Parametric Oscillator (OPO). The repetition rate of both pulses was 80 MHz. Both pulses were incident normal to the PhC surface and focused via an achromatic microscope objective (MO) down to 5 μm spot diameter, addressing ~7 periods of the PhC. In order to optimally couple the light into the FR, the probe beam was linearly polarized. We collected the retro reflected probe beam in the direction perpendicular to the PhC through the same microscope objective and focused it onto a single mode optical fibre (SMF) connected to an optical spectrum analyzer (OSA). A schematic representation of the set-up is

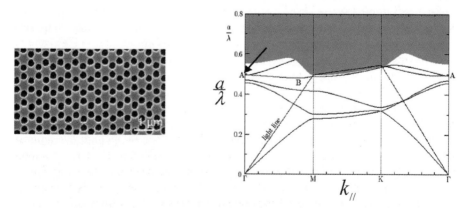

Fig. 5.6 A SEM picture of the fabricated 2D PhC slab (*left*); calculated band structure of our graphite lattice PhC (*right*). At high energy (*grey area*) the bands cannot be determined accurately due to the complexity of the interactions with the radiation modes

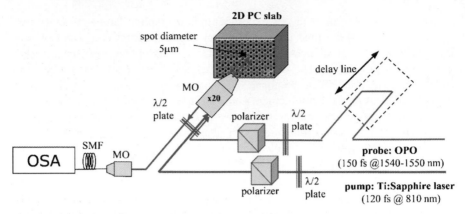

Fig. 5.7 Experimental pump and probe set-up

shown in Fig. 5.7. The pump is rejected prior to the injection into the fibre. The probe beam mean power was set at $0.2\,\mathrm{mW}$ ($1\,\mathrm{kW/cm^2}$ intensity). Since the sample is almost transparent around the probe wavelength, this intensity is sufficiently small to be unperturbative. The pump wavelength was chosen to be 810 nm in order to excite the electronic states high into the conduction bands.

When only the probe beam is incident onto the sample, a symmetric resonance, shown in Fig. 5.8a, appears at 1543.2 nm, very close to the expected wavelength predicted by the band structure calculation (see Fig. 5.6b). The shape of FR depends on the relative positions of the resonant 2D PhC discrete mode with respect to the continuum of radiation modes. In our structure, the resonant wavelength of the PhC coincides with a reflectivity minimum of the continuum determined by the $\lambda/2n$ Fabry–Perot. For this configuration, the coupling between the PhC and radiation modes leads to a symmetric resonance [19]. The FWHM of the linear resonance is 0.42 nm, corresponding to a high quality factor of $Q = 3700$. When the sample is excited only by the pump at the electronically resonant wavelength (810 nm), we obtain the laser effect for a threshold intensity of $4.5\,\mathrm{kW/cm^2}$. The laser intensity as a function of the pump intensity is shown in Fig. 5.8b.

When the sample is excited by the pump beam in the presence of the probe, the FR experiences a strong blue shift due to the pump induced nonlinear refractive index. The reflectivity spectra around the FR wavelength measured for different values of the pump power are shown in Fig. 5.9. Note that a shift due to thermal effect would be towards the red. A shift of one FWHM is obtained for a low ($0.13\,\mathrm{kW/cm^2}$) pump intensity. This shift is accompanied by a small decrease of reflectivity. As the pump beam is increased, the reflectivity increases and a strong shift, up to 4.5 nm, that is 10 times the FR FWHM, is observed! If the pump power is further increased up to the laser threshold value of $4.58\,\mathrm{kW/cm^2}$, the shift saturates and ends at a value of 8 nm, due to carrier depletion. The shift as a function of pump intensity is depicted in Fig. 5.10.

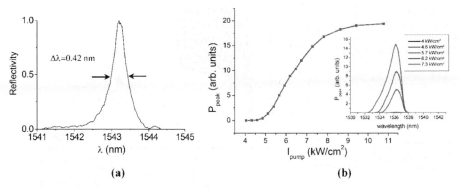

Fig. 5.8 (**a**) Linear resonance of the 2D PhC slab measured in reflection when only the probe signal is incident onto the sample. (**b**) Laser intensity as a function of the pump intensity. The *inset* shows several laser spectra corresponding to different pump intensities

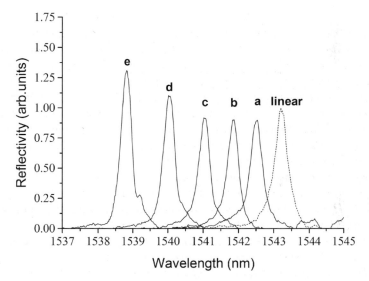

Fig. 5.9 Experimental reflectivity spectra around the FR wavelength for several pump powers. The *bold line* represents the linear reflectivity, when only the probe is incident on the sample. The *thin line* spectra correspond to the probe reflectivity at different pump intensities: $0.19\,\mathrm{kW/cm^2}$ (**a**), $0.43\,\mathrm{kW/cm^2}$ (**b**), $0.74\,\mathrm{kW/cm^2}$ (**c**), $1.13\,\mathrm{kW/cm^2}$ (**d**) and $1.78\,\mathrm{kW/cm^2}$ (**e**). All spectra are normalized to the maximum value of the linear reflectivity

Then we conducted measurements on the dynamics of the nonlinear response. We explored the possibility of basing a switching operation on the shifting of the wavelength of a PhC resonance via electronically resonant optical Kerr effect. A controlled temporal delay is introduced between pump and probe pulses by changing the probe optical path using a translation stage.

The dynamics of the shift is studied by measuring the reflectivity spectra for different values of the pump-probe delay time Δt. The external intensity of pump

Fig. 5.10 Nonlinear wavelength shift of the FR as a function of the pump intensity. The *square dots* correspond to the experimental measurements and the *solid curve* represents the wavelength shift predicted by our theoretical model, for the pump intensity used in the experiment

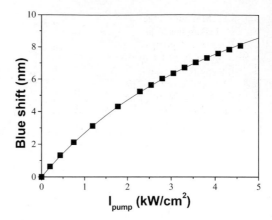

and probe are fixed at $1.6 \, \text{kW/cm}^2$ and $1 \, \text{kW/cm}^2$, respectively. In Fig. 5.11 the curve (a) represents the linear reflectivity, obtained for $\Delta t < 0$ i.e. when the probe arrives before the pump. At $\Delta t = 0$ a 2.6 nm blue shift of the resonance is observed (curve (b)). The onset time of the shift is extremely fast and is determined by the establishment of the stationary regime in the cavity ($\sim 1 \, \text{ps}$). The curves (c)–(j) correspond to measurements at increased pump-probe delays. It is clear that as Δt increases, the resonance shift diminishes due to carrier population relaxation and a consequent reduction in the induced nonlinear index. The total time that the resonance requires to return to its linear wavelength is of course directly related to the carrier population total lifetime τ.

Figure 5.12 shows the temporal evolution of the resonance shift for two different pump intensities. As expected, the maximum nonlinear shift increases with the pump intensity. The resonance shift versus delay time decays exponentially for all the pump intensities. The shift decay time τ, at $1/e$ from the maximum value it is a measure of the carrier population relaxation time in the quantum wells.

Fig. 5.11 Reflectivity spectra for different pump-probe time delays: curve (*a*) obtained at $\Delta t = -16 \, \text{ps}$ corresponds to the linear reflectivity, (*b*) $\Delta t = 0 \, \text{ps}$; (*c*) $\Delta t = 8 \, \text{ps}$; (*d*) $\Delta t = 16 \, \text{ps}$; (*e*) $\Delta t = 20 \, \text{ps}$; (*f*) $\Delta t = 32 \, \text{ps}$; (*g*) $\Delta t = 40 \, \text{ps}$; (*h*) $\Delta t = 56 \, \text{ps}$; (*i*) $\Delta t = 80 \, \text{ps}$ and (*j*) $\Delta t = 130 \, \text{ps}$. The pump and probe intensities were $1.6 \, \text{kW/cm}^2$ and $1 \, \text{kW/cm}^2$, respectively

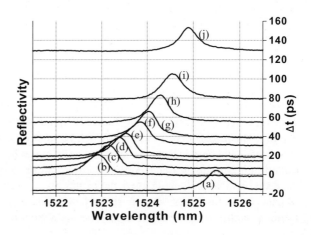

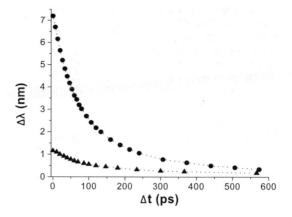

Fig. 5.12 Measured blue shift of the resonance as a function of the pump-probe time delay Δt for two different pump intensities: $I_{low} = 0.43\,\mathrm{kW/cm^2}$ (*triangle dots*) and $I_{high} = 5.6\,\mathrm{kW/cm^2}$ (*circular dots*)

The decrease in τ from 162 ps to 91 ps as the pump power increased from 1.1 to 7.4 kW/cm² is mainly due stimulated emission. Moreover, as observed in Fig. 5.12 the resonance never reaches its linear wavelength, because a certain number of carriers remain trapped at the etched interfaces and do not relax completely in the investigated Δt.

The fast initial change in the shift of the resonance depicted in Fig. 5.12 may be used very efficiently for operating a fast all-optical switching device. In such a device a shift of the order of one FWHM of the resonance would be enough, to switch with high contrast. For example, pumping the structure with an intensity of 1.17 kW/cm² shifts the PhC resonance by 1.6 nm with respect to its linear wavelength. A small supplementary pump intensity of 0.43 kW/cm² moves the central wavelength of the resonance by one FWHM (i.e. 0.6 nm) from its former position. The reflectivity spectra corresponding to these two values of the pump intensities are shown as curve (a) and (b) in Fig. 5.13a. We determine the rapidity of this shift by studying the time dependence of the differential reflectivity $\Delta R/R$ at $\lambda_1 = 1523.6$ nm and at $\lambda_2 = 1523$ nm as shown in Fig. 5.13b. A very large change, higher than 85%, is obtained in 20 ps. Further, for higher pump intensities a faster switching time can be obtained between two other wavelengths. A change in reflectivity of the order of 80% was observed with a time response as short as 7.3 ps. To simplify further, we can imagine the use of a CW laser at 810 nm to maintain the shifted resonance at λ_1. Then a switch may be operated between λ_1 and λ_2 separated by one FWHM by sending very low energy pump pulses.

The total time response (rise and recovery) of the resonance shorter than 10 ps can be obtained with reflectivity contrasts higher than 80%. These results confirm that an optically controlled nonlinear 2D PhC is an excellent platform for a versatile switching device. This may constitute one of the building blocs in the all-optical circuits envisaged with 2D PhCs.

The possibility of tuning photonic band-gaps at very short time scales by an external parameter opens the way to perform several functions, such as switching, routing, temporal demultiplexing, etc. Some "tuneable" PhCs based on non-optical

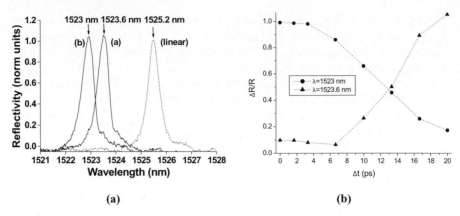

Fig. 5.13 (**a**) Reflectivity spectra around the PhC resonance wavelength for two different pump powers: $1.17\,\mathrm{kW/cm^2}$ for curve (*a*) and $1.6\,\mathrm{kW/cm^2}$ for curve (*b*). *Dashed line* represents the linear reflectivity. All spectra are normalized to the maximum of the linear reflectivity; (**b**) Differential reflectivity $\Delta R/R$ as function of the pump-probe delay at: 1523.6 nm (*triangular dots*) 1523 nm (*circle dots*)

processes using infiltrated liquid crystals, thermo-optic semiconductors, ferroelectric, ferromagnetic or II–VI materials have also been proposed. But the time scales of these processes are in the micro to millisecond range and they fall short of the requirements for certain telecommunication applications. Electronically, resonant optical third order nonlinear effect gives faster tuneability of the photonic bandgaps especially when the switching time is not based on the return to equilibrium of carriers and seems a more appropriate candidate for applications when picosecond round clock responses are required.

5.4 Conclusions

In conclusion, we have shown that semiconductor based 1D and 2D PhC structures can bring about second order and third order nonlinear optical phenomena that could be used for all-optical information transmission and processing. Specifically, we have investigated two different nonlinear phenomena. First, we have studied how quadratic dispersive effects brought about by an intense control pulse at frequency 2ω can change the transmittance and reflectance characteristics of the material structure for a signal beam at frequency ω. Several 1D configurations involving different materials and geometries have been investigated. We have found that the best efficiency can be achieved with a waveguide modulated transversely with alternating $\mathrm{Al_{30\%}GaAs}$ – air layers. In this case, control-beam intensities of the order of one or a few $\mathrm{GW/cm^2}$ would be necessary, for pulses of duration above 30 ps. Further improvements would probably require considering 2D or 3D structures.

Second, we have investigated third-order nonlinearities at the probe-field frequency brought about by carrier density changes in the conduction and valence bands induced by an optical control pulse. Here the control pulse can have any frequency connecting the valence and conduction bands and its intensity can be very low, below kW/cm^2, since the carrier-induced effects are strong, especially at points where the density of states defined by the PhC structure is large. The response time is short, of the order of 10 ps. Thus, 2D PhC slabs are shown to be an efficient platform to obtain enhanced nonlinear Kerr response. When the nonlinearity is turned on, the effective index increases and moves the photonic bands of the structure. Indeed, the nonlinear response can be engineered and substantially enhanced through tailoring the photonic modes and through spatial and spectral field localization. Under moderate pumping powers – today typically $100 \, W/cm^2$ – the PhC resonances can be dynamically manipulated at will and the spatial, spectral and/or temporal PhC response can be modified. Improvements on design and fabrication procedures should further lower the pumping powers by at least two orders of magnitude. These experimental demonstrations towards reconfigurable PhCs bring the prospect of all optical reconfigurable circuits a little closer to reality.

Numerous applications can be envisioned for highly nonlinear micro/nanostructured devices in the field of information and communication technologies. An all-optical approach for basic functions such as probe regeneration, switching and routing can significantly reduce the systems complexity. In the field of optical telecommunications, WDM-compatible microcavity-based devices [20] have been successfully used to demonstrate efficient 2R (1R is reamplify; 2R is 1R+reshape; 3R is 2R+retime) all-optical regeneration in a 40 Gbit/s long-haul transmission [21]. The benefit of this device was to increase the errorless transmission distance. Significant improvements are expected from the use of 2D PhCs in nonlinear devices due to their ability to obtain an increased photonic confinement. This should lead to a reduction of switching energy to $< kW/cm^2$, that would allow operation at still higher bit rates. For 3R all-optical regeneration, the existing technologies require the use of semiconductor optical amplifiers (SOA), which suffer from relatively high power consumption, and hence cannot easily be extrapolated to WDM systems. The combination of narrow resonance shifts and amplification could lead to the development of entirely novel switching gates with high switching contrast and low insertion losses, useful for 3R all-optical regeneration and routing or multiplexing in WDM systems.

Acknowledgements This work was partly financially supported by the COST Action P11: Physics of Linear, Nonlinear and Active Photonic Crystals, by the Spanish Ministerio de Educación y Ciencia through the project FIS2005-07931-C03-03 and by the Generalitat de Catalunya trough the project SGR2005 00457.

References

1. Yablonovitch E.: Inhibited Spontaneous Emission in Solid-State Physics and Electronics, *Phys. Rev. Lett.* 58, 2059–2062 (1987); John S.: "Strong localization of photons in certain disordered dielectric superlattices", *Phys. Rev. Lett.* 58, 2486–2489 (1987).
2. Sakoda K.: Enhanced light amplification due to group-velocity anomaly peculiar to two- and three-dimensional photonic crystals, *Opt. Express* 4, 167–176 (1999).

3. Monat C., Seassal C., Letartre X., Viktorovitch P., Regreny P., Gendry M., Rojo-Romeo P., Hollinger G., Jalaguier E., Pocas S. and Aspar B.: InP 2D photonic crystal microlasers on silicon wafer: room temperature operation at 1.55 μm, *Electron. Lett.* 37, 764–766 (2001).
4. Monat C., Seassal C., Letartre X., Regreny P., Rojo-Romeo P., Viktorovitch P., Le Vassor d'Yerville M., Cassagne D., Albert J.P., Jalaguier E., Pocas S. and Aspar B.: InP-based two-dimensional photonic crystal: in-plane Bloch mode laser, *Appl. Phys. Lett.* 81, 5102–5104 (2002); Sakoda K., Ohtaka K., Ueta T.: InP 2D photonic crystal microlasers on silicon wafer: room temperature operation at 1.55 μm, *Opt. Express* 4, 481–489 (1999).
5. Imada M., Noda S., Chutinan A. and Tokuda T.: Coherent two-dimensional lasing action in surface-emitting laser with triangular-lattice photonic crystal structure *Appl. Phys. Lett.* 75, 316–318, (1999); Mouette J., Seassal C., Letartre X., Rojo-Romeo P., Leclercq J.-L., Recgreny P., Viktorovitch P., Jalaguier E., Perreau P., Moriceau H.: Very low threshold vertical emitting laser operation in InP graphite photonic crystal slab on silicon *Electron. Lett* 39, 526–528 (2003).
6. Leonard S.W., van Driel H.M., Schilling J. and Wehrspohn R.B.: Ultrafast band-edge tuning of a two-dimensional silicon photonic crystal via free-carrier injection, Phys. Rev. B 66, 161102(R) (2002); Hu X., Zhang Q., Liu Y., Cheng B., Zhang D.: Ultrafast three-dimensional tunable photonic crystal, *Appl. Phys. Lett.* 83, 2518–2520 (2003).
7. Banaee M.G., Cowan A.R. and Young J.F.: Third-order nonlinear influence on the specular reflectivity of two-dimensional waveguide-based photonic crystals *J. Opt. Soc. Am. B* 19, 2224–2231 (2002).
8. Eggleton B.J., Slusher R.E., de Sterke C.M., Krug P.A. and Sipe J.E.: Bragg grating solitons, *Phys. Rev. Lett.* 76, 1627–1630 (1996).
9. Russell P.S.J., *Electon. Lett.* 29, 1228–1229 (1993).
10. Hagan D.J., Wang Z., Stegeman G., Van Stryland E.W., Sheik-Bahae M. and Assanto G., *Opt. Lett.* 19, 1305–1307 (1994).
11. Krumbügel M.A., Sweetser J.N., Fittinghoff D.N., DeLong K.W. and Trebino R.: *Opt. Lett.* 22, 245–247 (1997).
12. Asobe M., Yokohama I., Itoh H. and Kaino T., *Opt. Lett.* 22, 274–276 (1997).
13. Cojocaru C., Martorell J., Vilaseca R., Trull J. and Fazio E.: Active reflection via a phase-insensitive quadratic nonlinear interaction within a microcavity, *Appl. Phys. Lett.* 74, 504–506 (1999).
14. Cojocaru C., Martorell J., Díaz F. and Vilaseca R.: Actively induced transmission via a quadratic nonlinear optical interaction in a potassium titanyl phosphate microcavity, *Appl. Phys. Lett.* 79, 4479–4471 (2001).
15. Cojocaru C. and Martorell J.: Induced group and phase velocity change via a cascaded quadratic nonlinear interaction within a one-dimensional photonic crystal *J. Opt. Soc. Am. B* 19, 2141–2147 (2002).
16. Dumeige Y., Raineri F., Levenson A. and Letartre X., *Phys. Rev. E* 68, 066617 (2003).
17. Fano U.: Effects of configuration interaction on intensities and phase shift, *Phys. Rev.* 6, 1866 (1961).
18. Astratov V.N. et al.: Resonant coupling of near-infrared radiation to photonic band structure waveguides, *J. Lightwave Technol.* 17, 2050–2057 (1999).
19. Fan S.: Sharp asymmetric line shapes in side-coupled waveguide-cavity systems, *Appl. Phys. Lett.* 80, 908–910 (2002).
20. Mangeney J., Oudar J-L., Harmand J-C., Mériadec C., Patriarche G., Aubin G., Stelmakh N. and Lourtioz J-M.: Ultrafast saturable absorption in heavy-ion irradiated quantum well vertical cavity, *Appl. Phys. Lett.* 76, 1371–1373 (2000).
21. Rouvillain D., Brindel P., Seguineau F., Pierre L., Leclerc O., Choumane H., Aubin G. and Oudar J-L.: Optical 2R regenerator based on passive saturable absorber for 40 Gbit/s WDM long-haul transmissions, *Electron. Lett.* 38, 1113–1114 (2002).

6 Nonlinear Optics with Photonic-Crystal Fibres

Aleksei M. Zheltikov

Physics Department, International Laser Center, M.V. Lomonosov Moscow State University
Vorob'evy gory, 119992 Moscow, Russia

Abstract. Breakthroughs in photonic crystal fibre technologies push the development of a new class of fibre-optic frequency converters, broadband light sources, and short-pulse lasers. The frequency profile of dispersion and the spatial profile of electromagnetic field distribution in waveguide modes of photonic crystal fibres can be tailored by modifying the core and cladding design on a micro- and nanoscale, suggesting the ways of creating novel fibre-optic devices for highly efficient spectral and temporal transformation of laser pulses with pulse widths ranging from tens of nanoseconds to a few optical cycles (several femtoseconds) within a broad range of peak powers from hundreds of watts to several gigawatts. In new fibre lasers, microstructure fibres provide a precise balance of dispersion within a broad spectral range, allowing the creation of compact all-fibre sources of high-power ultrashort light pulses.

Key words: photonic-crystal fibers; nonlinear optics

6.1 Introduction: Photonic-Crystal Fibre Components for Advanced Optical Technologies

Photonic-crystal fibres (PCFs) [1–3], also referred to as microstructure, or holey, fibres, are optical waveguides of a new type. In PCFs, radiation can be transmitted through either a solid (Fig. 6.1a–e) or hollow (Fig. 6.1f,g) core, surrounded with a microstructured cladding, consisting of an array of cylindrical air holes running along the fibre axis. Such a microstructure is usually fabricated by drawing a preform composed of capillary tubes and solid silica rods.

Along with conventional waveguide regimes, provided by total internal reflection, PCFs under certain conditions can support guided modes of electromagnetic radiation due to the high reflectivity of their cladding within photonic band-gaps (PBGs) or regions of low densities of photonic states [4–7], as well as by the antiresonance mechanism of waveguiding [3, 8]. Such regimes can be supported by fibres with a hollow [5–7,9] or solid [10] core and a two-dimensionally periodic (photonic crystal) cladding. A high reflectivity provided by the PBGs in the transmission of such a cladding confines radiation in a hollow core, substantially reducing the loss, which is typical of hollow-core-guided modes in conventional, capillary-type

97

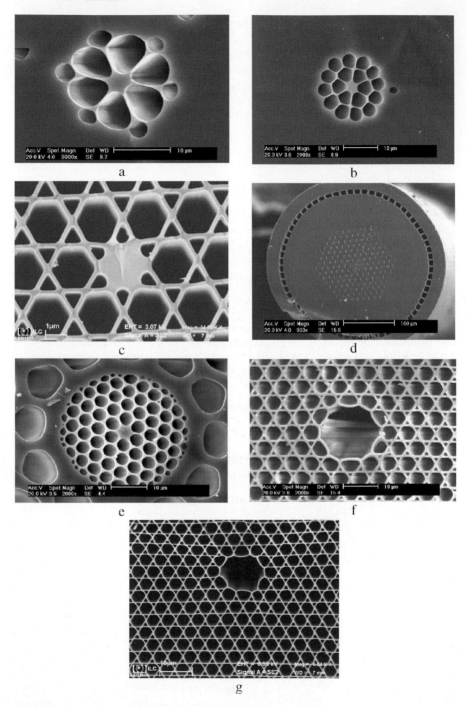

Fig. 6.1 Cross-section images of photonic-crystal fibres: (**a–c**) fibres with a high optical nonlinearity provided by a small fibre core and a high refractive-index contrast between the core and the cladding, (**d**) dual-cladding PCF, (**e**) large-mode-area PCF, and (**f,g**) hollow-core PCFs

hollow waveguides and which rapidly grow with a decrease in the diameter of the hollow core [11, 12].

Unique properties of PCFs open up new routes for a long-distance transmission of electromagnetic radiation [1–3], as well as for nonlinear-optical transformation of laser pulses [13]. As shown by Knight et al. [14], PCFs can support single-mode waveguiding within a remarkably broad frequency range. Photonic-crystal fibres offer new solutions for laser physics, nonlinear optics, and optical technologies, as they combine dispersion tuneability and a high degree of light-field confinement in the fibre core. Dispersion of such fibres is tailored by changing their core–cladding geometry [15, 16], while a strong light-field confinement is achieved due to the high refractive-index step between the core and the microstructure cladding [17]. Controlled dispersion of PCFs is the key to new solutions in optical telecommunications and ultrafast photonics. The high degree of light-field confinement, on the other hand, radically enhances the whole catalogue of nonlinear-optical processes and allows observation of new nonlinear-optical phenomena. Figure 6.1a–c display the cross-section views of PCFs with a high refractive-index step from the fibre core to the fibre cladding, controlled by the air-filling fraction of the microstructure cladding. The fibres of this type can strongly confine the electromagnetic field in the fibre core, providing high optical nonlinearities, thus radically enhancing nonlinear-optical interactions of light fields. Highly efficient fibre-format frequency converters of ultrashort light pulses [13] and PCF supercontinuum sources [18–22] based on highly nonlinear PCFs (Fig. 6.1a–c) are at the heart of advanced systems used in optical metrology [23–26], ultrafast optical science [27, 28], laser biomedicine [29], nonlinear spectroscopy [30–32], and nonlinear microscopy [33–35].

The possibility of dispersion tailoring makes PCFs valuable components for dispersion balance and dispersion compensation in fibre-optic laser oscillators intended to generate ultrashort light pulses with a high quality of temporal envelope. Lim et al. [36] have demonstrated an ytterbium-fibre laser source of 100-fs pulses with an energy of about 1 nJ with dispersion compensation based on a PCF instead of free-space diffraction gratings. A highly birefringent hollow-core PCF [37] provides a robust polarization-maintaining generation of 70-fs laser pulses with an energy of about 1 nJ in a fibre laser system [38]. Isomäki and Okhotnikov [39] have achieved dispersion balance in an ytterbium femtosecond fibre laser using an all-solid PBG fibre [10]. In contrast to silica–air index-guiding microstructure fibres, including silica and air holes, an all-solid PBG fibre guides light along a silica core surrounded with a two-dimensional periodic lattice of high-index glass inclusions. Dispersion tailoring and a high nonlinearity of small-core PCFs, on the other hand, allow efficient optical parametric oscillation and amplification due to the third-order optical nonlinearity of the fibre material [40–42]. Optical parametric oscillators based on PCFs can serve as efficient sources of correlated photon pairs [43, 44].

The maximum laser fluence in an optical system is limited by the laser damage of material of optical components. An increase in a fibre cross section is a standard strategy for increasing the energy of laser pulses delivered by fibre lasers. Standard large-core-area fibres are, however, multimode, making it difficult to achieve a high quality of the transverse beam profile. This difficulty can be resolved by using PCFs with small-diameter air holes in the cladding, which filter out high-order waveguide modes [14, 45]. This strategy can provide single-mode waveguiding even for large-core-area fibres [46, 47] (Fig. 6.1d). A dual-clad PCF design helps to confine the pump field in the microstructured cladding and to optimize a spatial overlap between the pump field and laser radiation. In this type of PCFs, the microstructured part of the fibre is isolated from the cladding by an array of large-diameter air holes (Fig. 6.1d). Large-mode-area ytterbium-doped PCFs [48, 49] are employed for the creation of high-power lasers [47, 50, 51]. Large-mode-area silica PCFs are also used for the compression of high-power subpicosecond laser pulses [52] and the generation of supercontinuum with an energy in excess of 1 μJ [53, 54].

Photonic-crystal fibre design presented in Fig. 6.1e is of special interest also for the development of novel fibre-optic sensors [55–57]. In sensors of this type, excitation radiation is delivered to an object along the fibre core. The inner part of the microstructured cladding features micrometer-diameter air holes and serves for a high-numerical-aperture collection of the scattered or fluorescent signal from the object, as well as for the fibre delivery of this signal to a detector. With such a scheme of sensing, a detector can be placed next to a radiation source [56, 57]. This fibre design is advantageous for sensing chemical and biological samples by means of one- and two-photon luminescence. A microstructured cladding of PCF can be also conveniently filled with a liquid-phase analyte. Radiation propagating along the fibre core will then induce luminescence of the analyte, allowing the detection of specific types of molecules from the minimal amount of analyte [57]. Such fibre sensors can be integrated into chemical and biological data libraries and data analyzers, including biochips, suggesting an attractive format for the readout and processing of the data stored in such devices.

The energy of laser pulses in fibre-optic devices can be radically increased through the use of hollow-core fibres. For standard, capillary-type fibres, however, the loss rapidly grows (as $\propto a^{-3}$) with a decrease in the core radius a [11, 12]. Because of this problem, such fibres cannot provide single-mode guiding or help to achieve high intensities for pulses with moderate peak powers. The loss of core-guided modes in hollow fibres can be radically reduced if the fibre has a two-dimensionally periodic (photonic crystal) cladding [5–7, 9] (Fig. 6.1f,g). A strong coupling of incident and reflected waves, occurring within a limited frequency range, called a photonic band-gap, leads to a high reflectivity of a periodically structured cladding, allowing low-loss guiding of light in a hollow fibre core. Hollow PCF compressors in fibre-laser systems [58, 59] allow the generation of output light pulses with a pulse width on the order of 100 fs in the megawatt range of peak powers.

Thus, PCFs play the key role in the development of novel fibre-laser sources of ultrashort light pulses and creation of fibre-format components for the control of such pulses. In what follows, we examine the physical mechanisms behind supercontinuum generation in such fibres, analyze various scenarios of spectral broadening and wavelength conversion, and discuss applications of PCF white-light sources and frequency converters in nonlinear spectroscopy and microscopy, as well as in optical metrology.

6.2 Supercontinuum Generation in Photonic-Crystal Fibres

Supercontinuum (SC) generation – a physical phenomenon leading to a dramatic spectral broadening of laser pulses propagating through a nonlinear medium – was first demonstrated in the early 1970s [60, 61] (see [62] for an overview of early experiments on supercontinuum generation). As a physical phenomenon, supercontinuum generation involves the whole catalog of classical nonlinear-optical effects, such as self- and cross-phase modulation, four-wave mixing, stimulated Raman scattering, solitonic phenomena and many others, which add up together to produce emission with extremely broad spectra, sometimes spanning over more than an octave.

Presently, more than three decades after its discovery, supercontinuum generation is still one of the most exciting topics in laser physics and nonlinear optics [20–22], the area where the high-field science in the most amazing way meets the physics of low-energy unamplified ultrashort pulses. The advent of photonic-crystal fibres, capable of generating supercontinuum emission (Fig. 6.2) with unamplified, nano- and even subnanojoule femtosecond pulses, has resulted in revolutionary changes in frequency metrology [23–26] opened new horizons in ultrafast science [27, 28, 63] and allowed the creation of novel wavelength-tunable and broadband fibre-optic sources for spectroscopic [30–32] and biomedical [29] applications. Rainbow of colours produced by a laser beam became an optical instrument and a practical tool.

The key concept underlying the latest breakthroughs in optical metrology involves using frequency combs [64–66] generated by mode-locked femtosecond lasers for the measurement of frequency intervals. Mode-locked femtosecond lasers deliver sequences of light pulses separated by a time interval T equal to the round-trip time of the laser cavity. In the frequency domain, such pulse trains are represented by equidistant frequency combs with a total spectral width determined by the time width of the laser pulses in the train and a frequency interval $\Delta\omega$ between

Fig. 6.2 Supercontinuum generation in a photonic-crystal fibre

the adjacent spectral components equal to $\Delta\omega = 2\pi/T$. Such a frequency comb can be calibrated against an atomic frequency standard and employed as a ruler measuring spectral and, hence, temporal and spatial intervals. Thus, while optical spectroscopy conventionally deals with wavelength measurements, the new concept in optical metrology involves the measurement of frequency intervals by using frequency rulers, improving the precision of optical measurements by many orders of magnitude and allowing the creation of a new generation of frequency standards and optical clocks [26]. The significance of this concept for optical metrology and high-precision laser spectroscopy has been highlighted by the statement of the Nobel Committee on the 2005 Nobel Prize in physics, which specifically mentions the frequency-comb measurement technique [67].

The idea of using mode-locked laser sources of short pulses for high-precision optical measurements was proposed in the late 1970s [68,69]. Hänsch's group were the first to experimentally demonstrate high-precision fine-structure measurements on atomic energy levels using frequency combs generated by picosecond mode-locked lasers. With the total width of the frequency comb being inversely proportional to the pulse width, picosecond pulses cannot provide a sufficiently broad range of measurements – picosecond frequency combs turn out to be too short. Sources of much shorter, femtosecond laser pulses were needed to make frequency combs a practical tool for optical metrology. The advent of PCFs provided an opportunity to directly link such frequency combs to atomic frequency standards.

Convenient, reliable, and compact solid-state femtosecond laser systems, which have become available in the early 1990s (see [70] for a review), can generate frequency combs with a bandwidth sufficient for practical applications in optical metrology and high-precision spectroscopy. The intermode interval $\Delta\omega$ can be locked to a radio-frequency reference source, such as an atomic cesium clock [26], for example. However, even with this task fulfilled, the frequency comb is still not fully linked to a reference frequency standard. The main source of difficulties is that the frequency of the nth spectral component in the frequency comb is not an exact multiple of the intermode interval $\Delta\omega$, but is given by $\omega_n = n\Delta\omega + \omega_0$, where ω_0 is the offset frequency. One of the physical factors giving rise to this frequency offset is the dispersion of optical elements inside the laser cavity, making the phase velocity up of laser radiation different from the group velocity ug of laser pulses. Because of the difference in up and ug, the phase of the light field (the carrier phase) is systematically shifted relative to the pulse envelope from one pulse to another in a pulse train generated by a mode-locked laser. As a result, a femtosecond mode-locked laser delivers a sequence of pulses separated by equal time intervals T, but having a nonzero offset $\Delta\varphi$ of the carrier phase relative to the envelope phase. Fourier transform of such a field gives a spectrum in the form of an equidistant comb of spectral components with frequencies $\omega_n = n\Delta\omega + \omega_0$, where $\omega_0 = \Delta\varphi/T$.

To make a frequency comb suitable as a ruler for high-precision frequency measurements, the frequency ω_0 needs to be measured and stabilized relative to an external frequency standard. This problem is solved through supercontinuum generation in a PCF [23–25] or a tapered fibre [71,72]. A fraction of a light beam produced by a femtosecond laser source of a frequency comb is coupled into a PCF. The remaining part of the laser beam is frequency-doubled in a nonlinear crystal. The key requirement to supercontinuum generation in a PCF is that the spectrum of the supercontinuum at the output of the fibre should span over an octave [26]. If this condition is satisfied, then, for any spectral component n picked from the low-frequency part of the frequency comb, the supercontinuum PCF output contains a frequency component $\omega_{2n} = 2n\Delta\omega + \omega_0$, close to the spectral component $2\omega_n = 2n\Delta\omega + 2\omega_0$ in the spectrum of the second harmonic from the nonlinear crystal. The beat signal produced by the second-harmonic field and the PCF output then contains a component $2\omega_n - \omega_{2n} = \omega_0$, allowing the offset frequency ω_0 to be measured and stabilized. This operation enables an accurate locking of all the spectral components in the femtosecond frequency comb with respect to a frequency standard using a single laser source.

Systems of optical metrology using frequency combs can measure frequency intervals with a relative accuracy at the level of 5×10^{-16} [26]. The method of measurement and stabilization of the carrier–envelope phase offset based on supercontinuum generation in a PCF suggests the possibility of phase control of few-cycle laser pulses [73, 74]. Light pulses with a stabilized carrier–envelope phase offset are necessary for a controlled above-threshold ionization [75], high-order optical harmonic generation [76], and the generation of attosecond pulses [77]. Due to its simplicity, compactness, and reliability, the frequency-comb technique is gaining a wide acceptance for the measurement of fundamental physical constants and creation of practical optical-clock systems [26]. Applications of frequency-comb methods for satellite navigation and a high-precision synchronization of optical networks are under consideration. Through high-order harmonic generation, femtosecond frequency combs are being extended to the UV and X-ray ranges [78].

6.3 Photonic-Crystal Fibre Sources for Nonlinear Raman Microspectroscopy

Through the past two decades, nonlinear-optical techniques have proven to be powerful tools for the investigation of ultrafast dynamics in physical, chemical, and biological systems [79–82]. Adaptively shaped and optimally time-ordered ultrashort laser pulses have given the key for understanding potential surfaces of complicated molecular systems, providing efficient means to control vibrational wave packets and steer chemical reactions [79, 83–87]. Since ultrafast nonlinear-optical methods of time-resolved studies and coherence control typically involve a resonant, sometimes multiphoton, optical excitation of a quantum system, wavelength-tunable sources of ultrashort light pulses have been in great demand in ultrafast optical science. Femtosecond dye lasers, as well as optical parametric oscillators and amplifiers have gained a wide recognition as practical and highly efficient light sources for the investigation of ultrafast processes in matter [70, 88], giving an access to a remarkably broad variety of atomic and molecular systems, including molecular aggregates [89, 90] and complexes of biological significance [91]. Femtosecond dye lasers and optical parametric oscillators and amplifiers, however, inevitably increase the cost of laser experiments and make the laser system more complicated, unwieldy, and difficult to align.

Fibre-optic technologies suggest a new, compact and cost-efficient way of adding wavelength tunability to a laser system. This novel concept of a wavelength-tunable source is based on highly efficient nonlinear-optical transformations of ultrashort laser pulses in a photonic-crystal fibre. Light sources based on PCFs have been recently shown to offer attractive fibre-optic solutions for simplification and compactization of nonlinear-optical microscopes [33] and spectrometers [32,92]. In particular, Paulsen et al. [33] have demonstrated microscopy based on coherent anti-Stokes Raman scattering (CARS) using a blue-shifted dispersive-wave emission from a PCF as a pump field and employing Ti:sapphire laser pulses as a Stokes field. Efficient frequency conversion and supercontinuum generation in PCFs have been shown to enhance the capabilities of chirped-pulse CARS [93] and coherent inverse Raman spectroscopy [94]. Efficient spectral broadening of ultrashort pulses in PCFs with carefully engineered dispersion profiles makes these fibres ideal light sources for pump–supercontinuum probe time- and frequency-resolved nonlinear-optical measurements [95]. Novel light sources based on frequency shifting in PCFs provide a useful tool for the measurement of second-order optical nonlinearities in organic materials [96] and offer interesting new options for multiplex CARS microscopy [34,97]. The capabilities of time-resolved four-wave-mixing spectroscopy with PCF sources can be enhanced by the pulse shaping of the frequency-shifted PCF output [98]. In recent experiments by Motzkus' group [99], a single-beam CARS approach [100–102] has been implemented in a new and elegant way through the generation of supercontinuum in a PCF followed by a compression and pulse shaping of this supercontinuum with the help of a pulse shaper based on a spatial light modulator.

Coherent anti-Stokes Raman scattering [103–106] involves two laser fields with frequencies ω_1 and ω_2, referred to as the pump and Stokes fields, providing a coherent excitation of a Raman-active mode in a medium under study with a frequency $\Omega = \omega_1 - \omega_2$. The third field with a frequency ω_3 is then scattered off these coherently excited Raman vibrations, giving rise to the anti-Stokes signal at the frequency $\omega_a = \omega_1 - \omega_2 + \omega_3$, which provides spectroscopic information on the system or serves to image Raman-active species in a medium. Sidorov-Biryukov et al. [31] have employed a specifically designed PCF to provide a wavelength-tunable Stokes field for two-colour time-resolved CARS with $\omega_1 = \omega_3$ and $\omega_a = 2\omega_1 - \omega_2$. In these experiments, a Ti:sapphire laser oscillator pumped with a second-harmonic output of a diode-pumped Nd:YVO$_4$ laser generated light pulses with a typical temporal width of about 30 fs, an energy up to 3 nJ, and a central wavelength of 800 nm at a pulse repetition rate of 100 MHz. The Ti:sapphire laser output was divided into two beams with a 70:30 beam splitter. The laser pulses of the first beam had an energy of about 2 nJ, providing pump and probe photons for CARS measurements. The Ti:sapphire laser pulse width increased up to 50 fs as the beam was transmitted through the delay line and focusing optics.

The laser pulses of the second beam from the beam splitter were coupled into the fundamental mode of a fused silica PCF with a core diameter of 1.8 μm

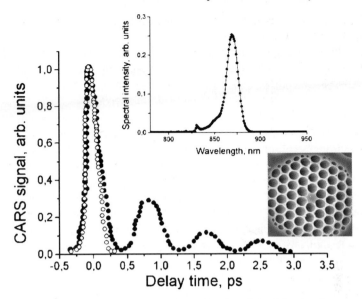

Fig. 6.3 (*Filled circles*) CARS signal measured as a function of the delay time between the probe pulse and the pump–Stokes pulse dyad. (*Open circles*) Cross-correlation of the red-shifted PCF output and a 50-fs Ti:sapphire laser pulse measured through sum-frequency generation in a BBO crystal. An SEm image of PCF is shown in *inset 1*. The spectrum of the red-shifted PCF output reflected off a diffraction grating is presented in *inset 2*

(shown in inset 1 to Fig. 6.3). For the fundamental mode of the fibre, the group-velocity dispersion (GVD) vanishes at 754 nm, providing anomalous dispersion for 800-nm Ti:sapphire laser pulses. The effective area of the fundamental mode at 800 nm is 2.1 μm^2, corresponding to the nonlinearity coefficient $\gamma \approx 110\,W^{-1}\,km^{-1}$. Ti:sapphire laser pulses tend to form optical solitons as they propagate through the PCF. These solitons undergo a continuous frequency down-shift due to the soliton self-frequency shift.

The central wavelength of the red-shifted soliton at the output of the fibre is controlled by the input peak power and the fibre length. With the peak power of input laser pulses increased from 1 to 6 kW, the central wavelength of the red-shifted soliton at the output of a 12-cm PCF is tuned from 840 to 930 nm. As the soliton is shifted toward longer wavelengths, its dispersion changes and effective mode area increases. With the input peak power ranging from 3 to 11 kW, the SSFS in the PCF with a length of 5–20 cm can provide the wavelength tunability range from 870 to 1180 nm for the red-shifted soliton at the output of the fibre. For the $\omega_a = 2\omega_1 - \omega_2$ CARS process with Ti:sapphire laser pulses used as a pump field (ω_1) and the red-shifted soliton from the PCF used as a Stokes field (ω_2), the demonstrated tunability range gives an access to the 1000–4025-cm^{-1} region of wavenumbers characteristic of fingerprint Raman transitions in a broad variety of molecular systems.

In contrast to dispersive waves, which sense the normal dispersion and tend to spread out in time, solitons are well-localized in the time domain, suggesting the way of achieving a femtosecond time resolution in CARS spectroscopy and microscopy. To assess the pulse width of the red-shifted PCF output, we mixed this signal with 50-fs 800-nm pulses of the Ti:sapphire laser in a 4-mm BBO crystal and measured the intensity of the resulting sum-frequency signal as a function of the delay time between the pulses. Open circles in Fig. 6.3 display such a cross-correlation trace for the red-shifted PCF output centered at $\lambda_s \approx 870$ nm (see inset 2 in Fig. 6.3). The pulse width of this soliton at the output of the fibre is estimated as 160 fs. The uncertainty in the time-domain localization of the red-shifted soliton PCF output, originating from the timing jitter of the soliton induced by fluctuations in the input peak power, was estimated as 30 fs.

To demonstrate time-resolved CARS with a Stokes field provided by the red-shifted solitonic output of a PCF, we performed time-delay CARS measurements on a doublet of Raman transitions in liquid-phase pyridine with Raman wave numbers $\Omega/2\pi c$ (c being the speed of light) of 988 and 1028 cm^{-1}. These transitions have close Raman cross sections, giving rise to well-pronounced quantum beats in the nonlinear response [107]. Experiments in [31] were performed with 50-fs pulses of 800-nm Ti:sapphire laser radiation as a pump (ω_1) and the 870-nm solitonic PCF output as a Stokes field (ω_2). The spectrum of the Stokes pulse reflected off a diffraction grating is presented in inset 2 to Fig. 6.3. The 870-nm solitonic PCF output gives an access to both 988- and 1028-cm^{-1} Raman modes, as the Stokes wavelengths corresponding to the exact $\Omega = \omega_1 - \omega_2$ Raman resonance with these Raman modes are equal to 868.7 and 871.7 nm, respectively. Time-delayed 50-fs Ti:sapphire laser pulses were employed to probe the Raman vibrations excited by the pump and Stokes fields in the folded CARS geometry, allowing the CARS signal to be spatially separated from the pump, Stokes, and probe fields.

In Fig. 6.3, we plot the intensity of the $\omega_a = 2\omega_1 - \omega_2$ CARS signal measured as a function of the delay time τ between the probe pulse and the pump–Stokes pulse dyad. The sharp peak observed in the region of small τ represents the nonresonant CARS signal related to electronic transitions in the medium. Since both the response time of the nonresonant part of cubic nonlinearity and the pulse width of the pump are much shorter than the duration of the Stokes pulse, the time response in the region of small τ (filled circles in Fig. 6.3) closely follows the cross-correlation of the Stokes and probe pulses (open circles). For delay times τ exceeding the duration of this cross-correlation, nonresonant CARS is suppressed, and the time-domain CARS response displays a characteristic oscillatory behavior, visualizing the quantum beats of the 988- and 1028-cm^{-1} Raman modes. Red-shifted soliton output provided by the PCF has been also applied to probe stretching C–H vibrations in ethanol ($\Omega/2\pi c \approx 2930$ cm^{-1}, $\lambda_s \approx 1.045\,\mu$m) and the 1210-cm^{-1} Raman resonance in toluene ($\lambda_s \approx 886$ nm). In both cases, the nonresonant background was efficiently suppressed in time-domain CARS, similar to the result of Fig. 6.3, by introducing a time delay between the probe pulse and the pump–Stokes pulse dyad.

Experiments discussed in this section thus demonstrate time-resolved CARS spectroscopy using a frequency-tunable soliton output of a silica PCF as a Stokes field. The wavelength tunability of the soliton PCF output within the 1000–4025-cm^{-1} wavenumber range is demonstrated, giving an access to fingerprint Raman transitions in a broad variety of molecular systems. A 160-fs red-shifted soliton output of a silica PCF was used as a Stokes field in our CARS experiments, allowing an efficient suppression of nonresonant background and making it possible to resolve quantum beats originating from close Raman-active modes of pyridine. These results suggest cost-efficient fibre-optic solutions for time-resolved CARS and allow the improvement in the sensitivity of PCF-based CARS spectroscopes and microscopes.

6.4 Photonic-Crystal Fibre Components for Time-Resolved Studies of Ultrafast Molecular Dynamics

In a recent experiment by Ivanov et al. [108], wavelength-tunable 100-fs pulses generated through the soliton self-frequency shift in a photonic-crystal fibre have been employed to visualize femtosecond coherence and population relaxation dynamics in molecular aggregates by means of time-resolved sum-frequency generation. The laser system used in those experiments consisted of a Cr^{4+}: forsterite master oscillator [109], a stretcher, an optical isolator, a regenerative amplifier, and a compressor. The master oscillator, pumped with an ytterbium fibre laser, generated 30-fs light pulses of 1.24-μm radiation at a repetition rate of 120 MHz. These pulses were amplified in a Nd:YLF-laser-pumped amplifier and recompressed to the 100-fs pulse duration with the maximum laser pulse energy up to 40 μJ at 1 kHz.

The Cr:forsterite laser system included four channels delivering pump and probe pulses for time-resolved nonlinear-optical measurements. The first channel generated amplified pulses of 1.24-μm Cr:forsterite laser radiation (curve 1 in Fig. 6.4) with a pulse energy up to 30 μJ and a pulse width $\theta_1 \approx 100$ fs. In the second channel, 1.24-μm amplified 100-fs Cr:forsterite laser pulses were coupled into a fused silica PCF with a core diameter of about 1.7 μm (see the inset in Fig. 6.4), providing zero group velocity dispersion at 730 nm. The laser pulses coupled into the PCF form solitons, which experience a continuous red shifting induced by the retarded fibre nonlinearity. This soliton frequency shift can be tuned by varying the input pulse energy and by changing the fibre length. A soliton with a pulse width $\theta_2 \approx 100$ fs and a central wavelength of 1.33 μm was generated using the PCF frequency shifter (curve 2 in Fig. 6.4) to provide a two-photon-absorption-resonant excitation of molecules studied in our experiments. In the third channel, unamplified Cr:forsterite laser pulses with a pulse energy of 5 nJ and pulse width $\theta_p \approx 30$ fs passed through an adjustable optical delay line. Finally, in the fourth channel a 1-mm-thick BBO crystal was used to generate second-harmonic pulses with a central wavelength of 620 nm, pulse width of about 90 fs, and the energy ranging from 10 to 80 nJ.

Experiments were performed with thin-film samples of J-type molecular aggregates [89] of thiacarbocyanine dye. The samples were prepared by spin-coating a 5×10^{-3} mol/l solution of this dye in a 2:2:1 mixture of acetonitrile, dichloroethane, and chloroform on a 1-μm-thick substrate. The thickness of the dye layer on the substrate was measured by the ellipsometric technique and was estimated as 30 nm. The absorption coefficient of the aggregate film at the wavelength of 633 nm

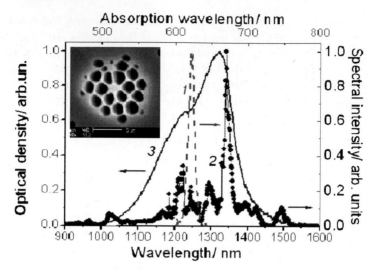

Fig. 6.4 The spectrum of the amplified fundamental-wavelength output of the Cr:forsterite laser (*1*), the spectrum of the frequency-shifted soliton output of the first-type photonic-crystal fibre (*2*), and the absorption spectrum of a thin-film sample of molecular aggregates (*3*). The *inset* shows an SEM image of the PCF

was $105\,\mathrm{cm}^{-1}$. Interactions between molecules in an aggregate give rise to collective electronic states delocalized over large chains of molecules, resulting in the formation of exciton energy bands [89, 110, 111]. Broad absorption bands, typical of isolated dye molecules, display narrowing and spectral shifting in the aggregate state. A well-pronounced peak at 660 nm in the absorption spectrum of our samples (curve 3 in Fig. 6.4) represents the excitonic absorption of *J* aggregates.

The method of time-resolved studies of coherent excitations in molecular aggregates adopted in [108] was based on sum-frequency generation (SFG) $\omega_{\mathrm{sf}} = 2\omega_1 + \omega_2$ (see the inset in Fig. 6.5) with a two-photon-absorption (TPA) resonance at the frequency $\omega_1 + \omega_2$. The pulses with carrier frequencies ω_1 and ω_2 have been delivered by the first and second channels of the above-described laser system. Two-photon-excited luminescence has been used to find the optimal frequency ω_2 of the solitonic PCF output for the maximum efficiency of the $\omega_1 + \omega_2$ TPA process. Such a maximum has been achieved with a soliton PCF output centered at $1.33\,\mu\mathrm{m}$ (curve 2 in Fig. 6.4). The measured dependences of the SFG signal on the excitation laser powers were consistent with the diagram shown in inset 2 to Fig. 6.5. The pulses with the frequencies ω_1 and ω_2 in our experiments could also reach the wing of the TPA spectrum of thiacarbocyanine monomers. However, due to the collective enhancement of optical nonlinearity in aggregates [110, 111], the SFG signal from aggregates in our experiments was at least an order of magnitude higher than the SFG signal from the monomers.

Time-delayed unamplified Cr:forsterite laser pulses with a pulse width $\theta_p \approx 30\,\mathrm{fs}$, delivered by the third channel, served to probe coherent excitations driven

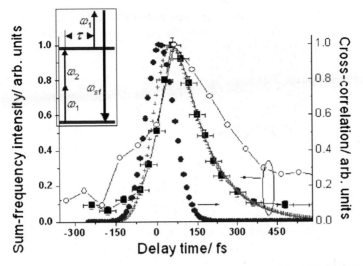

Fig. 6.5 Time-resolved SFG measurements with a solitonic output of the PCF. The *filled circles* show the power of the sum-frequency signal generated in a silica plate by amplified Cr:forsterite laser pulses, frequency-shifted pulses from the photonic-crystal fibre, and time-delayed unamplified pulses of the Cr:forsterite laser measured as a function of the delay time between the pulses (cross-correlation trace). *Rectangles* present the power P_{sf} of the sum-frequency signal from a molecular aggregate film measured as a function of the delay time τ between the probe pulse and the pulses providing a TPA-resonant coherent excitation of aggregates. *Crosses* show the fit of the experimental data with the use of expressions for P_{sf} and Q with the dephasing time $T_2 = 120$ fs and pulse widths $\theta_1 = \theta_2 = 100$ fs. *Open circles* represent the inverted dependence of the sum-frequency signal from a molecular aggregate film on the delay time between the excitation second-harmonic pulse and the light pulses involved in SFG. The inset presents a diagram of SFG $\omega_{sf} = 2\omega_1 + \omega_2$ with a TPA resonance at $\omega_1 + \omega_2$ and a time-delayed probe pulse

by the first two pulses. The power of the sum-frequency signal P_{sf} measured as a function of the delay time τ between the probe pulse and the pulses driving the coherent excitation of aggregates is then given by

$$P_{sf}(\tau) \propto \gamma^2 L^2 \int_{-\infty}^{\infty} E_p^2(\xi - \tau) Q^2(\xi) \, d\xi \tag{6.1}$$

where γ is the relevant nonlinear coefficient, L is the interaction length, $E_p(t)$ is the temporal envelope of the probe pulse, and Q is the amplitude of coherence induced in the medium by the TPA-resonant two-colour field with central frequencies ω_1 and ω_2. When the TPA-resonant field consists of two pulses with temporal envelopes $E_1(t)$ and $E_2(t)$, the amplitude Q can be represented as

$$Q(t) = \int_0^{\infty} \exp(-\eta/T_2) E_1(t - \eta) E_2(t - \eta) \, d\eta \tag{6.2}$$

where T_2 is the phase relaxation time. For two-photon-excited levels of aggregates accessed in our experiments, the main dephasing mechanism typically involves rapid intraband downward population relaxation towards the bottom of the exciton band.

In our experiments, the time-delayed pulse provides an ultrafast probe for coherently excited aggregates. With $\theta_1 \approx \theta_2 \approx 100\,\text{fs}$ and $\theta_3 \approx 30\,\text{fs}$, we have $(\theta_p/\theta_{1,2})^2 \approx 0.09$. In this regime, with the time T_2 being much less than the pulse width θ_p, measuring the dependence $P_{sf}(\tau)$ yields the temporal profile of the coherent response of aggregates. The time resolution is then determined by the cross-correlation

$$Y(\tau) = \int_{-\infty}^{\infty} E_1(\eta) E_2(\eta) E_p(\eta - \tau)\, d\eta . \tag{6.3}$$

This expression for $Y(\tau)$ with $\theta_1 = \theta_2 = 100\,\text{fs}$ and $\theta_p = 30\,\text{fs}$ provides a good fit for the cross-correlation trace (shown by the filled circles in Fig. 6.5) measured by mixing the pump, Stokes, and probe pulses in a thin silica plate with the beam interaction geometry identical to that used in experiments with aggregate samples. Rectangles in Fig. 6.5 present the power of the sum-frequency signal P_{sf} from the film with molecular aggregates measured as a function of the delay time τ. A nearly perfect fit for the measured $P_{sf}(\tau)$ dependence is obtained by using the above expressions for P_{sf} and Q with $\theta_1 = \theta_2 = 100\,\text{fs}$ and the phase relaxation time $T_2 = 120\,\text{fs}$ (crosses in Fig. 6.5). Notably, this phase relaxation time is much shorter than typical population relaxation time T_{ex} of one-exciton states of aggregates. To compare T_2 and T_{ex} for our sample, we measured nonlinear absorption spectra of molecular aggregates excited by the second-harmonic output of the Cr:forsterite laser, providing a resonant population transfer from the ground state to the one-exciton band. The supercontinuum output of PCF was then applied with a variable delay time δt relative to the second harmonic pulse. Differential absorption spectra measured in these experiments display well-pronounced minima at 665 nm, indicative of bleaching through pump-induced transitions between the ground state and the one-exciton band, and blue shifted peaks at 640 nm, originating from induced absorption due to transitions between one- and two-exciton bands. The amplitudes of induced-absorption and bleaching peaks in nonlinear absorption spectra decrease with the increase in the delay time δt, allowing the relaxation time for one-exciton states of aggregates to be estimated as $T_{ex} \approx 770\,\text{fs}$.

To demonstrate an ultrafast switching of optical nonlinearity of molecular aggregates, we used the above-described SFG process $\omega_{sf} = 2\omega_1 + \omega_2$ with the delay time τ chosen in such a way as to provide the maximum SFG efficiency. The nonlinear-optical response of the aggregate sample was modified by applying a 90-fs pulse of 620-nm second-harmonic output of the Cr:forsterite laser, depopulating the ground state of aggregates. Open circles in Fig. 6.5 display the inverted dependence of the sum-frequency signal on the delay time Δt between the excitation second-harmonic pulse and the light pulses involved in SFG. The maximum in this trace corresponds to the minimum in the sum-frequency signal, and the background corresponding to large $|\Delta t|$ has been subtracted. For small Δt, as can be seen from this dependence, the resonant excitation of aggregates substantially reduces the sum-frequency signal. As Δt increases, the initial level of the sum-frequency signal (corresponding to large negative Δt) is recovered. A monoexponential time dependence fails to provide an adequate fit for the experimentally measured trace. However, the results presented in Fig. 6.5 demonstrate a 75% recovery of the sum-frequency signal within a characteristic delay time of about 450 fs.

6.5 Spectral Transformation of Megawatt Femtosecond Laser Pulses in a Large-Mode-Area Photonic-Crystal Fibre

Small-core PCFs, typically operating in the regime of anomalous dispersion, show an excellent performance as supercontinuum sources ideally suited for nano- and subnanojoule input laser pulses. To accommodate higher input energies without the risk of laser-induced damage of the fibre material, PCFs with a larger core size are needed. Such large-mode-area (LMA) PCF components [46, 47] have been used for the creation of high-power fibre lasers [49, 50] and amplification of a short-pulse fibre laser output [51]. Südmeyer et al. [52] have demonstrated an efficient spectral broadening of a submegawatt, subpicosecond thin-disk laser output in LMA PCFs, enabling a compression of 810-fs thin-disk laser pulses to a 33-fs pulse duration. Genty et al. [112] have employed LMA PCFs to transform nanosecond pulses from a Q-switched Nd:YAG laser into a supercontinuum radiation.

Waveguide dispersion for large-core PCFs is typically weak, which limits the possibilities of fibre dispersion tailoring through fibre structure modifications. It becomes difficult, in particular, to achieve large shifts for the zero group-velocity dispersion (GVD) wavelength relative to the zero-GVD point in the bulk of the fibre material. As a result, the central wavelengths of many of the commonly used sources of high-energy ultrashort pulses are left in the regime of normal dispersion of silica large-mode-area PCF, making it difficult to generate broadband radiation through nonlinear-optical processes in the fibre. Mitrofanov et al. [113] have recently shown that a combination of Cr:forsterite laser sources with large-mode-area PCFs resolves the conflict between the mode area and dispersion, allowing high-peak-power laser output to be efficiently transformed into supercontinuum radiation. This strategy offers much promise for the creation of high-peak-power fibre-format sources of broadband radiation for spectroscopic, microscopic, biomedical, and micromachining applications, as well as for the creation of front-end component for laser sources designed to deliver extremely high-peak powers at their output.

Large-mode-area PCFs used in our experiments were made of fused silica using a standard stack-and-draw technology. For the fibre used in our experiments, the core diameter was about 22 μm with an effective mode area estimated as 380 μm^2 for the fundamental mode. The fibre core was surrounded with four rings of air holes (see the inset in Fig. 6.6) with a diameter $d \approx 4.0$ μm and a pitch $\Lambda \approx 11$ μm. Such a fibre could support higher order guided modes of 1.24-μm Cr:forsterite laser radiation used in our experiments. With an appropriate fibre alignment, however, a robust fundamental mode of 1.24-μm radiation was observed at the output of the fibre

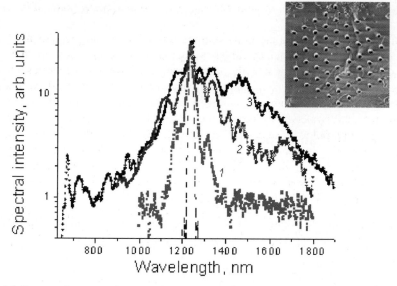

Fig. 6.6 Spectra of supercontinuum radiation from a 20-cm segment of a PCF with a mode area of $380\,\mu m^2$ (shown in the *inset*). The input pulse width is 300 fs. The energy of laser radiation coupled into the fibre is (*1*) 0.15 μJ, (*2*) 0.98 μJ, and (*3*) 1.3 μJ. The input spectrum of the laser pulse is shown by the *dashed line*

in the low-peak-power regime, when nonlinear-optical effects were negligible. The zero-GVD wavelength for the PCF employed in our experiments is $\lambda_z \approx 1.27\,\mu m$, and the fibre nonlinearity $\gamma = 2\pi n_2 (\lambda S)^{-1}$ (here, n_2 is the nonlinear refractive index of the fibre material, λ is the radiation wavelength, and S is the effective mode area) is about $0.4\,km^{-1}\,W^{-1}$ at $\lambda = 1.24\,\mu m$.

The laser system used in our experiments consisted of a Cr^{4+}:forsterite master oscillator, a stretcher, an optical isolator, a regenerative amplifier, and a compressor. The master oscillator, pumped with an ytterbium fibre laser, generated 30–60-fs light pulses of radiation with a central wavelength of 1.25 μm at a repetition rate of 120 MHz. These pulses were transmitted through a stretcher and an isolator, to be amplified in a Nd:YLF-laser-pumped amplifier and recompressed to a pulse width of 90–400 fs with a maximum pulse energy up to 30 μJ at 1 kHz. To avoid self-focusing-induced damage of the fibre, the laser output was stretched, through a compressor adjustment, to a pulse width of approximately 300 fs. The energy of amplified Cr:forsterite laser pulses delivered to the fibre input through an objective was about 3 μJ. The laser energy fluence on the input end of the fibre was thus about $0.79\,J/cm^2$, remaining well below the laser damage threshold. A maximum of 1.3 μJ was coupled into the fundamental mode of the fibre, corresponding to a peak power of $P \approx 4.3$ MW. Although the input peak power in our experiments was slightly higher than the lower-bound self-focusing threshold, a catastrophic self-focusing at this level of P was prevented by fibre dispersion [114], with a robust operation of

PCFs provided and no sign of laser damage observed over many hours of measurements.

The spectra of laser radiation transmitted through a 20-cm piece of the large-mode-area PCF are presented by curves 1–3 in Fig. 6.6 in comparison with the spectrum of the input field (shown by the dashed line). For lower input energies (curve 1 in Fig. 6.6), PCF output spectra exhibit moderate broadening with well-pronounced Stokes and anti-Stokes sidebands, falling in the range of anomalous and normal dispersion respectively. These spectral features are indicative of modulation instabilities of the pump field whose central wavelength lies close to the zero-GVD wavelength, thus facilitating phase matching for Stokes and anti-Stokes sideband generation. For higher input energies (curves 2 and 3 in Fig. 6.6), isolated spectral components originating from the pump field and its Stokes and anti-Stokes sidebands merge together, giving rise to a broadband spectrum at the output of the fibre. In this regime, the output spectra display a powerful long-wavelength wing, stretching up to 1800 nm for an input energy of 1.3 μJ (curve 3 in Fig. 6.6), which indicates the significance of soliton self-frequency shifting phenomena, induced by the retarded part of the fibre nonlinearity.

The visible part of the spectrum observed at the output of a 20-cm large-mode-area PCF is much less intense, carrying about 15% of the total energy of the fibre output, estimated as 1.15 μJ. As the short-wavelength part of the spectrum was generated in a mixture of guided modes, a considerable portion of the short-wavelength part of the spectrum (about 50%) was lost in our experiments in the course of propagation through leakage losses within the first 7–8 cm of the fibre, leading to a decrease in the total radiation energy from approximately 1.30 μJ (radiation energy coupled into the fibre) to about 1.15 μJ (radiation energy measured at the output of the 20-cm fibre).

To visualize the temporal envelope of the PCF output, we employed cross-correlation frequency-resolved optical gating (XFROG) technique [115, 116] by mixing the PCF output with the amplified fundamental-wavelength Cr:forsterite laser radiation through sum-frequency generation in an LBO crystal. The spectrally resolved sum-frequency signal from the nonlinear crystal was measured as a function of the delay time τ between the PCF output and the reference Cr:forsterite laser pulse, yielding sonograms of the light field at the output of the fibre. In Fig. 6.7, we present a typical XFROG trace measured for a 20-cm segment of PCF with an input pulse energy of 1.1 μJ. This trace features a distinct kink around the zero-GVD wavelength λ_z, separating regimes of normal and anomalous dispersion. For shorter wavelengths, $\lambda < \lambda_z$, because of the normal dispersion, high-frequency components are delayed in time with respect to spectral components with lower frequencies. For longer wavelength, $\lambda > \lambda_z$, well-resolved temporally isolated solitonic features are observed, indicating the significance of solitonic phenomena and the Raman-effect-induced soliton self-frequency shift in the enhancement of the long-wavelength part of the supercontinuum spectrum at the output of the fibre.

To quantify the enhancement provided by an LMA PCF for high-power supercontinuum generation relative to a bulk material, we compared the output spectra of the considered type of PCF with the spectral broadening of amplified Cr:forsterite

Fig. 6.7 An XFROG trace of the PCF output measured for an input pulse energy of 1.1 μJ. The fibre length is 20 cm. The *left-* and *right-*hand *ordinate axes* represent the wavelength of the sum-frequency signal λ_{SF} and the wavelength of the PCF output λ_{PCF}, respectively

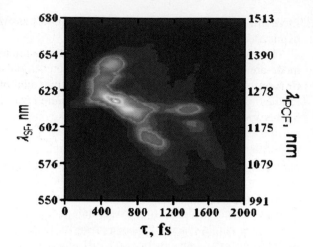

laser pulses attainable in the bulk of fused silica. In the latter experiment, an amplified 300-fs, 1.24-μm Cr:forsterite laser output (dashed curve 1 in Fig. 6.8) was focused inside a silica plate into a spot with a diameter of about 20 μm, close to the core diameter of the PCF used in our experiments. To achieve maximum spectral broadening, the energy of laser pulses in these experiments was set equal to 8 μJ, which was slightly below the damage threshold of the silica plate. The thickness of the silica plate was chosen equal to 2 cm, which substantially exceeded the effective interaction length (a few millimeters for the above-specified experimental parameters), so that a further increase in the plate thickness lead to no increase in the output spectral width. With these subcritical for a silica plate parameters of incident laser pulses, output radiation spectra (open circles in Fig. 6.8) remained substantially narrower than the spectra of Cr:forsterite laser pulses with an input

Fig. 6.8 Radiation spectra of the amplified Cr:forsterite laser output (*dashed line 1*), a focused 300-fs, 8-μJ Cr:forsterite laser pulse spectrally broadened in a 2-cm silica plate (*open circles*), and a 300-fs, 1.3-μJ Cr:forsterite laser pulse transmitted through a 20-cm segment of the large-mode area PCF (*filled circles*)

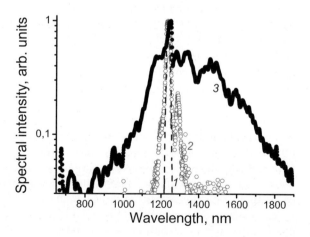

energy of $1.3\,\mu J$ and the same input pulse width transmitted through 20 cm of the large-mode-area PCF (filled circles in Fig. 6.8).

We have thus demonstrated microjoule supercontinuum generation using a large-mode-area PCF pumped by an amplified stretched-pulse output of a femtosecond Cr:forsterite laser. This result highlights the potential of large-mode-area PCFs as sources of high-peak power broadband radiation for spectroscopic, microscopic, biomedical, and micromachining applications.

6.6 Third-Harmonic Generation by Ultrashort Laser Pulses

Third-harmonic generation (THG) is one of the basic nonlinear-optical processes, which has been intensely studied and employed for numerous applications since the early days of nonlinear optics [117–119]. The seminal work by Miles and Harris [120] has demonstrated a tremendous potential of direct THG related to the cubic optical nonlinearity $\chi^{(3)}$ of gases for efficient frequency conversion of laser radiation and for the diagnostics of the gas phase. Solid-state strategies of frequency conversion, on the other hand, mainly rely on the quadratic nonlinearity $\chi^{(2)}$ of noncentrosymmetric crystals, with frequency tripling conventionally implemented through cascaded second-order nonlinear-optical processes, phase matched by crystal anisotropy [121, 122] or a periodic poling of nonlinear materials [123].

Controlled dispersion of guided modes and large interaction lengths provided by PCFs result in a radical enhancement of nonlinear-optical processes. Third-order nonlinear-optical processes enhanced in PCFs offer a useful alternative to frequency-conversion schemes using $\chi^{(2)}$ nonlinear crystals. Highly efficient THG has been recently observed in fused silica [124–130] and multicomponent-glass PCFs [131], as well as in tapered fibres [132]. These experiments not only demonstrated the significance of THG for efficient, guided-wave frequency tripling of femtosecond laser pulses, but also revealed several new interesting nonlinear-optical phenomena. The third-harmonic signal has been shown to display an asymmetric spectral broadening [131–133] or even a substantial frequency shift [134–137]. The sign and the absolute value of the third-harmonic frequency shift, observed in many PCF experiments, are controlled by the phase- and group-index mismatch for the interacting pair of pump and third-harmonic modes. The possibility to tune the frequency of the main spectral peak in the spectrum of the third harmonic by varying the group-velocity mismatch is a unique property of THG-type processes, which is not typical of standard parametric FWM processes, where the first-order dispersion terms cancel out of the balance of the field momenta. New regimes of THG will be identified with no signal produced at the central frequency of the third harmonic, $3\omega_0$, and with the pump energy efficiently converted to spectrally isolated narrowband frequency components, which can be tuned within a spectral range of several tens of terahertz from the $3\omega_0$ frequency.

As shown by Fedotov et al. [130] soliton regimes of pulse propagation in optical fibres and in PCFs, in particular, offer interesting new options for third-harmonic generation. In those experiments, solitons excited in a PCF by unamplified fem-

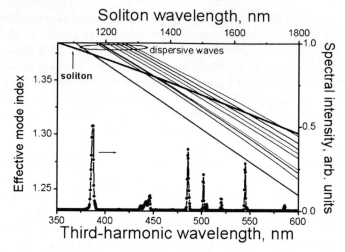

Fig. 6.9 Spectrum of the third harmonic (*filled circles* connected by a *solid line*) generated in a 30-cm PCF by 2-nJ 120-fs Cr:forsterite laser pulses. *Solid lines* labeled as dispersive waves show the effective mode indices n_m of high-order PCF modes as functions of radiation wavelength in the short-wavelength (third-harmonic) spectral range (the *lower abscissa axis*). The *bold line* shows the effective mode index n_{sol} of the soliton pump in the fundamental PCF mode as a function of the pump wavelength (the *upper abscissa axis*). Phase matching is achieved at the wavelengths λ where the *bold line*, representing $n_{sol}(\lambda)$, *crosses* one of the *solid lines*, representing $n_m(\lambda/3)$

tosecond pulses of a Cr:forsterite laser swept over the spectral range from 1.25 to 1.63 μm, scanning through a manifold of THG phase-matching resonances with third-harmonic dispersive waves in PCF modes. As a result, intense third-harmonic peaks were observed to build up in the range of wavelengths from 370 to 550 nm at the output of the fibre (Fig. 6.9), making PCF a convenient fibre-format multi-frequency source of short-wavelength radiation. Time-resolved fluorescence measurements with photoexcitation provided by the third-harmonic PCF output [138] demonstrate the high potential of PCF sources for an ultrafast photoexcitation of fluorescent molecular systems in physics, chemistry, and biology.

6.7 Hollow Photonic-Crystal Fibres for the Delivery and Nonlinear-Optical Transformation of High-Peak-Power Ultrashort Pulses

Through the past few years, hollow-core photonic-crystal fibres (PCFs) [1–7] have emerged as an interesting novel type of optical waveguides, offering much promise for high-field physics and nonlinear optics. Diffraction-induced radiation losses, typical of hollow waveguides [11, 12], can be substantially reduced in hollow PCFs due to the high reflectivity of a two-dimensional periodic cladding within photonic band-gaps (PBGs) or a cladding structure providing an antiresonance with the modes guided in the hollow core region of a waveguide [1, 8]. Hollow PCFs offer a fibre format of high-power beam delivery in materials processing and related technologies [139–141] and suggest novel solutions for the development of advanced fibre-laser sources of ultrashort pulses [38, 58, 59, 142], fibre microendoscopes [143, 144], and other fibre-based components for biomedical applications [145]. Fibres of this type open new horizons in optical science by allowing guided-wave nonlinear-optical interactions of nondiverging laser beams with transverse sizes of a few microns – a unique regime of nonlinear optics that could not be accessed with previously known optical waveguides [146]. Large propagation lengths, phase-matching management through PCF structure engineering, and high peak powers of laser fields attainable with small-core hollow PCFs suggest attractive strategies toward a radical enhancement of a variety of nonlinear-optical processes, such as stimulated Raman scattering [147, 148], off-resonance four-wave mixing [149, 150], and coherent anti-Stokes Raman scattering [151, 152]. The spatial self-action of intense ultrashort laser pulses gives rise to interesting waveguiding regimes in hollow PCFs below the beam blowup threshold [153].

In the regime of anomalous dispersion, the temporal self-action of laser pulses guided in the hollow core of PCFs can lead to the formation of solitons. While in standard optical fibres, the peak power of an individual (fundamental) soliton with a typical pulse width of 100 fs is usually on the order of a hundred of watts, the peak power of solitons that can be produced in hollow PCFs can be as high as a few megawatts [154]. Soliton phenomena in hollow PCFs permit compression of high-peak-power laser pulses [155], allow the creation of wavelength-tunable high-peak-power fibre sources for nonlinear spectroscopy [156], and enable attractive regimes of long-disance transmission for high-power ultrashort laser pulses [157]. Skryabin et al. [158] have predicted the existence of novel two-colour solitons in hollow PCFs filled with a Raman-active gas.

Ivanov et al. [156] have demonstrated that hollow PCFs make the fibre format of beam delivery fully compatible with the requirements of CARS microscopy and bioimaging. The laser system used in those experiments consisted of a Cr^{4+}: forsterite master oscillator, a stretcher, an optical isolator, a regenerative amplifier, and a compressor. The master oscillator, pumped with a fibre ytterbium laser, generated 30–60-fs light pulses of radiation with a wavelength of 1.23–1.25 μm at a repetition rate of 120 MHz. These pulses were amplified in a Nd:YLF-laser-pumped amplifier and recompressed to the 170-fs pulse duration with the maximum laser pulse energy up to 30 μJ at 1 kHz. A 1-mm-thick BBO crystal was used for second-harmonic generation. Second-harmonic pulses with an energy ranging from 1 to 1000 nJ and a central wavelength of 618 nm (their spectrum is shown by the dotted line in Fig. 6.10) were coupled into a hollow PCF with a core diameter of 12 μm and a cross-section structure shown in inset 1 to Fig. 6.10. The fibre was fabricated from S-93 soft glass and provided a transmission band from 0.60 to 0.66 μm. With no true closed PBG achieved in the considered wavelength range, the leakage loss of the air-guided modes was quite high (3.1 dB/m in the peak of transmission).

Since the central wavelength of the input light pulses falls in the range of anomalous dispersion (the wavelength of zero group-velocity dispersion (GVD) for our PCF is $\lambda_z \approx 614$ nm), the light pulses tend to generate optical solitons as they

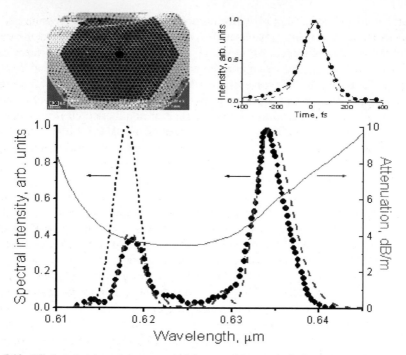

Fig. 6.10 *Filled circles* show the spectrum of the second-harmonic Cr:forsterite laser pulse transmitted through a hollow PCF (shown in *inset 1*) with a length of 50 cm filled with an atmospheric-pressure air. The input pulse (its spectrum shown by the *dotted line*) has a central wavelength of 618 nm and an initial energy of 300 nJ. The *dashed line* represents the numerical solution of the generalized nonlinear Schrödinger equation. Also shown is the wavelength dependence of PCF attenuation. *Inset 2* displays the temporal envelope of the blue-shifted soliton PCF output reconstructed from XFROG (*filled circles*) and calculated by solving the GNSE (*dashed line*)

propagate through the fibre. For the solitons defined as solutions to the nonlinear Schrödinger equation (NSE), the peak power P_s of the fundamental soliton with a pulse width τ_s is given by

$$P_s = 0.079\lambda^3 DS_{\text{eff}}\tau_s^{-2}(cn_2)^{-1} \tag{6.4}$$

where D is the GVD, λ is the radiation wavelength, S_{eff} is the effective mode area, c is the speed of light in vacuum, and n_2 is the nonlinear refractive index of the gas filling the core of the PCF. With $\lambda = 618$ nm, $D \approx 70$ ps/(nm km), $S_{\text{eff}} \approx 100\,\mu\text{m}^2$, $n_2 \approx 3 \times 10^{-19}\,\text{cm}^2/\text{W}$ (for atmospheric-pressure air), we arrive at an estimate $P_s \approx 1.4$ MW for the peak power of the fundamental NSE soliton with a pulse width $\tau_s \approx 100$ fs.

The spectral and temporal parameters of the PCF output were measured in our experiments by using cross-correlation frequency-resolved optical gating (XFROG). To this end, the pulse transmitted through the PCF was mixed through sum-frequency generation with a 618-nm 90-fs second-harmonic output of the Cr:forsterite laser in a BBO crystal. The spectrum of the PCF output was additionally measured using an Ocean Optics spectrometer. In agreement with the above estimate for the peak power of the fundamental NSE soliton, we observed solitonic phenomena for 90-fs 618-nm input pulses with energies exceeding 180 nJ. Figure 6.10 displays a typical spectrum and temporal envelope of the PCF output, measured for an input pulse width of 90 fs and an input pulse energy of 250 nJ. In contrast to low-peak-power input pulses, which tend to spread out in time as they propagate through the fibre due to the fibre dispersion, the soliton PCF output remains localized in the time domain (inset 2 in Fig. 6.10). The pulse width of the PCF output is estimated as 120 fs, and its energy is 130 nJ, corresponding to a peak power of 1.1 MW. The retarded (Raman) part of optical nonlinearity of the gas filling the PCF core down-shifts the soliton frequency – phenomenon know as the soliton self-frequency shift (SSFS). Similar to solitons in standard, solid-core silica fibres, the SSFS of megawatt solitons is controlled by the peak power of input pulses, as well as by the fibre length and dispersion. In hollow PCFs, the SSFS can be additionally controlled by varying the pressure and the content of the gas filling the fibre core – an option that is unavailable with conventional fibre technologies. We verified the above arguments by numerically solving the generalized nonlinear Schrödinger equation (GNSE) for high-peak-power pulses propagating in a gas-filled hollow PCF. These computations provide a nearly perfect fit for the spectrum (cf. the filled circles and the dashed line in Fig. 6.10) and the temporal envelope (inset 2 in Fig. 6.10) of the hollow-PCF output in our experiments. The high leakage loss of our PCF results in a dissipation of soliton energy, eventually inducing a decay of solitons, thus limiting the attainable soliton frequency shift. We therefore expect that the tunability range of PCF soliton frequency shifters of high-power laser pulses can be further expanded by optimizing the PCF structure for the low-loss waveguiding.

The frequency-shifted megawatt soliton output of the hollow PCF was employed as a Stokes pulse in time-resolved two-colour CARS measurements (inset 1 in Fig. 6.11). In two-colour CARS [104–106], the pump and Stokes fields with frequencies ω_1 and ω_2 coherently excite Raman-active molecular vibrations with a frequency $\Omega = \omega_1 - \omega_2$. The probe field with the frequency ω_1 is then scattered off the coherence induced in the medium by the pump and Stokes fields, giving rise to an anti-Stokes signal at the frequency $\omega_a = 2\omega_1 - \omega_2$. With the 90-fs 618-nm second-harmonic output of the Cr:forsterite laser used as a pump field (ω_1), the frequency shift of megawatt solitons in the hollow PCF was adjusted in such a way as to tune the frequency difference $\omega_1 - \omega_2$ to a resonance with the Raman-active tetrahedron A_{1g} vibration of carbon tetrachloride molecules ($\Omega/(2\pi c) \approx 459 \, \text{cm}^{-1}$). Another 90-fs second-harmonic pulse of the Cr:forsterite laser, applied with a delay time τ with respect to the pump and Stokes pulses, was employed as a probe field in our experiments. In the noncollinear CARS geometry used in our experiments, the anti-Stokes signal can be conveniently separated in space from the pump, Stokes, and probe fields (inset 1 in Fig. 6.11).

The filled circles in Fig. 6.11 present the power of the anti-Stokes signal P_a from the liquid-phase CCl_4 sample measured as a function of the delay time τ. For small τ, the $P_a(\tau)$ dependence is dominated by the nonresonant part of the CARS

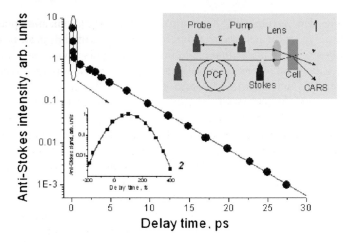

Fig. 6.11 *Filled circles* represent the intensity of the anti-Stokes signal from a CCl_4 cell measured as a function of the delay time τ between the probe pulse and the pump–Stokes pulse dyad. The *solid line* shows the $\exp(-2\tau/T_2)$ dependence. *Inset 1* sketches a diagram of the CARS setup. The *filled circles* in *inset 2* show a close-up of the $P_a(\tau)$ dependence measured for small τ. The *solid line* in this inset represents the cross-correlation function for the pump, Stokes, and probe pulses

signal. This section of the $P_a(\tau)$ dependence (inset 2 in Fig. 6.11) recovers the cross-correlation trace of the pump, Stokes, and probe pulses. For delay times τ exceeding the pulse widths of the input light fields, the $P_a(\tau)$ dependence visualizes the exponential decay of the Raman-resonant response, originating from the dephasing of coherent vibrations of CCl_4. The high peak power of the Stokes pulse provided by the soliton output of the hollow PCF provides a broad dynamic range of CARS signal detection (about four decades in our experiments), allowing time-resolved CARS measurements within the range of delay times from tens of femtoseconds up to tens of picoseconds. In agreement with the earlier work [159], the decay of the coherent response of the 459-cm^{-1} CCl_4 vibration (filled circles in Fig. 6.11) is well approximated by an exponential $\exp(-2\tau/T_2)$ with the phase relaxation time $T_2/2 \approx 4\,\mathrm{ps}$ (the solid line in Fig. 6.11).

Experiments presented in [156] show that, with hollow-core PCFs, the fibre format of beam delivery becomes fully compatible with the requirements of CARS microspectroscopy. In the experiments presented in this work, hollow PCFs have been used for the delivery and soliton frequency shifting of 2.8-MW femtosecond pulses with an input central wavelength of 618 nm. The frequency-shifted megawatt soliton output of hollow PCFs is shown to be ideally suited as a Stokes field for coherent Raman spectroscopy and imaging, as well as for time-resolved CARS measurements. The high peak power of the Stokes pulse provided by the soliton output of the hollow PCF provides a dynamic range of CARS signal detection of about four decades, allowing time-resolved CARS studies of ultrafast relaxation processes on time scales from tens of femtoseconds up to tens of picoseconds.

As recently demonstrated by Serebryannikov et al. [159], ionization phenomena can substantially modify the soliton propagation dynamics of high-peak-power laser pulses in hollow-core PCFs. Numerical solution of the pulse-evolution equation for a high-peak-power laser field in an ionizing gas medium reveals two qualitatively different scenarios of soliton evolution in a hollow PCF controlled by the ionization potential I_p of the gas filling the fibre core. Hollow PCFs filled with high-I_p gases are shown to allow formation of gigawatt soliton features, which remain stable over large propagation distances, with their spectrum undergoing a continuous red shift due to the retarded nonlinearity of the fibre cladding. In hollow PCFs filled with low-I_p gases, the ionization-induced change in the refractive index of the gas leads to a blue shifting of soliton transients, pushing their spectrum beyond the point of zero group-velocity dispersion, thus preventing formation of stable high-peak-power solitons. Hollow waveguides capable of providing soliton transmission regimes for gigawatt laser pulses suggest attractive solutions for long-distance transmission of high-power optical signals, creation of wavelength-tunable sources of high-peak-power ultrashort light pulses, fibre-format beam delivery in materials microprocessing, as well as the development of fibre endoscopes and fibre components for laser surgery, vessel photodisruption, laser ophthalmology, and optical histology.

Recent experiments [160] have demonstrated that the ionization-induced change in the refractive index of a gas can substantially blue-shift megawatt light pulses transmitted through hollow PCFs (Fig. 6.12). Given the type of a fibre and the sort

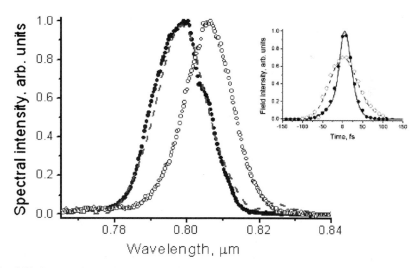

Fig. 6.12 Spectra of amplified Ti:sapphire laser pulses measured at the input (*open circles*) and at the output (*filled circles*) of the hollow-core PCF with a length of 5 cm. The initial pulse width is 60 fs. The input pulse intensity is 5×10^{13} W/cm^2. The *dashed line* shows the results of numerical simulations. The *inset* shows the pulse shape of the PCF output: (*solid and dashed lines*) results of numerical simulations and (*filled and open circles*) reconstruction from cross-correlation measurements. The input pulse intensity is (*open circles and dashed line*) 3.0×10^{12} W/cm^2 and (*filled circle and solid line*) 5.0×10^{13} W/cm^2

of gas filling the fibre core, the sign of the frequency shift of the laser field and its rate can thus be controlled by the input laser peak power, offering attractive solutions for the development of high-peak-power wavelength-tunable fibre-format sources of ultrashort pulses, as well as for the processing and regeneration of high-peak power optical signals.

6.8 Conclusion

Photonic-crystal fibres play a progressively significant role in the creation of compact and efficient fibre-optic systems for the generation and control of ultrashort light pulses. Dispersion and field-profile tailoring is the key advantage of photonic-crystal fibres, which allows a high-precision dispersion balance to be achieved within a broad spectral range, enabling the creation of novel types of fibre-optic sources of ultrashort light pulses. Methods of nano-optics help to tailor dispersion profiles of PCF modes, providing a highly efficient frequency conversion of femtosecond laser pulses and enabling wavelength-tunable generation of broadband radiation. Special strategies of micro- and nanostructuring of the core and the cladding of optical fibres help to realize an efficient spectral and temporal transformation of laser pulses with input pulse widths from tens of nanoseconds down to several field cycles within the range of peak powers from hundreds of watts up to several gigawatts. Hollow-core photonic-crystal fibres supporting soliton transmission regimes for megawatt pulses open up new possibilities in laser biomedicine and optical technologies.

Advanced fibre-optic technologies suggest an attractive format of frequency-tunable ultrashort-pulse sources for time-resolved spectroscopy and microscopy. In experiments discussed in this paper, frequency-tunable ultrashort pulses generated through the soliton self-frequency shift in a photonic-crystal fibre with a special dispersion profile were used for time-resolved studies of coherence and population dynamics in individual molecules and molecular aggregates using nonlinear Raman and sum-frequency generation spectroscopy. Time-resolved CARS spectroscopy has been implemented using a wavelength-tunable short-pulse soliton output of solid- and hollow-core PCFs as a Stokes pulse. Hollow-core PCFs make the fibre format of beam delivery fully compatible with the requirements of CARS microspectroscopy. The frequency-shifted megawatt soliton output of hollow PCFs has been shown to be ideally suited as a Stokes field for coherent Raman spectroscopy and imaging, as well as for time-resolved CARS measurements.

Acknowledgements I am pleased to gratefully acknowledge fruitful collaboration and illuminating discussions with A.B. Fedotov, D.A. Sidorov-Biryukov, E.E. Serebryannikov, I.V. Fedotov, and A.A. Podshivalov (Physics Department, M.V. Lomonosov Moscow State University); A.A. Ivanov and M.V. Alfimov (Center of Photochemistry, Russian Academy of Sciences); A. Baltuška and E. Wintner (Technological University of Vienna); F. Krausz and A. Apolonsky (Max Planck Institute for Quantum Optics, Garching); D. von der Linde and A.P. Tarasevitch (University of Essen–Duisburg); P.S.J. Russell (University of Erlangen–Nuremberg), J.C. Knight (University of Bath); Yu.N. Kondrat'ev, V.S. Shevandin, K.V. Dukel'skii, and A.V. Khokhlov (S.I. Vavilov State Optical Institute, St. Petersburg); V.I. Beloglazov, N.B. Skibina, and A.V. Shcherbakov (Institute of

Technology and Equipment for Glass Structures, Saratov); D. Chorvat, I. Bugar, D. Lorenc, and F. Uherek (International Laser Center, Bratislava); C. Sibilia ("La Sapienza" University of Rome), R. Miles and M. Shneider (Princeton University); Ching-yue Wang, Minglie Hu, and Yanfeng Li (Tianjin University).

This study was supported in part by the Russian Foundation for Basic Research (projects 06-02-16880, 07-02-12175, and 07-02-91215), the Russian Federal Research and Technology Program, and Award no. RUP2-2695 from the U.S. Civilian Research & Development Foundation (CRDF). Experiments and calculations presented in this chapter were partially performed under COST P11 action.

References

1. Russell P.S.J.: "Photonic Crystal Fibres" Science **299**, 358–362 (2003).
2. Knight J.C.: Nature **424**, 847 (2003).
3. Russell P.S.J.: J. Lightwave Technol. **24**, 4729 (2006).
4. Knight J.C., Broeng J., Birks T.A., and Russell P.S.J.: Science **282**, 1476, (1998).
5. Cregan R.F., Mangan B.J., Knight J.C., Birks T.A., Russell P.S.J., Roberts P.J., and Allan D.A.: Science **285**, 1537 (1999).
6. Smith C.M., Venkataraman N., Gallagher M.T., Muller D., West J.A., Borrelli N.F., Allan D.C., and Koch K.W.: Nature **424**, 657 (2003).
7. Konorov S.O., Fedotov A.B., Kolevatova O.A., Beloglazov V.I., Skibina N.B., Shcherbakov A.V., and Zheltikov A.M.: JETP Lett. **76**, 341 (2002).
8. Litchinitser N.M., Abeeluck A.K., Headley C., and Eggleton B.J.: Opt. Lett. **27**, 1592 (2002).
9. Zheltikov A.M.: Phys. Uspekhi **47**, 1205 (2004).
10. Luan F., George A.K., Hedley T.D., Pearce G.J., Bird D.M., Knight J.C., and Russell P.S.J.: Opt. Lett. **29**, 2369 (2004).
11. Marcatili E.A.J., Schmeltzer R.A.: Bell Syst. Tech. J. **43**, 1783 (1964)
12. Adams M.J.: An Introduction to Optical Waveguides, Wiley: New York (1981).
13. Zheltikov A.M.: Phys. Uspekhi **47**, 69 (2004).
14. Knight J.C., Birks T.A., Russell P.S.J., and Atkin D.M.: Opt. Lett. **21**, 1547 (1996).
15. Ferrando A., Silvestre E., Miret J.J., and Andres P.: Opt. Lett. **25**, 790 (2000).
16. Reeves W.H., Skryabin D.V., Biancalana F., Knight J.C., Russell P.S.J., Omenetto F.G., Efimov A., and Taylor A.J.: Nature **424**, 511 (2003).
17. Fedotov A.B., Zheltikov A.M., Tarasevitch A.P., and von der Linde D.: Appl. Phys. B **73**, 181 (2001).
18. Ranka J.K., Windeler R.S., and Stentz A.J.: Opt. Lett. **25**, 25 (2000).
19. Wadsworth W.J., Ortigosa-Blanch A., Knight J.C., Birks T.A., Mann T.P.M., and Russell P.S.J.: J. Opt. Soc. Am. B **19**, 2148 (2002).
20. Zheltikov A.M. (ed): Supercontinuum Generation, Special issue of Applied Physics B **77**, nos. 2/3 (2003)
21. Zheltikov A.M.: Phys. Uspekhi **49**, 605 (2006).
22. Dudley J.M., Genty G., and Coen S.: Rev. Mod. Phys. **78**, 1135 (2006).
23. Jones D.J., Diddams S.A., Ranka J.K., Stentz A., Windeler R.S., Hall J.L., and Cundiff S.T.: Science **288**, 635 (2000).
24. Holzwarth R., Udem T., Hänsch T.W., Knight J.C., Wadsworth W.J., and Russell P.S.J.: Phys. Rev. Lett. **85**, 2264 (2000).
25. Diddams S.A., Jones D.J., Ye J., Cundiff S.T., Hall J.L., Ranka J.K., Windeler R.S., Holzwarth R., Udem T., and Hänsch T.W.: Phys. Rev. Lett. **84**, 5102 (2000).
26. Udem T., Holzwarth R., and Hänsch T.W.: Nature **416**, 233 (2002).
27. Serebryannikov E.E., Zheltikov A.M., Ishii N., Teisset C.Y., Köhler S., Fuji T., Metzger T., Krausz F., and Baltuška A.: Phys. Rev. E **72**, 056603 (2005).

28. Teisset C.Y., Ishii N., Fuji T., Metzger T., Köhler S., Holzwarth R., Baltuska A., Zheltikov A.M., and Krausz F.: Opt. Express **13**, 6550 (2005).
29. Hartl I., Li X. D., Chudoba C., Rhanta R.K., Ko T.H., Fujimoto J.G., Ranka J.K., and Windeler R.S.: Opt. Lett. **26**, 608 (2001).
30. Konorov S.O., Akimov D.A., Serebryannikov E.E., Ivanov A.A., Alfimov M.V., and Zheltikov A.M.: Phys. Rev. E **70**, 057601 (2004)
31. Sidorov-Biryukov D.A., Serebryannikov E.E., and Zheltikov A.M.: Opt. Lett. **31**, 2323 (2006).
32. Zheltikov A.M. and Raman J.: Spectrosc. **38**, 1052 (2007).
33. Paulsen H.N., Hilligsøe K.M., Thøgersen J., Keiding S.R., and Larsen J.J.: Opt. Lett. **28**, 1123 (2003).
34. Kano H. and Hamaguchi H.: Opt. Express **13**, 1322 (2005).
35. von Vacano B., Wohlleben W., and Motzkus M.: Opt. Lett. **31**, 413 (2006).
36. Lim H., Ilday F.Ö., and Wise F.W.: Opt. Express **10**, 1497 (2002).
37. Chen X., Venkataraman M., Li, N., Gallagher M.T., Wood W.A., Crowley A.M., Carberry J.P., Zenteno L.A., and Koch K.W.: Opt. Express **12**, 3888 (2004).
38. Lim H., Chong A., and Wise F.W.: Opt. Express **13**, 3460 (2005).
39. Isomäki A. and Okhotnikov O.G.: Opt. Express **14**, 4368 (2006).
40. Sharping J.E., Fiorentino M., Kumar P., and Windeler R.S.: Opt. Lett. **27**, 1675 (2002).
41. de Matos C.J.S., Taylor J.R., and Hansen K.P.: Opt. Lett. **29**, 983 (2004).
42. Deng Y., Lin Q., Lu F., Agrawal G.P., and Knox W.H.: Opt. Lett. **30**, 1234 (2005).
43. Sharping J.E., Chen J., Li X., Kumar P., and Windeler R.S.: Opt. Express **12**, 3086 (2004).
44. Rarity J.G., Fulconis J., Duligall J., Wadsworth W.J., and Russell P.S.J.: Opt. Express **13**, 534 (2005).
45. Birks T.A., Knight J.C., and Russell P.S.J.: Opt. Lett. **22**, 961 (1997).
46. Knight J.C., Birks T.A., Cregan R.F., Russell P.S.J., and de Sandro J.P.: Electron. Lett. **34**, 1347 (1998).
47. Furusawa K., Malinowski A., Price J., Monro T., Sahu J., Nilsson J., and Richardson D.: Opt. Express **9**, 714 (2001).
48. Wadsworth W.J., Knight J.C., Reeves W.H., and Russell P.S.J.: Electron. Lett. **36**, 1452 (2000).
49. Furusawa K., Monro T.M., Petropoulos P., and Richardson D.J.: Electron. Lett. **37**, 560 (2001).
50. Wadsworth W., Percival R., Bouwmans G., Knight J., and Russell P.: Opt. Express **11**, 48 (2003).
51. Limpert J., Schreiber T., Nolte S., Zellmer H., Tünnermann T., Iliew R., Lederer F., Broeng J., Vienne G., Petersson A., and Jakobsen C.: Opt. Express **11**, 818 (2003).
52. Südmeyer T., Brunner F., Innerhofer E., Paschotta R., Furusawa K., Baggett J.C., Monro T.M., Richardson D.J., and Keller U.: Opt. Lett. **28**, 1951 (2003).
53. Mitrofanov A.V., Ivanov A.A., Alfimov M.V., Podshivalov A.A., and Zheltikov A.M.: Opt. Commun. **280**, 453 (2007).
54. Mitrokhin V.P., Ivanov A.A., Fedotov A.B., Alfimov M.V., Dukel'skii K.V., Khokhlov A.V., Shevandin V.S., Kondrat'ev Yu.N., Podshivalov A.A., and Zheltikov A.M.: Laser Phys. Lett. **4**, 529 (2007).
55. Pickrell G., Peng W., and Wang A.: Opt. Lett. **29**, 1476 (2004).
56. Jensen J.B., Pedersen L.H., Hoiby P.E., Nielsen L.B., Hansen T.P., Folkenberg J.R., Riishede J., Noordegraaf D., Nielsen K., Carlsen A., and Bjarklev A.: Opt. Lett. **29**, 1974 (2004).
57. Konorov S.O., Zheltikov A.M., and Scalora M.: Opt. Express **13**, 3454 (2005).
58. Limpert J., Schreiber T., Nolte S., Zellmer H., and Tünnermann A.: Opt. Express **11**, 3332 (2003).
59. de Matos C.J.S., Popov S.V., Rulkov A.B., Taylor J.R., Broeng J., Hansen T.P., and Gapontsev V.P.: Phys. Rev. Lett. **93**, 103901 (2004).
60. Alfano R.R., Shapiro S.L.: Phys. Rev. Lett. **24**, 584 (1970).
61. Alfano R.R., Shapiro S.L.: Phys. Rev. Lett. **24**, 592 (1970).

62. R. Alfano (ed.) "The Supercontinuum Laser Source", Springer, Berlin (1989).
63. Baltuska A., Fuji T., and Kobayashi T.: Opt. Lett. **27**, 1241 (2002).
64. Reichert J., Holzwarth R., Udem Th., and Hänsch T.W.: Opt. Commun. **172**, 59 (1999).
65. Udem T., Reichert J., Holzwarth R., and Hänsch T.W.: Opt. Lett. **24**, 881 (1999).
66. Udem T., Reichert J., Holzwarth R., and Hänsch T.W.: Phys. Rev. Lett. **82**, 3568 (1999).
67. http://nobelprize.org/physics/laureates/2005/index.html
68. Eckstein J.N., Ferguson A.I., and Hänsch T.W.: Phys. Rev. Lett. **40**, 847 (1978).
69. Baklanov Y.V. and Chebotayev V.P.: Appl. Phys. **12**, 97 (1977).
70. Brabec T. and Krausz F.: Rev. Mod. Phys. **72**, 545 (2000).
71. Birks T.A., Wadsworth W.J., and Russell P.S.J.: Opt. Lett. **25**, 1415 (2000).
72. Bagayev S.N., Dmitriyev A.K., Chepurov S.V., Dychkov A.S., Klementyev V.M., Kolker D.B., Kuznetsov S.A., Matyugin Y.A., Okhapkin M.V., Pivtsov V.S., Skvortsov M.N., Za-kharyash V.F., Birks T.A., Wadsworth W.J., Russell P.S.J., and Zheltikov A.M.: Laser Phys. **11**, 1270 (2001).
73. Apolonski A., Poppe A., Tempea G., Spielmann C., Udem T., Holzwarth R., Hänsch T.W., and Krausz F.: Phys. Rev. Lett. **85**, 740 (2000).
74. Telle H.R., Steinmeyer G., Dunlop A.E., Sutter D.H., and Keller U.: Appl. Phys. B **69**, 327 (1999).
75. Paulus G.G., Grasbon F., Walther H., Nisoli M., Stagira S., Priori E., and De Silvestri, S.: Nature **414**, 182 (2001).
76. Drescher M., Hentschel M., Kienberger R., Tempea G., Spielmann Ch., Reider G.A., Corkum P.B., and Krausz F.: Science **291**, 1923 (2001).
77. Baltuska A., Udem T., Uiberacker M., Hentschel M., Goulielmakis E., Gohle C., Holzwarth R., Yakovlev V.S., Scrinzi A., Hänsch T.W., and Krausz F.: Nature **421**, 611 (2003).
78. Gohle C., Udem T., Herrmann M., Rauschenberger J., Holzwarth R., Schuessler H.A., Krausz F., and Hänsch T.W.: Nature **436**, 234 (2005).
79. Zewail A.H., Femtochemistry – Ultrafast Dynamics of the Chemical Bond, World Scientific, Singapore, Vols. I and II (1994).
80. Sundström V (ed.), Femtochemistry & Femtobiology, World Scientific, Singapore (1997).
81. Zheltikov A.M., L'Huillier A., and Krausz F., "Nonlinear Optics", in Träger F. (ed), Handbook of Lasers and Optics, pp. 157–248. Springer, New York (2007).
82. Radi P. and Zheltikov A.M. (eds), Proceedings of the European Conference on Nonlinear Optical Spectroscopy (ECONOS), Oxford, UK, Special Issue of J. Raman Spectrosc. **37**, 6 (2006).
83. Warren W.S., Rabitz H., and Dahleh M.: Science **259**, 1581 (1993).
84. Assion A., Baumert T., Bergt M., Brixner T., Kiefer B., Seyfried V., Strehle M., and Gerber G.: Science **282**, 919 (1998).
85. Rabitz H., de Vivie-Riedle R., Motzkus M., and Kompa K.: Science **288**, 824 (2000).
86. Brixner T., Damrauer N.H., Niklaus P., and Gerber G.: Nature **414**, 57 (2001).
87. Levis R.J., Menkir G.M., and Rabitz H.: Science **292**, 709 (2000).
88. Kärtner F.X. (ed.), Few-Cycle Laser Pulse Generation and Its Applications, Springer, Berlin (2004).
89. Kobayashi T. (ed.), J-Aggregates, World Scientific, Singapore (1996).
90. Akimov D.A., Ivanov A.A., Alfimov M.V., Grabchak E.P., Shtykova A.A., Petrov A.N., Podshivalov A.A., and Zheltikov A.M.: J. Raman Spectrosc. **34**, 1007 (2003).
91. Herek J.L., Wohlleben W., Cogdell R.J., Zeidler D., and Motzkus M.: Nature **417**, 533 (2002).
92. Akimov D.A., Ivanov A.A., Alfimov M.V., and Zheltikov A.M.: Vibrational Spectrosc. **42**, 33 (2006).
93. Konorov S.O., Akimov D.A., Ivanov A.A., Alfimov M.V., and Zheltikov A.M.: Appl. Phys. B **78**, 565 (2004).
94. Kano H. and Hamaguchi H.: Opt. Lett. **28**, 2360 (2003).
95. Zheltikov A.M.: JETP **100**, 833 (2005).
96. Konorov S.O., Akimov D.A., Ivanov A.A., Alfimov M.V., Yakimanskii A.V., and Zheltikov A.M.: Chem. Phys. Lett. **405**, 310 (2005).

97. Andresen E.R., Paulsen H.N., Birkedal V., Thøgersen J., and Keiding S.R.: J. Opt. Soc. Am. B **22**, 1934 (2005).
98. Linik Y.M., Ivanov A.A., Akimov D.A., Alfimov M.V., Siebert T., Kiefer W., and Zheltikov A.M.: J. Raman Spectrosc. **37**, 705 (2006).
99. von Vacano B., Wohlleben W., and Motzkus M.: Opt. Lett. **31**, 413 (2006).
100. Dudovich N., Oron D., and Silberberg Y.: Nature **418**, 512 (2002).
101. Dudovich N., Oron D., and Silberberg Y.: Chem J. Phys. **118**, 9208 (2003).
102. Oron D., Dudovich N., and Silberberg Y.: Phys. Rev. Lett. **89**, 273001 (2002).
103. Eesley G.L., Coherent Raman Spectroscopy Pergamon, Oxford (1981).
104. Eckbreth A.C., Laser Diagnostics for Combustion Temperature and Species Abacus, Cambridge, MA (1988).
105. Maker P.D. and Terhune R.W.: Phys. Rev. **137**, A801 (1965).
106. Druet S.A.J. and Taran J.-P., In: Moss, T.S., Stenholm, S. (eds.), Progress in quantum electronics. Vol. 7, p. 1. Pergamon Press Oxford (1981).
107. Leonhardt R., Holzapfel W., Zinth W., and Kaiser W.: Chem. Phys. Lett. **133**, 373 (1987).
108. Ivanov A.A., Alfimov M.V., and Zheltikov A.M.: Opt. Lett. **31**, 3330 (2006).
109. Ivanov A.A., Alfimov M.V., and Zheltikov A.M.: Physics Uspekhi, **47**(7), 743 (2004).
110. Spano F.C. and Mukamel S.: Phys. Rev. A **40**, 5783 (1989).
111. Knoester J.: Phys. Rev. A **47**, 2083 (1993).
112. Genty G., Ritari T., and Ludvigsen H.: Opt. Express **13**, 8625 (2005).
113. Mitrofanov A.V., Ivanov A.A., Podshivalov A.A., Alfimov M.V., and Zheltikov A.M.: JETP Lett. **85**, 231 (2007).
114. Gaeta A.L.: Phys. Rev. Lett. **84**, 3582 (2000).
115. Linden S., Kuhl J., and Giessen H.: Opt. Lett. **24**, 569 (1999).
116. Gu X., Xu L., Kimmel M., Zeek E., O'Shea P., Shreenath A.P., Trebino R., and Windeler R.S.: Opt. Lett. **27**, 1174 (2002).
117. Shen Y.R., The principles of nonlinear optics, New York, Wiley (1984).
118. Bloembergen N., Nonlinear Optics, New York, Benjamin (1964).
119. Akhmanov S.A. and Khokhlov R.V., Problems of nonlinear optics, Moscow, VINITI, 1964 [in Russian]; Engl. transl. New York, Gordon & Breach (1972).
120. Miles R.B. and Harris S.E.: IEEE J. Quantum Electron. **9**, 470 (1973).
121. Maker P.D., Terhune R.W., Nisenoff M., and Savage C.M.: Phys. Rev. Lett. **8**, 21 (1962).
122. Giordmaine J.A.: Phys. Rev. Lett. **8**, 19 (1962).
123. Feier M.M., Magel G.A., Jundt D.H., and Byer R.L.: IEEE J. Quantum Electron. **28**, 2631 (1992).
124. Ranka J.K., Windeler R.S., and Stentz A.J.: Opt. Lett. **25**, 796 (2000).
125. Omenetto F.G., Taylor A.J., Moores M.D., Arriaga J., Knight J.C., Wadsworth W.J., and Russell P.S.J.: Opt. Lett. **26**, 1158 (2001).
126. Omenetto F.G., Efimov A., Taylor A.J., Knight J.C., Wadsworth W.J. and Russell P.S.J.: Opt. Express **11**, 61 (2003).
127. Efimov A., Taylor A.J., Omenetto F.G., Knight J.C., Wadsworth W.J., and Russell P.S.J.: Opt. Express **11**, 910 (2003).
128. Efimov A., Taylor A.J., Omenetto F.G., Knight J.C., Wadsworth W.J., and Russell P.S.J.: Opt. Express **11**, 2567 (2003).
129. Ivanov A.A., Lorenc D., Bugar I., Uherek F., Serebryannikov E.E., Konorov S.O., Alfimov M.V., Chorvat D., and Zheltikov A.M.: Phys. Rev. E **73**, 016610 (2006).
130. Fedotov A.B., Voronin A.A., Serebryannikov E.E., Fedotov I.V., Mitrofanov A.V., Ivanov A.A., Sidorov-Biryukov D.A., and Zheltikov A.M.: Phys. Rev. E **75**, 016614 (2007).
131. Naumov A.N., Fedotov A.B., Zheltikov A.M., Yakovlev V.V., Mel'nikov L.A., Beloglazov V.I., Skibina N.B., and Shcherbakov A.V.: J. Opt. Soc. Am. B **19**, 2183 (2002).
132. Akimov D.A., Ivanov A.A., Naumov A.N., Kolevatova O.A., Alfimov M.V., Birks T.A., Wadsworth W.J., Russell P.S.J., Podshivalov A.A., and Zheltikov A.M.: Appl. Phys. B **76**, 515 (2003).
133. Naumov A.N. and Zheltikov A.M.: Opt. Express **10**, 122 (2002).

134. Zheltikov A.M.: Phys. Rev. A **72**, 043812 (2005).
135. Zheltikov A.M.: J. Opt. Soc. Am. B **22**, 2263 (2005).
136. Serebryannikov E.E., Fedotov A.B., Zheltikov A.M., Ivanov A.A., Alfimov M.V., Beloglazov V.I., Skibina N.B., Skryabin D.V., Yulin A.V., and Knight J.C.: J. Opt. Soc. Am. B **23**, 1975 (2006).
137. Konorov S.O., Ivanov A.A., Alfimov M.V., and Zheltikov A.M.: Appl. Phys. B **81**, 219 (2005).
138. Konorov S., Ivanov A., Ivanov D., Alfimov M., and Zheltikov A.: Opt. Express **13**, 5682 (2005).
139. Konorov S.O., Fedotov A.B., Kolevatova O.A., Beloglazov V.I., Skibina N.B., Shcherbakov A.V., Wintner E., and Zheltikov A.M.: J. Phys. D, Appl. Phys. **36**, 1375 (2003).
140. Shephard J.D., Jones J.D.C., Hand D.P., Bouwmans G., Knight J.C., Russell P.S.J., and Mangan B.J.: Opt. Express **12**, 717 (2004).
141. Stakhiv A., Gilber R., Kopecek H., Zheltikov A.M., and Wintner E.: Laser Phys. **14**, 738 (2004).
142. Chen X., Li M., Venkataraman N., Gallagher M.T., Wood W.A., Crowley A.M., Carberry J.P., Zenteno L.A., and Koch K.W.: Opt. Express **12**, 3888 (2004).
143. Flusberg B., Jung J., Cocker E., Anderson E., and Schnitzer M.: Opt. Lett. **30**, 2272 (2005).
144. Fu L., Jain A., Xie H., Cranfield C., and Gu M.: Opt. Express **14**, 1027 (2006).
145. Konorov S.O., Fedotov A.B., Mitrokhin V.P., Beloglazov V.I., Skibina N.B., Shcherbakov A.V., Wintner E., Scalora M., and Zheltikov A.M.: Appl. Opt. **43**, 2251 (2004).
146. Zheltikov A.M.: Nature Materials **4**, 265 (2005).
147. Benabid F., Knight J.C., Antonopoulos G., and Russell P.S.J.: Science **298**, 399 (2002).
148. Benabid F., Couny F., Knight J.C., Birks T.A., and Russell P.S.J.: Nature **434**, 488 (2005).
149. Konorov S.O., Fedotov A.B., and Zheltikov A.M.: Opt. Lett. **28**, 1448 (2003).
150. Konorov S.O., Akimov D.A., Serebryannikov E.E., Ivanov A.A., Alfimov M.V., and Zheltikov A.M.: Phys. Rev. E **70**, 066625 (2004).
151. Fedotov A.B., Konorov S.O., Mitrokhin V.P., Serebryannikov E.E., and Zheltikov A.M.: Phys. Rev. A **70**, 045802 (2004).
152. Konorov S.O., Serebryannikov E.E., Fedotov A.B., Miles R.B., and Zheltikov A.M.: Phys. Rev. E **71**, 057603 (2005).
153. Konorov S.O., Zheltikov A.M., Zhou P., Tarasevitch A.P., and von der Linde D.: Opt. Lett. **29**, 1521 (2004).
154. Ouzounov D.G., Ahmad F.R., Müller D., Venkataraman N., Gallagher M.T., Thomas M.G., Silcox J., Koch K.W., and Gaeta A.L.: Science **301**, 1702 (2003).
155. Ouzounov D.G., Hensley C.J., Gaeta A.L., Venkateraman N., Gallagher M.T., and Koch K.W.: Opt. Express **13**, 6153 (2005).
156. Ivanov A.A., Podshivalov A.A., and Zheltikov A.M.: Opt. Lett. **31**, 3318 (2006).
157. Luan F., Knight J.C., Russell P.S.J., Campbell S., Xiao D., Reid D.T., Mangan B.J., Williams D.P., and Roberts P.J.: Opt. Express **12**, 835 (2004).
158. Skryabin D.V., Biancalana F., Bird D.M., and Benabid F.: Phys. Rev. Lett. **93**, 143907 (2004).
159. Serebryannikov E.E. and Zheltikov A.M.: Phys. Rev. A **76**, 013820 (2007).
160. Fedotov A.B., Serebryannikov E.E., and Zheltikov A.M.: Phys, Rev. A **76**, 053811 (2007).

Part III
Technology, Integration
an Active Photonic Crystals

7 Photonic Crystal and Photonic Band-Gap Structures for Light Extraction and Emission Control

Richard M. De La Rue

Optoelectronics Research Group, Department of Electronics and Electrical Engineering
University of Glasgow, Glasgow G12 8QQ, Scotland, U.K.

Abstract. Research into photonic crystal (PhC) and photonic band-gap (PBG) structures has been motivated, from the start, by their possible use in controlling, modifying and enhancing the light emission process from high refractive index solid materials. This chapter considers the possible role of such structures when incorporated into semiconductor diode based light-emitting devices. Both light-emitting diodes (LEDs) and lasers will be considered. In order to provide a proper framework for discussion and analysis, space is devoted to the historical development of III–V semiconductor based LEDs – and to competing alternative approaches that have been demonstrated for enhanced light extraction. The possible advantages of photonic quasi-crystal (PQC) structures over regularly periodic photon crystal structures for advanced LED designs are also considered. Photonic crystal structures potentially provide major enhancements in the performance of laser diodes (LDs) – and progress towards this performance enhancement will be reviewed.

Key words: photonic crystals; photonic quasi-crystals; light emitting diodes; semiconductor lasers

7.1 Introduction

Light emitting diodes (LEDs) and semiconductor laser diodes (SLDs) are all-pervasive in the diverse applications of modern photonics technology. Many billions of LEDs and on the order of a billion SLDs are produced per year. In aggregate, all of these light sources are responsible for the consumption of a considerable amount of power. But, at the same time, the intrinsic efficiency with which electrical power can be converted into light in III–V semiconductor-based devices is encouraging for their exploitation in large-scale applications such as illumination, display devices and automotive lighting systems. Important challenges remain if the performance of the light emitting structures is to be maximized. These challenges include the creation, using low-cost mass-production techniques, of structures that can reduce the threshold electrical current of a semiconductor laser diode (SLD) by providing more-or-less omni-directional feedback to the source region. Such a structure may be considered as a form of generalized distributed feed-back (DFB) grating or distributed Bragg reflector (DBR) mirror. For light-emitting diodes, the challenge, again with the need for low-cost mass-production, is to maximize the extraction of the light that is generated within the device, with the strong additional requirement that this extraction should be through a single surface, with a well-defined beam pattern and direction.

Photonic crystals (PhCs) and, with sufficiently strong feedback, photonic bandgap (PBG) structures have been identified as a potential route to solution of the difficulties underlying these challenges.

7.2 Basic Background

The use of the words 'photonic crystal' has a history that goes back to the authors of two seminal papers that were published within a few weeks of each other in the year 1987, i.e. just over twenty years ago [1, 2]. The interested reader may wish to consult the more recent article by one of those authors, Eli Yablonovitch [3] – but that article is primarily concerned with defining what kinds of structure may legitimately be called photonic crystals and what structures should be regarded as lying outside this definition. For the moment, we shall work with the definition that *a photonic crystal is a structure that is periodic in two or three space dimensions –* and that the photonic (or optical) property that exhibits this periodic variation is most typically the refractive index, which should change, up-and-down, by a 'large' amount as the position in the periodic 'lattice' is changed. It is characteristic of periodically varying component properties that they lead, for propagating waves, to the existence of frequency ranges where *stop-band* behaviour occurs.

We may now make the significant jump implied by introducing the phrase 'photonic band-gap' – and say that a photonic crystal (PhC) exhibits *full photonic bandgap* (full PBG) behaviour when the propagation of light through the photonic crystal is *forbidden* for all directions of propagation and optical polarisation. If it is desired to achieve a full PBG situation in a medium comprised of two distinct materials, it is appropriate to use a guideline value for the refractive index contrast of 3-to-1 or more, together with an appropriate distribution of refractive index that includes a substantial proportion of each constituent material. Because of the optical anisotropy that is characteristic of photonic crystal media, the spectral range (i.e. bandwidth) over which stop-band behaviour occurs varies with the propagation direction – and with the polarisation-state of the light. In 1975, Byko [4] recognised that the operation of a laser could not merely be controlled by inserting a periodic structure into the gain medium, such as the distributed feedback (DFB) grating structures that have become the norm for semiconductor lasers in large-capacity fibre-optical communications systems, but that generalising the DFB concept to all three space dimensions could lead to a more-or-less complete suppression of spontaneous emission. Given that spontaneous emission and stimulated emission unavoidably accompany each other, but that the relationship between the two is determined by the environment in which the emission process takes place, suppression of spontaneous emission can lead to a reduction in the threshold intensity for the pumping process that drives emission. At this point, it seems appropriate to quote Bykov directly:

Control of the spontaneous emission and particularly its suppression may be important in lasers. For example, the active medium of a laser may have a three-dimensional periodic structure. Let us assume that this structure has such anisotropic properties that at the transition frequency of a molecule there is a narrow cone of directions in which the propagation of electromagnetic waves is allowed, whereas all the other directions are forbidden. Then the laser threshold of this medium (in the allowed direction) should be much lower than that of a medium without a periodic structure because the spontaneous emission will be suppressed for the majority of directions and it is the spontaneous emission that has a strong influence on the laser threshold.

The use of the word 'molecule' in the present context may be understood as meaning the basic building-block of the gain medium. In a semiconductor medium, the individual molecule may be regarded as formed by a single hole-electron pair, possibly in the form of an exciton.

In the case of a semiconductor diode, the creation of light (i.e. of photons) is due, in appropriate circumstances, to the *radiative* recombination of holes and electrons that takes place at the more-or-less direct interface (i.e. the 'junction') between the regions that are doped to provide holes or electrons for the transport of current into the junction. With the correct choice of semiconducting materials and the right kind of detailed structure, e.g. a III–V semiconductor-based quantum-well hetero-structure, the light emission process can be very efficient – and almost all hole-electron recombination events then lead to the generation of a photon. In the context of semiconductor lasers and light-emitting diodes (LEDs), the critical questions that follow this possibility of very efficient generation of light are, for LEDs, how well can the generated photons be extracted from the device and sent in the right direction – and, for lasers, what are the values of the extraction efficiency and the threshold current for laser action? The quality of the output light beam, in terms of divergence, uniformity and stability, is also important. The preceding considerations provide a substantial part – but by no means all – of the motivation for research on photonic crystals and the exploitation of photonic band-gap (PBG) behaviour.

7.3 Why III–V Semiconductors?

The earlier mention of efficient *radiative* recombination between holes and electrons brings us almost automatically to the III–V semiconductors. In the case of the basic binary III–V compounds gallium arsenide (GaAs) and indium phosphide (InP), an obvious reason for their use in devices that produce light through current-injection electroluminescence is that they are direct *electronic* band-gap semiconductors. Moreover, the ternary and quaternary compounds that can be formed as thin heteroepitaxial layers on single-crystal wafers of the binary III–V compounds are also direct band-gap semiconductors over usefully wide ranges of their possible compositions – but not all. Examples include aluminium gallium arsenide ($Al_xGa_{1-x}As$)

and indium gallium arsenide phosphide ($In_xGa_{1-x}As_yP_{1-y}$), also known as 'phosphorus quaternary or 'P-Q'.

In semiconductors with direct band-gap properties, the requirements for energy and momentum conservation in the processes that take place are relatively easily satisfied. Because the momentum of the photon that is to be produced is very small in comparison with the possible momentum values of 'massive' particles like electron and holes, the energy conservation requirement must be satisfied by the recombination of holes and electrons at negligible velocities, and therefore with very small momentum values. The direct transition point in ω–k space, with a minimum in the conduction band (energy) occurring directly above the maximum in the valence band and at the origin of momentum space, is just the right location for the process required to give efficiently radiative emission.

In contrast, in the indirect band-gap semiconductors, momentum conservation requires (using particle language) the involvement of *phonons*. In different situations that can occur in the same semiconductor – and at the same time – the phonon involved may be either an emitted or an absorbed photon, depending on whether energy is extracted or, more typically, absorbed by the crystal lattice. Phonons are quantized lattice vibrations – and the random distribution of phonon energies and momenta that are associated with lattice vibrations relates directly to the temperature of the lattice. Most obviously in metals, a substantial part of the energy and momentum that determine the temperature of the material may be directly associated with the behaviour of the characteristically free electrons. But semiconductors also have free carriers present in them, so a fraction of the thermal energy of the semiconductor is due to these free carriers and is controlled, for a typical situation, by the doping of the semiconductor.

The end-result of the nature of the hole-electron recombination process in *indirect* band-gap semiconductors is that the probability of generating a photon is typically much less than in a *direct* band-gap semiconductor, so the light emission process is intrinsically much less efficient. This result is particularly obvious for the case of the most important indirect band-gap semiconductor, i.e. silicon. Notwithstanding current research efforts towards the creation of the 'silicon laser', the direct band-gap of the typical III–V semiconductors makes them still the most obvious base for efficient electrically pumped sources of photons. But there is an interesting partial exception to the argument just presented concerning light emission in indirect band-gap semiconductors – and that is the case of the III–V semiconductor gallium phosphide (GaP).

Gallium phosphide (GaP) has occupied a central role in the history of the light-emitting diode (LED). This central role is due simply to the fact that the vast majority of the LEDs that have been produced emit light in the visible part of the spectrum – and to the fact that, among binary compound semiconductors, GaP has the best combination of properties so far demonstrated for light emission from a binary semiconductor, over much of the visible part of the spectrum. Gallium phosphide is transparent throughout a large part of the visible spectrum, not including the blue. In reality, the epitaxial device structures involved in currently produced LEDs are hetero-structures that typically involve ternary or more complex compound alloys.

Depending on the emission wavelength required, the substrate wafers used may be made of either gallium arsenide (GaAs) or of gallium phosphide (GaP) – in order to better accommodate the degree of lattice mismatch associated with the composition and heterogeneity of the materials involved. To spell it out in slightly more detail, the substrate on which the epitaxial heterostructure is grown, with various alloy compositions of the form $GaAs_xP_{1-x}/GaAs_yP_{1-y}$, may well be single-crystal GaP – but, for operation at the longer wavelength end of the visible spectrum, the substrate used may instead preferably be gallium arsenide (GaAs), because of its better lattice match.

For the shorter wavelength region of the spectrum available from GaP LEDs, i.e. for green light emission, it is appropriate to use $GaAs_xP_{1-x}$ alloy compositions that involve a relatively large fraction of phosphorus – implying that the light emitting material has an indirect electronic band-gap. The key to the successful use of gallium-phosphide for green LEDs is that it is possible to treat it in such a way that it behaves substantially like a direct band-gap semiconductor. For practical purposes, GaP can be made into a sufficiently good current-injection electroluminescent semiconductor by using special doping processes (involving, for example, the implantation of nitrogen atoms) that create a high density of *states in the (electronic) band-gap*. Phonons are still involved in the emission process, but their role is to help in populating the levels created by the implanted dopants, so that direct transitions can occur between these states and the valence band. This process for generating light in an indirect band-gap semiconductor ('gallium phosphide') is clearly not as efficient as in a direct band-gap III–V semiconductor such as gallium arsenide (where close to 100% internal quantum efficiency is possible experimentally). But the efficiency of the light generation process in suitably doped gallium phosphide LEDs has nevertheless been sufficient (a few percent) for it to be used in a large fraction of the light emitting diodes produced so far.

In a moderately recent paper on more advanced LED technology for light production in the green to infra-red part of the spectrum, R.M. Fletcher [5] has both reviewed the history of the development of gallium-phosphide based LEDs and identified techniques that have enabled, for instance, red LEDs to be produced with nearly 100% internal quantum efficiency and as much as 15% external efficiency. For greater internal efficiency, the use of an $Al_xGa_{1-x}As$-based double-heterostructure with suitable compositions enables direct band-gap emission. Substrate transparency, implying greater extraction efficiency, is obtainable through the use of a ternary $Al_yGa_{1-y}As$ substrate with a large enough value for the Al-fraction. Fletcher also describes how further performance improvements result from the use of $Al_xIn_yGa_zP_{1-x-y-z}$ epitaxial hetero-structures on GaAs substrates. In particular LEDs can be produced that have superior efficiency throughout the whole spectrum from the near infra-red to the green. It should be remarked that, so far, we have made no mention of LEDs that can produce light in the blue part of the visible spectrum – or in the ultra-violet spectrum.

In the present context of light emitting devices based on III–V semiconductors, it is appropriate to mention the binary semiconductor gallium nitride (GaN), a material that *does* have a direct electronic band-gap. Gallium nitride now provides the crucial basis for LEDs that produce light in the blue part of the visible spectrum (e.g. at a wavelength of 465 nm), as well as further into the short wavelength part of the spectrum, i.e. the blue-violet (405 nm) and ultra-violet (down to 350 nm). Again, it is important that quantum well hetero-structures can be realised within epitaxial material that has a direct electronic band-gap, e.g. structures that contain gallium indium nitride ($Ga_xIn_{1-x}N$) quantum wells. The best internal quantum efficiency (and wall-plug efficiency) values obtained so far in gallium nitride LEDs fall well short of what is possible (in the red and infra-red) with aluminium gallium arsenide. But further improvement in performance is to be expected from advances in the epitaxial growth processes for nitride semiconductors, allied to adoption of substrates that provide a better intrinsic match to the lattice constant of the active material and adjacent regions. The quality and reliability already achieved with gallium nitride based blue LEDs has led to a massive upsurge in the use of LEDs as the primary light source for large-volume applications such as computer and TV displays, automotive lighting (white light sources), road signs and a variety of other applications in illumination. It is now probable that both incandescent and fluorescent light sources will largely be displaced by LED-based solid light sources in many areas of application. The short review paper by Kovacs et al. [6] provides a very useful and wide-ranging perspective of the state-of-the-art in LED development a few years ago, with both a historical perspective and immediate relevance to on-going applications issues such as white light generation and the use of organic LEDs (OLEDs) in displays. One aspect of the construction of LEDs for visible wavelengths is that the use of a GaAs substrate may be desirable from the point of view of obtaining lattice-matched growth, e.g. with an AlGaInP active layer, as shown schematically in Fig. 7.1(a), but the substrate must be transparent to the emitted light in order for the LED to allow light extraction through it. A wafer-bonding process in which the GaAs substrate is removed and substituted by a GaP substrate can produce the desired situation – and can lead to a much more efficient device, particularly when the truncated pyramid configuration [7] is used. This type of structure will be considered in a little more detail later in this chapter. Figure 7.1(b) therefore shows an identical epitaxial structure, but with the GaP substrate substituted that is required if the emitted light is to be able to pass through the substrate. Figure 7.1(c) and (d) shows schematic diagrams of blue LEDs [6] grown on a semiconducting silicon carbide (SiC) substrate and an insulating sapphire substrate, respectively. The possible advantages of the semiconducting substrate, in terms of convenient electrical transport through the active epitaxial structure do not appear, so far, to have outweighed the cost advantages of using sapphire substrates.

7.4 Efficient Extraction of Light from LEDs

As has already been implied earlier in this chapter, an important engineering problem in the design and construction of light emitting diodes (LEDs) is that of maximising the fraction of the total light generated within the device that is actually extracted from the device. In asking the question: 'Where does the light go – or

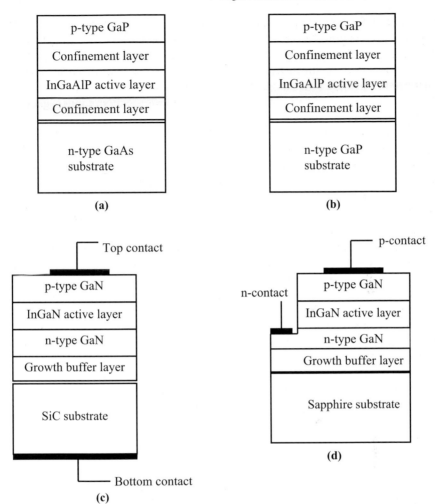

Fig. 7.1 (**a**) Schematic of high quantum efficiency red-light emitting epitaxial structure grown on GaAs substrate; (**b**) the same light emitting structure as in (**a**), but transported on to a conductive GaP structure; (**c**) simplified schematic of GaN based blue LED grown on semiconducting SiC substrate; (**d**) GaN based blue LED grown on insulating single-crystal sapphire substrate

fail to go?', one important factor that has been successfully addressed in a practical sense is the impact of the shape of the small chip of semiconductor in which the light is being generated, together with the issue of light absorption, e.g. because the electronic band-gap of the substrate on which the LED epitaxial heterostructure has been grown may be smaller than the photon energy of the emitted light.

Photonic crystal structures offer the possibility of enhancing the efficiency of the light extraction process from both LEDs and LDs – and of adding a substantial measure of beam control in terms of the direction and shape of the emitted light beam. But before we address the possible benefits of using photonic crystal structures in light emitters it is appropriate to review some of the alternatives that have been investigated more-or-less extensively and successfully.

A basic reason for the difficulty of extracting light effectively from an LED is illustrated in Fig. 7.2(a). For simplicity, the LED is shown in cross-section – and it has the form of a cuboidal box, i.e. it has a rectangular cross-section. Light is generated in the relatively thin active junction region of the LED by radiative recombination of injected holes and electrons. The light is predominantly produced by spontaneous emission and has low coherence. The emitted light radiates from the region where it is generated over a wide range of directions. For the particular case where only light that escapes from the single top surface of the wafer section is regarded as useful, it is then found that total internal reflection, in accordance with Snell's law, begins to occur at angles that deviate only slightly from the normal to the surface – because of the high refractive index ($\sim$3) of the semiconducting material from which the LED has been made. Taking account of the 3D nature of the problem, so that what matters is the solid angular range of the light escaping from one face, in relation to a total possible angular range available of 4π steradians, simple geometrical calculations show that only about three or four per cent of the internally generated light is usefully extracted through a single top surface, depending on the refractive index of the LED material.

One way to enhancement that has been used successfully in the LED light-extraction 'game' is to roughen the surface. As long ago as 1993, the photonic crystal/band-gap pioneer, E. Yablonovitch, together with several co-workers [7], demonstrated that one approach that could produce a substantial enhancement in the extraction efficiency of an LED was to roughen its surface in a suitable manner. In that work, the estimated enhancement obtained was a factor of three – giving a value of 30% for the total light extraction efficiency. More recent papers [8, 9] indicate that various roughening processes can produce about a factor of two increase in the amount of light that is extracted. Figure 7.2(b) shows how roughening works,

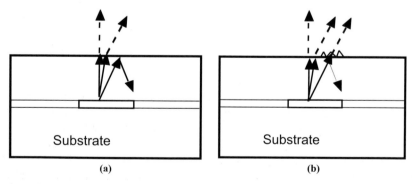

(a)   (b)

Fig. 7.2 (a) TIR occurring for light emitted from within rectangular LED box, due to sufficiently off-normal incidence on the surface; (b) sharply-angular roughened surface provides additional light extraction

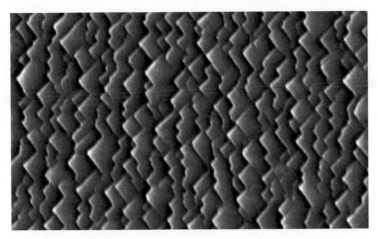

Fig. 7.3 Micrometre to sub-micrometre scale, sharply angular, roughened surface. (After reference [8])

in a simple schematic way. Figure 7.3 is an electron micrograph of a suitable roughened surface produced by a wet-etching process that has crystallographic orientation dependence.

When the substrate of an LED is transparent to the light emitted from the active region – and also if a useful part of the light generated in the active region is guided with small enough losses along the epitaxial structure to the edge of the device die, as much as six times more light can escape from the LED than from the single top surface, making extraction efficiencies greater than 20% a practical possibility. The issue then becomes that of re-directing the escaping light into the range of directions required for a specific application. An external (micro)-mirror system, possibly one that uses coated plastic mouldings, can provide a sufficiently efficient and directional beam for many applications.

With a transparent substrate and the ability to re-shape substantial parts of the LED chip to desirable shapes that may look like truncated pyramids or cones, it becomes possible to extract more than 50% of the light generated within the LED [10]. In this work, Krames and co-workers removed the gallium arsenide (GaAs) substrate on which the electroluminescent epitaxial structure was originally grown because, although necessary for growth of the best red-light emitting material, the substrate absorbs all of the red light that enters it. Having removed the GaAs substrate, a gallium phosphide (GaP) substrate was then fused onto the epitaxial structure before application of the angled sawing processes that produced the desired truncated pyramid geometry. In comparison with reference devices, the external efficiency of the devices produced showed an increase by about a factor of 1.4 – to 55% at the peak wavelength. Hundreds of milliWatts of output light were demonstrated – and conversion efficiencies as large as 100 lumen of light per watt of electrical input were shown to be possible. Before moving on to the topic of photonic crystal light extractors, we shall also mention the review by Craford et al. [11].

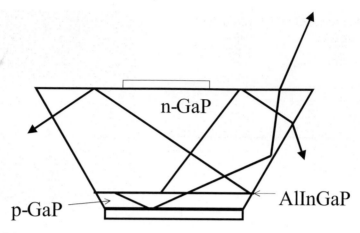

Fig. 7.4 Schematic representation of LED with transparent substrate and truncated pyramid geometry, showing emitted light rays escaping after one or, possibly two, reflective 'bounces' at the surfaces of the device. (After reference [10])

The way in which a truncated pyramidally or conically shaped geometry can help light to escape from an LED, after emission from an active region that is located close to one face of the device, is shown schematically in Fig. 7.4. Most of the LED structure is formed by the transparent substrate section – onto which the light emitting epitaxial structure has been wafer-bonded.

7.5 Photonic Crystal Light Extraction Structures

As we have already said, photonic crystal structures offer the possibility of enhancing the efficiency of the light extraction process – and of adding a substantial measure of beam control in terms of the direction and shape of the emitted light beam. The construction of the LED may be of a more advanced type, such as a micro-cavity formed either by periodic mirrors grown epitaxially as part of the LED – or by using a metallic reflective coating – in combination, possibly, with substrate thinning. The detailed nature and the area distribution and location of the photonic crystal structure added on to the LED may then be somewhat different. The photonic crystal structure may be required specifically to interact strongly with light that propagates outwards from the centre of a finite-size cylindrical region that relies primarily on micro-cavity principles for enhanced light extraction. The outwardly propagating light may be guided because of the refractive index distribution that is associated with the doped epitaxial hetero-structure diode that becomes, through current injected under forward bias, the source of the luminescence. The photonic crystal structure should then penetrate down to and through the active region of the LED. For instance, in the work of Rattier and co-workers [12], this approach has been adopted in an ingenious fashion, with the background issue of possible non-radiative recombination being addressed by organizing a mesa geometry and an all

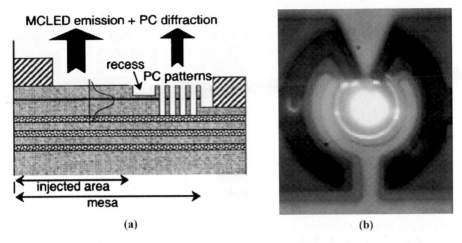

Fig. 7.5 (a) Schematic of advanced design of photonic crystal LED. The part shown may be considered as a half cross-section of a circularly symmetric (cylindrical) device, with the rotational symmetry axis to the left-hand side of the sketch. (After reference [12]); (b) plan view of operating micro-cavity/PhC LED – showing a distinct region of micro-cavity luminescence, surrounded by an arc of PhC coupled light

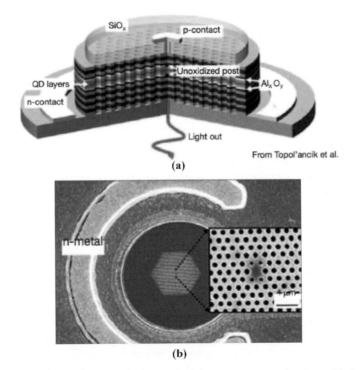

Fig. 7.6 (a) Schematic of mesa-geometry light-emitting diode incorporating a 2D PhC based micro-cavity; (b) plan view image showing hexagonal PhC region, with expanded view of central section showing micro-cavity region

top-contact arrangement – so that carrier recombination was not required to occur in regions in which the photonic crystal structure, with its accompanying fabrication-process induced damage, had been added to the device structure.

A somewhat similar device structure is described in reference [13] – and Fig. 7.6(a) and (b) closely resemble diagrams from that paper. This device used an epitaxially grown DBR bottom mirror and a region of 2D PhC around a small area of in-plane defect microcavity. An important feature of the structure is that the PhC was etched through the active region in order to obtain strongly interactive emission behaviour. Although this device was successfully operated at quite low drive current levels, its electrical to light conversion efficiency was quite small, because of non-radiative recombination associated with etch-damage on the inside surfaces of the PhC holes in the region where they intersect with the active quantum wells. It is also now widely accepted that significant improvement, in terms of reduction in non-radiative recombination, results from using an active region based on quantum dots rather on quantum wells.

A further point of interest is that the title of reference [13] clearly indicates that the device was a laser, albeit one with a very low efficiency. Indeed, the article, reference [14], for which the author of the present chapter was a co-author was written in the belief that a laser had been demonstrated.

7.6 Photonic Crystal (PhC) and Photonic Quasi-Crystal (PQC) Structures Incorporated into Large Band-Gap Nitride LEDs

The motivation for the statement of the photonic band-gap (PBG) concept given by both Bykov [4] and Yablonovitch [1] was that sufficiently strong and omni-directional periodic feedback would make it possible to reduce the threshold for lasing in an appropriate wavelength range, neglecting technological difficulties. But quite early on in the development of the concepts and technology of photonic crystals, it was also recognized that photonic crystal structuring might enhance the light extraction efficiency of LEDs. In a sense, a two-dimensionally periodic photonic crystal on the surface of a light emitting diode is simply a more complicated form of coupling grating.

Photonic quasi-crystals (PQCs)s have been identified by a number of workers [15–18] as of interest, in part because of the higher level of rotational symmetry that they can provide by comparison with the restricted choice of symmetry offered by fully regular photonic crystal lattices. The benefits of using photonic quasi-crystal structures and specifically of using Archimedean tiling for both efficient light extraction and emitted light beam shaping were subsequently demonstrated by Rattier and co-workers [17]. A notable feature of the LED devices in this last case is that, as in earlier work [12], the photonic crystal light extraction structure occupied a thin circular ring region around a mesa structure – and the PQC structure was deeply etched, so as to penetrate through the active region. This active region acted additionally as a waveguide for a substantial part of the light generated in the LED –

and the PQC structure was an effective coupler-out for that guided light because of its deep penetration.

Figure 7.7(a) shows a simple photonic quasi-crystal pattern formed out of holes grouped locally in square and triangular coordinated arrangements. Figure 7.7(b) shows an actual etched PQC pattern etched into the surface of a gallium nitride LED, realised with the objective of enhancing the light-extraction efficiency and beam quality of the LED. Elegant alternative PQC patterns from the web-site of Mesophotonics [19] are shown in Fig. 7.7(c) and (d).

An important difficulty in using photonic crystal structures as an integral part of light-emitting semiconductor devices arises from the need to fabricate the structure in epitaxial diode material, by means of a lithographically defined etching process. Electron-beam lithography (EBL) can define the pattern required with sufficient precision – and the subsequent processes of development and pattern transfer into the material structure via dry-etching are capable of being carried out *en masse*. Focused ion-beam etching (FIBE) is attractive because of its adequate precision for

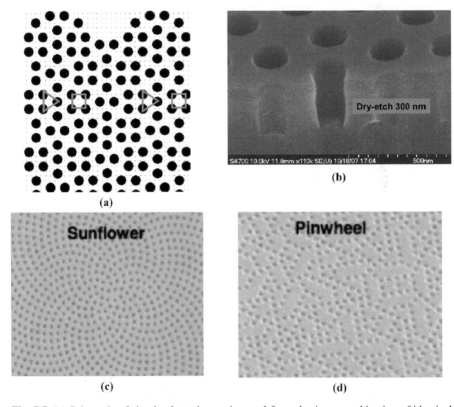

(a)

(b)

(c)

(d)

Fig. 7.7 (**a**) Schematic of simple photonic quasi-crystal formed using a combination of identical single square and identical single triangular hole formations; (**b**) Scanning electron micrograph showing hole-type 2D PhC etched into surface of GaN light emitting diode; (**c**) Sunflower photonic quasi-crystal pattern; (**d**) Pin-wheel photonic quasi-crystal pattern

some purposes – and because it involves a single process step. But the maximum area typically involved when writing with sufficient precision may be very small, e.g. less than $100 \times 100\,\mu m$. The need for a fully mass-scale process for fabricating 2D photonic crystals indicates a requirement for processes such as deep ultra-violet (DUV) lithography and nano-imprint lithography (NIL).

7.7 Emission Control and Lasing

It is perhaps appropriate to remember at this point that both the seminal work of Yablonovitch [1] and the earlier recognition by Bykov [4] of the possibilities for suppression of spontaneous emission through almost total distributed feedback were motivated by the desire to reduce the threshold pump power (more specifically the threshold value of the injection current in the case of electroluminescent devices) required to achieve the situation where emission primarily takes the form of stimulated emission, i.e. laser action. The DFB and DBR lasers that are familiar to the fibre-optical communications engineering community intrinsically involve the use of periodic grating structures, although controlled deviations from simple periodicity are also routinely involved. Possibly the simplest deviation from periodicity is the incorporation of one or two steps in the grating, effectively changing the positions of large sections of the grating with respect to other sections. Such shifts are called '$\lambda/4$' or 'double '$\lambda/8$' shifts when chosen with the magnitude that gives simple optimal behaviour.

In 'standard model' DFB and DBR lasers with one-dimensionally distributed periodic structures embedded in a stripe waveguide formed by a compound heterostructure, the feedback that is the key to changing the light emission from being predominantly through spontaneous emission to being predominantly through stimulated emission only operates initially on a small fraction of the total spontaneous emission. Most of the spontaneous emission is not confined to the waveguide region and either escapes out of the device completely without causing a stimulated emission event or is absorbed somewhere in the laser, e.g. in the substrate if that has a smaller electronic band-gap than the emitted photon energy. As the drive current is increased, progressively more stimulated and guided photons are produced – and the situation where stimulated photons generate other stimulated photons in significant quantities is reached. The DFB or DBR grating is, at the same time, selectively filtering the spectrum of the light that will actually be emitted through either end of the waveguide along which it propagates. The build-up of stimulated emission becomes so large that the fraction of the total emission that is escaping or being absorbed becomes relatively small, even though it may also increase in absolute terms.

It is clearly the case that distributed feedback that is produced by a strong 2D-periodic grating (i.e. a 2D photonic crystal) has the potential to substantially reduce the threshold current for lasing, because of the greater angular range of light that can be grabbed and involved in guided-wave stimulated emission. As with standard DFB and DBR lasers realised in hetero-epitaxial semiconductor diodes, 2D PhCs can be used to form a cavity and produce defect states in the photonic band-gap (or stop-band) produced by their periodic array of partial reflectors. Such cavities may, for instance, be produced simply by 'filling-in' one of the holes that form the 2D PhC lattice. As with the basic DFB or DBR laser, the balance of effects that lead to the combination of feedback and emission in a partial reflection situation – and to lasing – may occur at a band-edge, resulting in 'band-edge lasing'. In practical 2D photonic crystal lasers, proper choice of the quantum-well thicknesses or the quantum-dot size distribution, in the active region, can ensure that only one of the available band-edges is correctly positioned at the photon energy value that is appropriate for involvement in the lasing process.

This section of our chapter will finish with a rather rapid and selective annotated bibliography. Arguably what will be most remarkable in this review is that progress towards the 'obvious' goal of an efficient and low-threshold current, injection-electroluminescent, PhC laser has been quite slow. As a first example, there is the early publication by O'Brien and co-workers [20], involving possibly the first serious attempt to use a deeply-etched 2D photonic crystal structure as a (DBR) mirror in a semiconductor diode laser. Although the lasers certainly worked, they were relatively wide gain stripe structures with a quoted best threshold current of 110 mA – and the output spectrum indicated that lasing action was on the second level of the single quantum well GRINSCH material. More recently, some of the same authors have described another version of the photonic crystal semiconductor injection laser, emphasizing operation over a broad area [21]. This version of the photonic crystal laser has used a quite different form of PhC lattice, with much weaker reflectivity, because the holes do not penetrate as far as the active quantum well layers, and a rectangular lattice aligned to a waveguide channel. It was found that, with the correct choice of the two lattice parameters, stripe laser operation in a single transverse-mode could be obtained with a $1/e2$ width of 25 μm. Spectrally controlled operation with such a wide stripe would be useful for obtaining high output-power operation with good beam quality.

Several groups have pursued the objective of making compact in-plane emitting lasers with strong PhC type DBR mirrors, e.g. [22]. This paper is notable for its decisive demonstration of the reduced threshold current produced by use of deeply etched DBR mirrors, although only in a 1D format. Serious efforts have been made to produce injection current electroluminescent 2D PhC lasers using the strong confinement provided by a suspended membrane waveguide, with a central pillar providing both local support and a path for the injection of the device drive current [23]. The threshold current obtained with such lasers was a mere 260 μA. While indeed small, this threshold current value is still significantly larger than has been achieved in other compact geometry semiconductor lasers. The demonstrated quantum efficiency of the PhC laser was apparently very small, although the light output value estimates may be lower than the total actually generated. It seems reasonable to assume that there was in fact a substantial level of non-radiative carrier recombination at the etched surfaces of the PhC structure. The issue of surface recombination has been addressed quite recently, but in optically pumped semiconductor quantum well lasers, again in membrane format [24]. The well-known sulphide-based passivation process has been shown to produce a dramatic (factor of 4) reduction in the threshold pump power of the PhC lasers.

7.8 Conclusions

Photonic crystal (PhC) structures and photonic band-gap (PBG) principles may be applied widely in the design and operation of both LEDs and LDs. Potentially major advantages result from the demonstrated enhancement of light extraction efficiency and beam quality in LEDs, while there are real possibilities for better control of modal properties and reduction of the lasing threshold in laser diodes. Although there is much promise, photonic crystal based devices must compete with alternative approaches that can also provide major improvements in light emitting device performance. Further advances in device fabrication technology, e.g. the development of mass-production techniques and of completely effective and reliable passivation processes, are still required if the photonic crystal approach is to deliver its full potential.

Acknowledgements Research on the topics of this chapter carried out by the author, together with co-workers, has been supported by the European Community through the COST P11, ePIXnet and SMILED projects. Support for current work on the PQLDI project by the U.K. Department for Business, Enterprise and Regulatory Reform (BERR) is also acknowledged.

References

1. Yablonovitch E., 'Inhibited Spontaneous Emission in Solid-State Physics and Electronics', *Phys. Rev. Letts.* 15(20), 2059–2062, 18th May (1987).
2. John S., 'Strong Localisation of Photons in Certain Disordered Dielectric Superlattices', *Phys. Rev. Letts.* 15(23), 2486–2489, 8th June (1987).
3. Yablonovitch E., 'Photonic Crystals: What's in a Name?', *Optics and Photonics News* 18(3), 12–13, March (2007).
4. Bykov V.P., 'Spontaneous emission from a medium with a band spectrum', *Sov. J. Quant. Electron.* 4(7), 861–871, Jan (1975).
5. Fletcher R.M., Kuo C., Osentowski T.D., Yu J.G. and Robbins V.M., 'High-efficiency aluminum indium gallium phosphide light-emitting diodes', *Hewlett Packard Journal*, Aug (1993).
6. Kovac J., Peternai L. and Lengyel O., 'Advanced light emitting diodes structures for optoelectronic applications', *Thin Solid Films* 433, 22–26, (2003).
7. Schnitzer I., Yablonovitch E., Caneau. C, Gmitter T.J. and Scherer A., '30% external quantum efficiency from surface textured, thin-film light-emitting diodes', *Appl. Phys. Lett.* 63, 2174–2176, (1993).
8. Lee Y.J., Lu T.C., Kuo H.C., Wang S.C., Hsu T.C., Hsieh M.H., Jou M.J. and Lee B.J., 'Nano-roughening of n-side surface of AlGaInP-based LEDs for increasing extraction efficiency', *Materials Science and Engineering* B, 138, 157–160, (2007).
9. Fujii T., David A., Gao Y., Iza M., DenBaars S.P., Hu E.L., Weisbuch C. and Nakamura S., 'Cone-shaped surface GaN-based light-emitting diodes', *Phys. Stat. Sol* (c), 2(7), 2836–2840, (2005).

10. Krames M.R., Ochiai-Holcomb M., Höfler G.E., Carter-Coman C., Chen E.I., Tan I.-H., Grillot P., Gardner N.F., Gardner H.F., Chui H.C., Huang J.-W., Stockman S.A., Kish F.A., Craford M.G., Tan T.S., Kocot C.P., Hueschen M., Posselt J., Loh B., Sasser G. and Collins D., 'High-power truncated-inverted-pyramid $(Al_xGa_{1-x})_{0.5}In_{0.5}P/GaP$ light-emitting diodes exhibiting >50% external quantum efficiency', *Appl. Phys. Lett.* 75, 2365–2367, (1999).
11. Craford M.G., Holonyak N. and Kish F., 'In Pursuit of the Ultimate Lamp', *Scientific American* 284, 62–67, Feb (2001).
12. Rattier M., Krauss T.F., Carlin J.F., Stanley R., Oesterle U., R. Houdré, Smith C.J.M., De La Rue R.M., Benisty H. and Weisbuch C., 'High extraction efficiency, laterally injected, light emitting diodes combining microcavities and photonic crystals', *Opt. Quant. Electron* 34(1-3): 79–89, Jan (2002).
13. Zhou W.D., Sabarinathan J., Kochman B., Berg E., Qasaimeh O, Pang S. and Bhattacharya P., 'Electrically injected single-defect photonic bandgap surface-emitting laser at room temperature', *Electronics Letters* 36(18), 1541–1542, 31st Aug (2000).
14. De La Rue R. and Smith C., 'On the threshold of success', *Nature* 408, 653–656, 7th Dec (2000).
15. Kaliteevski M.A., Brand S., Abram R.A., Krauss T.F., De La Rue R.M. and Millar P., 'Two-dimensional Penrose-tiled photonic quasicrystals: diffraction of light and fractal density of modes', *Journal Of Modern Optics* 47(11), 1771–1778, Sept (2000).
16. Zoorob M.E., Charlton M.D.B., Parker G.J., Baumberg J.J., and Netti M.C., 'Complete photonic bandgaps in 12-fold symmetric quasi-crystals', *Nature* 404, 740–743, (2000).
17. Rattier M., Benisty H., Schwoob E., Weisbuch C., Krauss T.F., Smith C.J.M., Houdré R. and Oesterle U., 'Omnidirectional and Compact Guided Light Extraction from Archimedean Photonic Lattices', *Appl. Phys. Lett.* 83(7), 1283–1285, 18 Aug (2003).
18. David A., Fujii T., Matioli E., Sharma R., Nakamura S., DenBaars S.P. and Weisbuch C., 'GaN light-emitting diodes with Archimedean lattice photonic crystals', *Appl. Phys. Letts.* 88, 073510 (2006).
19. Mesophotonics website at: http://www.mesophotonics.com/
20. O'Brien J., Painter O., Lee R., Cheng C.C., Yariv A., and Scherer A., 'Lasers incorporating 2D photonic bandgap mirrors', *Electronics Letters* 32(24), 2243–2244, 21st November (1996).
21. Zhu L., Chak P., Poon J.K.S., DeRose G.A., Yariv A. and Scherer A., 'Electrically-pumped, broad-area, single-mode photonic crystal lasers', *Optics Express* 15(10), 5966–5975, 14th May (2007).
22. Raffaele L., De La Rue R.M., Roberts J.S. and Krauss T.F., 'Edge-emitting semiconductor microlasers with ultrashort-cavity and dry-etched high-reflectivity photonic microstructure mirrors', *IEEE Photon. Tech. Lett.* 13(3), 176–178, March (2001).
23. Park H.G., Kim S.H., Kwon S.H., Ju Y.G., Yang J.K., Baek J.H., Kim S.B., and Lee Y.H., 'Electrically driven single-cell photonic crystal laser', *Science* 305, 1444–1447, (2004).
24. Englund D., Altug H. and Vuckovic J., 'Low-threshold surface-passivated photonic crystal nanocavity laser', *Appl. Phys. Letts.* 91, 071124, (2007).

8 Silicon-Based Photonic Crystals and Nanowires

Bozena Jaskorzynska and Lech Wosinski

Royal Institute of Technology (KTH), Electrum 229, 164 40 Kista, Sweden

Abstract. This chapter highlights issues related to dense photonic integration based on silicon platform and reviews two alternatives to achieve this goal within the diffraction limit; photonic nanowires and photonic crystal waveguides. Examples of the device concepts and demonstrators, as well as the fabrication techniques for passive Si based mesostructures, are presented. Promising prospects and recent breakthroughs in heterogeneous integration of silicon with optically active are indicated.

Key words: Photonic crystal waveguides; photonic nanowires; photonic integration; silicon photonics; nanotechnology

8.1 Towards Dense Photonic Integration

A breathtaking but not too speculative view is that the miniaturization and large-scale integration of photonic components can have a similar impact to that experienced in electronic components in the 1960s and later. Photonics influences many aspect of our life and is one of the most important technologies for the 21st century [1]. However, in spite of tremendous progress in photonic components today's integrated optical products are very big in comparison with electronics. And it cannot be justified by photons be orders of magnitude larger than electrons, because the size of the available products is still far from reaching the fundamental diffraction limit of the half-wavelength, which for typical dielectric materials amounts to a few hundred nanometres. The main reason is that waveguides used in today's photonic devices have low refractive index contrast between the core and the cladding, which makes the components large and does not allow sharp bends necessary for dense integration. For example, in silica glass waveguides the bending radius at which the bending loss is acceptable is ca. 30 mm. In super-high contrast semiconductor waveguides, called photonic nanowires, the bending radius can be reduced to few micrometers, as it has been demonstrated [2] for Silicon-On-Insulator (SOI) technology, and is illustrated in Fig. 8.1.

Photonic crystals (PhCs) also offer very sharp bends, and when made in SOI or InP membrane PhC waveguides are theoretically lossless, but their propaga-

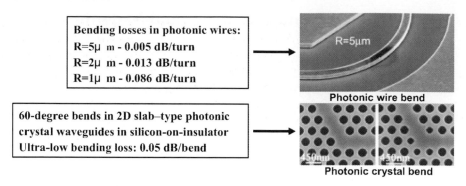

| Bending losses in photonic wires: |
| R=5μ m - 0.005 dB/turn |
| R=2μ m - 0.013 dB/turn |
| R=1μ m - 0.086 dB/turn |

Photonic wire bend

| 60-degree bends in 2D slab–type photonic crystal waveguides in silicon-on-insulator Ultra-low bending loss: 0.05 dB/bend |

Photonic crystal bend

Fig. 8.1 Bending losses in photonic crystal waveguides and photonic nanowires. (Ref. [2], picture reproduced from http://domino.research.ibm.com/comm/research_projects.nsf/pages/photonics.projects.html)

tion losses due to fabrication inaccuracies are still typically significantly larger (14 dB/cm) [3, 4] than those demonstrated (2.4 dB/cm) in SOI photonic nanowires [4].

The main source of propagation loss in photonic nanowires is light scattering from the sidewall roughness which is difficult to eliminate with the present fabrication techniques. Nevertheless, losses as low as 0.8 dB/cm for a single-mode waveguide of a 500 nm width have been demonstrated with roughness reduction by oxidation smoothing [5]. Most recently (January 2008) the research group at the University of Glasgow obtained losses of 0.92 dB/cm in a 500 nm wide photonic nanowire using highly selective one-step etching with hydrogen silsesquioxane as an electron-beam resist, and without any post-processing [6]. In general, the smaller is the waveguide cross-section the larger is the detrimental effect of the sidewall roughness [7].

Out-of-plane scattering losses in 2D slab photonic crystal waveguides may have several origins. They may be due to insufficient etching depth, non vertical etch profile, or other etching imperfections, which can be remedy by further improvement of the fabrication processes. O'Faolain et al. [8] obtained propagation losses 4.1 dB/cm in a single line defect (W1) waveguide in Si membrane with very smooth and vertical hole sidewalls. However, waveguides in low contrast PhC slabs as e.g. InP based heterostructures, suffer from the intrinsic loss due to unavoidable coupling to radiation modes. Here losses of the order of 18 dB/cm have been achieved in InP/InGaAsP heterostructures [9]. A detailed explanation of the power leakage mechanism and how it can be used to advantage in the so called 2.5D microphotonics approach, can be found in Chap. 2 of this book.

The high refractive index difference between silicon and silica or air allows for reduction of a mode size to about 0.1 μm², which is the same order of magnitude as used in VLSI electronics. On the other hand this leads to a very large modal mismatch for light in/out-coupling to optical fibre, giving 30 dB losses for an ordinary butt-coupling. There are several approaches for mode conversion used by different groups based on: adiabatic tapers [10], vertical coupling with surface gratings etched directly in silicon [11] or in thin Au overlays [12], inverted tapers [13], and others. Usually, coupling losses obtained with these techniques are of the order of 1 dB per coupling. Using two-step converter consisting of adiabatic and inverted tapers Vlasov's group obtained coupling losses as low as 0.5 dB [14].

Both photonic crystals and nanowires satisfy two of the prerequisites for dense integration. They provide strong light confinement in a waveguide core and sharp, low-loss bends. The requirement for low loss propagation, however, is not yet fulfilled and the problem is more severe for photonic crystals.

On the other hand, PhCs exhibit incomparably richer, unconventional dispersive properties that can be tailored by design to achieve novel or improved signal processing functions. Their recognized benefits and potential applications include: lossless frequency selective and omnidirectional mirrors [15], dispersion compensators [16], defect cavities with strongly enhanced quality factor [17] for "zero-threshold" lasers, narrow band filters [18,19], high efficiency and directionality surface light emitting diodes [20]. There has been much interest recently in utilizing those of the most peculiar dispersion related effects in photonic crystals. These are; negative refraction [21–26] with the implied sub-wavelength light focussing and novel imaging possibilities [27], super-prism effect [28] for orders of magnitude increase of wavelength resolution [29], slow modes [30,31] for optical time-delay [32] and enhancement of nonlinear interactions, or optical memory.

Currently, combination of photonic crystals and photonic nanowires is considered as the most viable alternative for densely integrated, highly functional circuits. For example an interesting solution for microcavities with high quality factor and high transmission has been recently shown [33], where single row photonic crystal structure with tapered PhC mirrors was embedded in a photonic nanowire.

However, also other approaches are being explored. Several ultra small components have been demonstrated based on a new concept of a slot waveguide, where light is highly confined in the air core [34, 35]. Another approach for even further size reduction (below the diffraction limit) relies on coupling of electromagnetic waves to electrons at metal-dielectric interfaces. Such so-called surface plasmon polaritons have however, excessively large damping loss (1.2 dB/μm) [36] at optical frequencies, which is today a fundamental limitation and a serious drawback, especially when large scale integration is considered. Nevertheless, the concept attracts a considerable interest because of the novel effect it offers, compatibility with silicon photonics, and the size comparable to the smallest metal wires used for electrical interconnects. To overcome the loss hurdle ways to amplify surface plasmons polaritons [37], and new types of engineered artificial materials are intensively sought.

Silicon has been the main material used in electronics industry for the past 40 years. Silicon photonics, although having its roots in the late 1980s due to pioneering works of Soref [38], only recently attracted a great deal of attention [39]. However, because of its indirect band-gap silicon is a very poor light emitter. It also lacks a linear electro-optic effect due to its centro-symmetric crystal structure. Therefore, Si has been mainly used as a substrate material for passive components offering a potential of cost effective optoelectronic integration. Today silicon is not only the main candidate for optical interconnects in the future VLSI electronics, but also intensive efforts are devoted to add active functions by heterogeneous integration with III–V semiconductors or other optically active or externally controlled materials, such as e. g. electro-optic polymers, or liquid crystals. For integration with III–Vs two viable alternatives are being investigated: the monolithic integration, where III–V material is directly grown on Si platform [40], and the hybrid approach with technology promising high volume and low cost assembling, where III–V materials or structures are (adhesively) bonded onto Si/SOI wafer. The latter is best illustrated by the recent breakthrough demonstrations of electrically pump Si hybrid lasers [41,42].

The idea of high contrast photonics is not new. However, it is only now that enabling nanotechnologies reached the stage to produce such small components with sufficiently high accuracy. The feature size of photonic nanowires and photonic crystals is a few hundred nanometres which is an order of magnitude larger than the required fabrication accuracy of a few tens nanometre. Therefore, adequately referring to the fabrication as nanotechnology one sometimes refers to those structures as nanophotonic components. Also more accurately, one calls them wavelength-scale devices or less often; mesophotonic components or mesostructures. Mesophotonics deals with optical components sized below the wavelength of light. Integrated circuits combining conventional electronics and mesophotonic components could represent ultimate limit for optoelectronic miniaturization.

8.2 Device Concepts and Demonstrators

In this section we present examples of our designs and demonstrators to illustrate the feasibility of photonic nanowires for device miniaturization and benefits offered by photonic crystal dispersion.

8.2.1 The Smallest AWG Demultiplexer

Due to strong light confinement in Si nanowires the size reduction of the functional integrated circuits can reach several orders of magnitude in comparison to standard integrated optics in silica-on-silicon technology. Based on amorphous silicon (α-Si) nanowires we have demonstrated the smallest to date Arrayed Waveguide Grating (AWG) demultiplexers (Fig. 8.2) [43]. The total size of the device is $40 \times 50\,\mu m$ and it is 30,000 times smaller than a conventional AWG in silica-on-silicon.

In Fig. 8.3 we visualize the size reduction obtained by application of SOI nanowire technology. Size of the small square corresponding to 1600 SOI nanowire based AWGs should be compared with that of the 4″ Si wafer on which one can see a series of conventional AWGs in silica-on-silicon.

Amorphous silicon (α-Si) is an interesting material and the technology gives more flexibility in nanowire structures, in particular allowing for fabrication of multilayer structures. Propagation loss of α-Si has been obtained to be $1.5\,dB/cm$ and refractive index at $1.55\,\mu m$ to be 3.63.

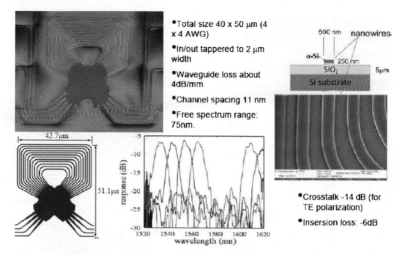

Fig. 8.2 Compact AWG demultiplexer based on amorphous silicon nanowires. *Top left*: SEM picture of the fabricated structure. *Right*: cross section of the nanowire and SEM picture of a few nanowires. *Bottom left*: sketch of the structure. *Middle*: spectral response of the fabricated AWG

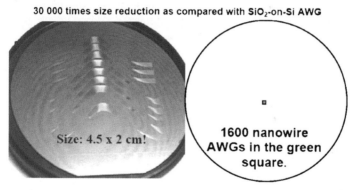

Fig. 8.3 Size comparison of conventional SiO_2/Si AWGs (*left*) and 2×2 mm sample that can contain 1600 nanowire AWGs (*right*)

8.2.2 Negative Refraction for Polarization Splitting

One of the most intriguing dispersive properties of photonic crystals is that at some frequency regimes they can refract light as if they had a negative refractive index, i.e. the angles of the incident and the refracted beam have opposite signs [21]. We utilized this to realize one of the first practical applications of this phenomenon, a polarization beam splitter (PBS) for the optical communication window at 1.5 μm [26]. The splitter consists of a PhC slab, an input waveguide and two output waveguides, as it is depicted in Fig. 8.4 (upper view). The PhC slab is formed by 15 rows of Si pillars, and the surface normal to the PhC slab is along ΓM direction. PhC is designed such that only TE polarization experiences the negative refraction, while the TM polarized beam is positively refracted.

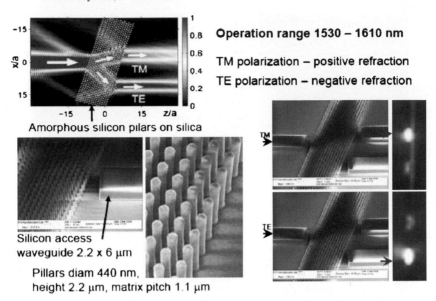

Operation range 1530 – 1610 nm

TM polarization – positive refraction

TE polarization – negative refraction

Amorphous silicon pilars on silica

Silicon access
waveguide 2.2 x 6 μm

Pillars diam 440 nm,
height 2.2 μm, matrix pitch 1.1 μm

Fig. 8.4 Demonstration of the polarization beam splitter. *Upper view*: Principle of operation. *Lower views*: SEM of the sample (*left*), and the experimental evidence (*right*) for the polarization splitting

The design of the PBS is based on 2D band structure calculations. Figure 8.5 shows the band structures and equal-frequency contours (EFCs) of a triangular lattice of infinitely long silicon pillars in air. The band structures and EFCs are both calculated with the plane-wave method. We can see from Fig. 8.5(e) and (f) that with increasing frequency the EFC moves inward for the TE polarization, whereas it moves outward for the TM polarization. This implies that the corresponding group velocities have opposite signs (determined by the gradient of the EFCs). Hence, the TE and TM polarized beams, respectively, experience negative and positive refraction within the same frequency range of $0.672 < a/\lambda < 0.732$, where a is the lattice constant (pillar matrix pitch).

The fabricated beam splitter has a total size of only $20 \times 20 \, \mu m$. The splitting angle of the two polarizations is $\sim 60°$, and the two beams are separated with a distance of $\sim 15 \, \mu m$ at the exit of the PhC slab. The working wavelength covers the whole spectral range 1530–1610 nm of the Amplified Spontaneous Emission (ASE) Erbium source. From the picture taken with IR camera (lower, right view in Fig. 8.4) one can see that TM polarized input beam mainly exits from the upper arm, whereas the TE beam goes to the lower arm. The extinction ratio is 15 dB in the whole spectral range of ASE.

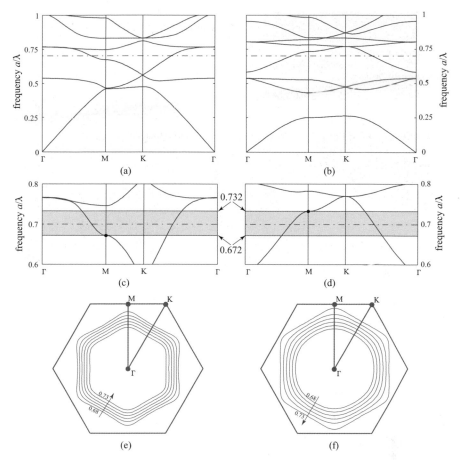

Fig. 8.5 (**a–d**) Band structures of a triangular lattice of silicon pillars in air with diameter $d = 0.4a$ (*a* is the lattice constant). *Dash-dotted lines* indicate the frequency $a/\lambda = 0.7$; (**c,d**) is the zoom-in view of (**a,b**) around $a/\lambda = 0.7$. (**e,f**) EFC contours of the present structure. The *arrows* indicate the direction of frequency increase (i.e. the direction of the group velocity). (**a,c,e**) for TE polarization, (**b,d,f**) for TM polarization

8.2.3 Channel Drop Filter in a Two-Dimensional Triangular Photonic Crystal

Add-drop filters based on resonant operation of two parallel waveguides and a cavity resonator system offer very narrow bandwidths at very short device lengths. Operation of such filters relies on the phase matching between the waveguides and the cavity system. This in general requires special design care in order to achieve so called accidental degeneracy of the cavity system modes, so that both drop and add functions can be realised at the same frequency.

We have designed such a filter in a two-dimensional photonic crystal with a triangular lattice of air holes [44]. The filter consists of two defect waveguides obtained

Fig. 8.6 Design of an add-drop filter in 2D photonic crystal. The size of the four holes (marked with the *solid circles*) is decreased to enforce the degeneracy between the cavity modes

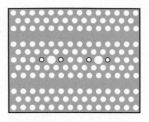

Fig. 8.7 Simulated field distribution of the drop filter depicted in Fig. 8.6

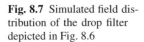

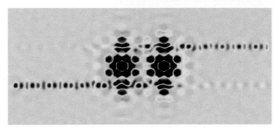

by removing a row of the holes, and two defect cavities formed by two bigger holes. The accidental degeneracy was enforced by decreasing the size of the neighbouring four holes marked with the solid circles, as it is shown in Fig. 8.6.

In contrast to the earlier designs [18], ours does not involve either inclusion of additional materials or an extra small feature size. The simulated field distribution for a drop operation is shown in Fig. 8.7.

For the wavelength 1.54 μm and the refractive index 3.32 the bandwidth of the theoretically lossless filter is 1.1 nm and its length is ca 10 μm.

8.2.4 Narrow Band Directional Coupler Filter with 1D PhC Arm

We have utilized [19] strong periodicity induced dispersion of a **B**ragg **R**eflection **W**aveguide (BRW) often referred to as 1D photonic crystal, to obtain a filter of a sub-nanometre bandwidth while sacrificing the ultra small size offered by 2D PhC based components. The filter has geometry of a directional coupler depicted in Fig. 8.8 (left), and is 1.7 mm long. The details of the cross-section are shown in Fig. 8.8 (right).

The upper input arm (bar) is a channel waveguide with a Ge:SiO$_2$ core buried in a silica glass. It is single mode in a wide wavelength range around 1.55 μm. The lower arm (cross) is a BRW with an amorphous silicon core. The Bragg Reflector claddings are made of alternating layers of amorphous silicon (0.15 μm) and silica (0.25 μm) deposited with PECVD (Plasma Enhanced Vapour Deposition). The "defect" core is a 0.28 μm silicon layer. The whole, vertically etched structure is buried in a silica glass.

The dispersion slopes of the two arms (Fig. 8.9 left) strongly differ, which is a prerequisite for a narrow band response. The measured bandwidth is 0.3 nm in a 1.7 mm long device.

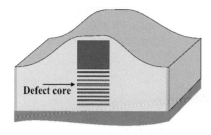
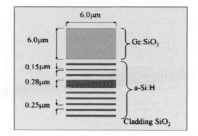

Fig. 8.8 Geometry (*left*) and the cross section (*right*) of the directional coupler filter with the conventional input arm and the BRW drop arm

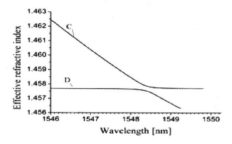
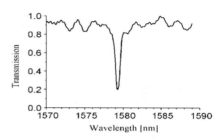

Fig. 8.9 *Left*: Dispersion of the coupler. *C* Bragg Reflection Waveguide, *D* conventional arm; *Right*: Measured spectrum of the transmission through the conventional-waveguide arm

8.2.5 Widely Tuneable Directional Coupler Filters with 1D Photonic Crystal

Here we give two examples of our designs illustrating hybrid approach to make silicon structures electrically tuneable at optical wavelengths.

The first is a modification of a filter presented in the previous section. We showed [45] that by replacing a silicon core of the BRW with a commercially available smectic liquid crystal (LC), as it is shown in Fig. 8.10 (left), ca. 100 nm range of continuous tuning is theoretically predicted under applying electric field of less than $\pm 5\,V/\mu m$. It can also be seen in Fig. 8.8 (right) that the slope of the dispersion very weakly depends on the operation wavelength, keeping the bandwidth variations over 1510–1589 nm below 6%. The drawback is that replacing the high index Si core with the low index liquid crystal degraded the theoretical (bandwidth × length) product from (0.3×2.2 mm) to (1.2×7.4 mm), and the device became significantly larger.

The other design [46] provides a miniaturized filter tuneable over 100 nm, however not continuously but in a mode-hop operation. The filter consists of a microdisc resonator evanescently coupled to two straight photonic nanowires embedded in a liquid crystal, as it can be seen in Fig. 8.11 (left). To the right the theoretically predicted mode-hopping tuneability under the applied voltage is shown.

The hybrid approach allows tailoring the desired properties by both a smart design of a Si host structure and the choice of an incorporated material.

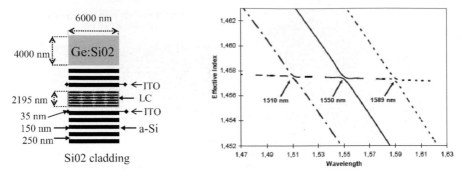

Fig. 8.10 Tuanble directional coupler filter with a BRW arm infiltrated with liquid crystal. *Left*: Filter cross-section. *Right*: Supermode dispersion at the off-state (1550 nm) and the extremes of the tuning range for $\Delta n = \mp 0.006 (\Delta E = \mp 5\,V/\mu m)$

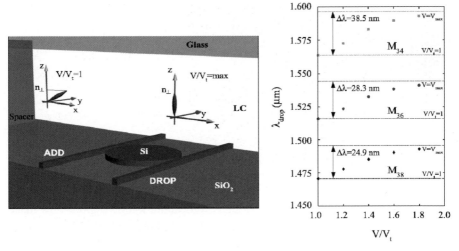

Fig. 8.11 Tuneable Add-Drop filter embedded in a LC cell consists of a microdisc resonator coupled with two nanowires. *Left*: Sketch of the device. *Right*: Three regimes of continuous tuneability – each corresponding to a particular mode (M_{38}, M_{36}, M_{34}) of the microdisc resonator. The applied voltage (*abscissa*) is normalized with its threshold values V_t for a corresponding operation mode

8.2.6 Vertical Waveguide-PhC Cavity Coupler

This design (Fig. 8.12 left) is an example of combining PhCs and photonic nanowires. It is a wavelength selective filter based on the vertical coupling between the photonic nanowire and a single hole missing cavity in a suspended silicon photonic crystal membrane [47]. PhC has lattice constant $a = 480\,nm$ and hole diameter $d = 289\,nm$. The holes close to the cavity have been properly designed to boost the Q value. The distance between the membrane and the waveguide is 250 nm. The intrinsic Q is estimated to be 500 and the rejection bandwidth is approximately 10 nm.

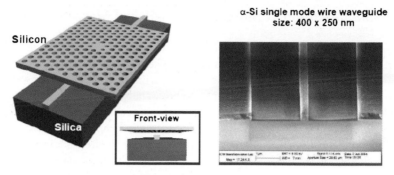

α-Si single mode wire waveguide size: 400 x 250 nm

Silicon

Front-view

Silica

Fig. 8.12 *Left*: Design of the filter based on vertical coupling between a silicon nanowire waveguide and a cavity in a suspended silicon photonic crystal membrane [47]. *Right*: Fabricated nanowire waveguide passing under the membrane

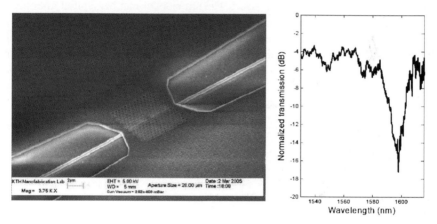

Fig. 8.13 SEM picture of the fabricated device and its normalized transmission spectrum [47]

From the light passing through the nanowire waveguide only the wavelength which is in resonance with the cavity is reflected back, whereas the other wavelengths are not affected.

Here fabrication consists of two steps of silicon deposition to obtain two-Si-level structure (which is not possible with an ordinary SOI wafer). At the first step a silicon nanowire is formed with height of 250 nm (Fig. 8.12 right). Next the wafer is spin-coated with a thick layer of polymer to planarize the structure, which is then thinned with oxygen plasma to a demanded distance between the waveguide and the cavity. In the second step, on top of such formed sacrificial layer, another 250 nm of α-Si is deposited and 2D photonic crystal with a designed cavity is e-beam patterned. The structure is etched in an ICP reactor and finally the membrane is released by oxygen ashing of the sacrificial polymer. Figure 8.13 shows the fabricated structure (left) and its normalized transmission spectrum (right).

8.3 Nanofabrication Technology for Si-Based Mesostructures

This section describes fabrication technology for realizing passive, high-contrast planar mesostructures in silicon-silica material system. Typical fabrication steps are explained and exemplified by a more detailed description of the approach we apply.

In the most common approach, where commercially available "sandwich" SOI wafers are used as the basis, the main fabrication steps are lithography and etching. Since the feature size of the components is usually below 1 μm, standard optical lithography has insufficient resolution for this application. Therefore, the deep UV lithography and electron beam (e-beam) lithography are typically employed. Deep UV lithography applies masks as in the case of optical lithography. Although the masks are here very expensive, they allow for wafer-scale manufacturing with CMOS standard processing. The resolution of 248 nm deep UV lithography is about 200 nm. For even smaller feature size a 193 nm stepper can be applied giving 90–100 nm resolution. For more demanding applications immersion lithography and extreme UV lithography can be used, but both are still not easily available, extremely expensive and just at the beginning of commercialization. For these applications e-beam lithography (EBL) is practically the only solution today, despite it is slow and not suitable for mass production. The performance of the fabricated mesostructures critically depends on the etching quality. In particular, on the achievable feature-size and aspect ratio, the etch profile smoothness and vertical shape. Therefore, only the best quality etching processes give satisfactory results. For both silica and silicon chemically assisted dry etching is normally applied. Among different solutions probably the best choice is Inductively Coupled Plasma (ICP) Reactive Ion Etching (RIE). For each material and anticipated profile chemistry and process parameters need to be carefully chosen and optimized.

Instead of using commercially available crystalline silicon-on-insulator (SOI) wafers we deposit both silica (SiO_2) and amorphous silicon (α-Si) layers using PECVD and subsequently etched with ICP Reactive Ion Etching. Although, the losses of α-Si are typically slightly higher than those of its crystalline counterpart, this approach gives us flexibility to freely adjust the layers thicknesses and to some extend, their refractive indices. Thus it makes possible to fabricate more complex structures including multi-layer components (e.g. Bragg reflectors), vertical cavities etc. We have developed technology for low loss amorphous silicon and recently demonstrated several ultra compact components based on photonic nanowires and photonic crystals. Some of them are shown in the previous section.

8.3.1 Plasma Enhanced Chemical Vapour Deposition (PECVD) of Amorphous Si and SiO$_2$

For optoelectronic applications silica-on-silicon (Si/SiO_2) and amorphous silicon (α-Si) waveguides of high optical quality are of critical importance. Two main prerequisites for obtaining good quality waveguides are: 1) high quality of the deposited films, and 2) smooth etch sidewalls to minimize scattering loss for channel waveguides.

For conventional Si/SiO_2 channel waveguides typical thicknesses of the deposited three-layer films is around 30 μm, which is much larger than that used in microelectronics. In the case of α-Si nanowire waveguides film thickness is comparable to the microelectronic feature size, i.e. 1 μm and below. The films should be homogeneous, with uniform thickness and refractive index over the deposition area, and have very good transmission for the optical communication wavelengths 1.3–1.6 μm. It should also be possible to add dopants to modify the refractive index and/or some other parameters such as UV photosensitivity, melting temperature, introducing active ions and other species.

8.3.1.1 PECVD Deposition Mechanisms

The main mechanism is based on Chemical Vapour Deposition (CVD), where the substrate placed in the reaction chamber is exposed to gas precursors comprising elements of the desired film. In the classical CVD thermal energy is used to generate the ionized species that subsequently react and deposit on the substrate surfaces forming the film. The volatile byproducts are removed from the chamber. This is a high temperature and high pressure process.

In Plasma Enhanced CVD (PECVD) process instead of using thermal energy the plasma energy is supplied by an external radio frequency (RF) source in the form of an electric or magnetic field. It accelerates the electrons and ions in the chamber for the generation of the reacting species. Both high temperature and high pressure requirements are released. PECVD allows deposition at temperature even below 150 °C, which is a grate advantage for integration with semiconductors having lower melting temperature than silicon.

As the plasma reactions are very complex and dependent on many parameters and variables, the whole plasma process is still not fully understood. The fabrication is often established on an empirical basis and then further optimization of the process can be based on models developed due to analysis and understanding of the empirical data. A set of reactions has been found that helps to understand the process of chemical dissociation of molecules and surface reactions leading to the building of stable layers. There are different gas phase precursors containing silicon and oxygen that are in different PECVD systems used to deposit SiO_2 and α-Si. The silicon precursors can be SiH_4, $SiCl_4$, SiF_4, and the organometallic TEOS, which diminishes cusping and fills the gaps better because of the higher surface mobility of the reactants. The oxygen precursors are O_2, N_2O, CO_2, and other oxygen containing gases. Depending on the chosen precursors the gas delivery system can be different because many of the hydrides react with oxygen at room temperature and must be delivered separately to the gas chamber to avoid spontaneous reactions and massive particle formation. Most common PECVD reactors are based on SiH_4 and N_2O chemistry that allows a common mix gas inlet to the chamber, as it is shown Fig. 8.14, and in case of SiO_2 provides the best uniformity of the deposited layers.

The process starts with the primary initial electron-impact reactions between electron and reactant gases to form ions and radical reactive species. In very simplified form plasma deposition of silica can consist of four steps:

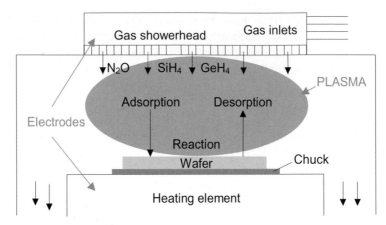

Fig. 8.14 Parallel plate capacitively coupled reactor for Plasma Enhanced Chemical Vapour Deposition of amorphous Si and SiO_2

1. generation of radicals and ions by the plasma
2. oxygen reacts with SiH_4 (GeH_4) to form silanols
3. ions and silanols are adsorbed at the surface
4. rearrangement, surface reactions, deposition

The reactive species and reaction products can be incorporated into the deposited films or re-emitted from surface back to the gas phase. The optimisation here goes towards stoichiometric composition of deposited material that should be obtained during the formation of the films and not after, during the post processing annealing as it takes place in standard PECVD processing. Ion bombardment plays a significant role in this process. Surface reactions critically influence film properties such as: density, stress, defects and impurity incorporation. Ion bombardment can modify these characteristics as well as contribute to gap-filling and surface planarization.

8.3.1.2 Deposition of Amorphous Silicon

Amorphous silicon deposited in low temperatures by sputtering or evaporation has poor optical qualities, due to the high density of localized states in the mobility band-gap, originated from either strained or dangling bonds in the silicon network. In PECVD deposition when using silane as the only gaseous precursor, the incorporation of hydrogen atoms in the film saturates the silicon dangling bonds. (PECVD) technique is used with very low RF power at 13.56 MHz. Low power prevents ion bombardment that generates defects rather than assists the deposition. Relatively low temperature between 200 °C and 300 °C and other process parameters are optimized to incorporate a right amount of hydrogen, that should passivate most of the dangling bonds, as the excess of hydrogen can increase the material porosity and increase scattering losses.

In both cases, silica and amorphous silicon deposition, instead of using thermal annealing to decrease absorption at 1.51 μm we have carefully optimized parameters of the deposition process to obtain high optical quality material directly during the deposition without any post process annealing. Moreover, this material is deposited in low temperatures, between 250 °C and 300 °C that allows for integration with other, temperature sensitive materials and components as it was pointed out earlier.

8.3.2 Electron Beam Lithography

EBL technique uses electron beam to form patterns on a surface covered by a thin layer of a special photoresist. The advantage over conventional lithography is that electrons are orders of magnitude smaller than photons, so features far below the diffraction limit of light can be patterned. The electrons accelerate from an electron gun with energies typically from 1 keV to 100 keV depending on applications and instruments. The electron beam is scanning across the surface to be patterned point by point. This makes the method slow and expensive. It can take tens of minutes or even several hours to scan over the desired structure.

Moreover, special care has to be taken to avoid stitching errors. For large structures the pattern is divided into several writing fields, the largest areas, the deflection unit can scan over at a specific resolution. When changing the writing field, the accuracy of the line continuity, kept by a laser interferometer, is called a stitching error. For highest accuracy patterning the writing field is usually $100 \times 100 \mu m$ and the resolution for an isolated line is 10–30 nm, whereas for patterns it is 20–50 nm. The resolution is not only dependent on the acceleration energy and quality of the focusing optics, but also on the photoresist and how large dose of the secondary and scattered electrons the photoresist can tolerate before being exposed.

The stages of the patterning process for a photonic crystal structure including etching, are shown in Fig. 8.15.

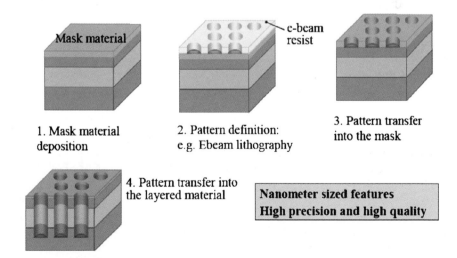

1. Mask material deposition

2. Pattern definition: e.g. Ebeam lithography

3. Pattern transfer into the mask

4. Pattern transfer into the layered material

Nanometer sized features
High precision and high quality

Fig. 8.15 E-beam patterning and etching steps for photonic crystals

8.3.3 Etching

In etching process one removes layers of patterned material to obtain desired depth profile. Wet chemical etching is isotropic and can be very selective if the etching solution is properly chosen. Dry chemical etching uses gaseous reactants instead of liquids and usually gives better control of the etching process (see Fig. 8.16a). Chemical plasma etching can more effectively etch the desired material as plasma energy ionizes the gas and produces radicals that react with the material to be removed and form volatile products which are pumped away. Sputter etching is a technique that applies physical bombardment, usually high energy argon ions, to remove the unwanted material. This technique is highly anisotropic, but least selective (see Fig. 8.16b)). Ion beam etching and milling are similar physical etching techniques. Finally Reactive Ion Etching (RIE) usually contains two main processes that contribute to material removement:

1. Chemical etching, where the generated reactive species react with the target material and remove it in form of volatile products. This kind of etching is similar to wet etching, it can be highly material selective, but is isotropic.
2. Physical etching (sputtering) where ions with high kinetic energies impact on the target and reject the material mechanically. The main advantages of this process are that it acts on all materials and can be highly anisotropic.

RIE is a technique, where quite a large number of parameters is to be optimized for the best etch quality. One of the most important issues is to properly adjust proportions of the physical and chemical contributions. The chemical etching has good resolution, good selectivity, and high etching rates. However the etching is isotropic. The physical part gives more vertical profiles (anisotropy) but has low rate and selectivity.

8.3.3.1 Etching of Si and SiO₂

High selectivity and/or high-aspect-ratio etching requires high-density plasma processing. For this purpose, we use Inductively Coupled Plasma (ICP) tool. It provides the ion density up to 10^{12} cm^{-3} which is up to a factor of 100 higher than in the ordinary RIE generator. It also operates at lower pressures; a few mT in comparison to

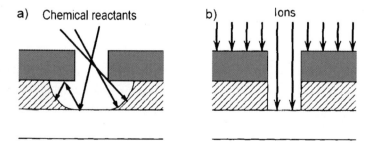

Fig. 8.16 Dry etching mechanisms, typical features for **a)** chemical and **b)** physical etching

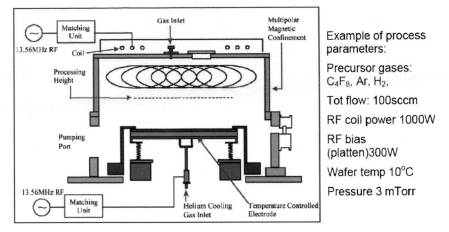

Example of process parameters:

Precursor gases: C_4F_8, Ar, H_2,

Tot flow: 100sccm

RF coil power 1000W

RF bias (platten)300W

Wafer temp 10°C

Pressure 3 mTorr

Fig. 8.17 Inductively Coupled Plasma etcher

several hundred mT. Efficient excitation of plasma and hence an enhanced discharge is achieved by inductive coupling of RF power via a matching unit and coil assembly (Fig. 8.17). The advantage of this configuration is the possibility of independent control of the energy of ions bombarding the substrate and the plasma density. Bias of the substrate gives the accelerating voltage to the ions bombarding and sputtering the target, whereas the amount of ions and reactive species (intensity of chemical etching) is controlled independently by adjustment of plasma density (ion flux) that depends on the coil power.

Additionally, ICP processing uses gas chemistries that supply inhibitors (polymer materials to deposit on sidewalls) to protect vertical sidewalls from chemical etching. In this way vertical walls are not affected at all, whereas the horizontal surfaces are etched both chemically and physically, which results in the desired highly anisotropic etching. This is followed by a short passivation step using C_4F_8, and the steps are alternately repeated. This technique differs from usually applied one, where SF_6 is used during the etch step to etch Si isotropically. For high-aspect ratio etching of mesostructures we use an Advanced Oxide Etcher (AOE), which is an improved version of ICP. Additional multipolar magnetic confinement at the chamber sidewalls is used to concentrate the plasma and make the ions access the substrate more vertically. Figure 8.18 shows examples of our etching results obtained with AOE: photonic crystal holes, photonic crystal pillars, and a plasmonic waveguide.

For optical applications it is very important to keep the sidewalls roughness on the level of single nm RMS that contribute to low scattering losses of fabricated devices.

Fig. 8.18 Fluorocarbon based etching of the stack a-Si/silica (*left*), silicon PhC pillars (*middle*), slot plasmonic waveguide (*right*)

8.4 Conclusions

High-index contrast mesostructures, in particular photonic crystals and photonic nanowires, have great potential for miniaturizing photonic components down to the diffraction limit, which is the prerequisite for densely integrated photonic circuits and optoelectronic integration. While photonic nanowires are simpler and lend themselves for device miniaturization, PhCs offer in addition a wide array of new functionalities due to their unconventional dispersion properties. The main issue is too high propagation loss of the demonstrated components. However, considering the rapid technological advances one can expect this problem to be soon resolved.

Heterogeneous integration of silicon mesostructures with optically active and externally controlled materials proves to be viable alternative for integrated photonics. Recently demonstrated heterogeneous integration of silicon with III/V semiconductors is a milestone towards functional silicon photonics and optoelectronic integration, where Si technology would allow adding optical functions to the existing electronics.

Acknowledgements The reviewed work was included in COST P11 Action, and the European Networks of Excellence: ePIXnet and PHOREMOST. The support from the Swedish Research Council (VR), the Swedish Foundation for Strategic Research (SSF), is gratefully acknowledged.

References

1. "Photonics for the 21st century", http://web13.vdi.net-build.de/pdf/visionpaperPh21.pdf
2. Vlasov Y. and McNab S., "Losses in single-mode silicon-on-insulator strip waveguides and bends", Opt. Express **12**, 1622–1631 (2004)
3. Settle M., Salib M., Michaeli A., and Krauss T.F., "Low loss silicon on insulator photonic crystal waveguides made by 193 nm optical lithography", Opt. Express **14**, 2440–2445 (2006)
4. Bogaerts W., Baets R., Dumon P., Wiaux V., Beckx S., Taillaert D., Luyssaert B., Campenhout J.V., Bienstman P., and Thourhout D.V., "Nanophotonic Waveguides in Silicon-on-Insulator Fabricated With CMOS Technology", J. Lightwave Technol. **23**, 401–412 (2005)

5. Lee K.K., Lim D.R., and Kimerling L.C., "Fabrication of ultralow-loss Si/SiO$_2$ waveguides by roughness reduction", Opt. Lett. **26**, 1888–1890 (2001)
6. Gnan M., Thomas S., Macintyre D.S., De La Rue R.M. and Sorel M., "Fabrication of low-loss photonic wires in silicon-on-insulator using hydrogen silsesquioxane electron-beam resist", Electron. Lett. **44**, 115–116 (2008)
7. Grillot F., Vivien L., Laval S., and Cassan E., "Propagation Loss in Single-Mode Ultrasmall Square Silicon-on-Insulator Optical Waveguides", J. Lightwave Technol. **24**, 891–896 (2006)
8. O'Faolain L., Yuan X., McIntyre D., Thoms S., Chong H., De La Rue R.M., and Krauss T.F., "Low-loss propagation in photonic crystal waveguides" Electron. Lett. **42**, 1454–1455 (2006)
9. Kotlyar M.V., Karle T., Settle M.D., O'Faolain L., and Krauss T.F., "Low-loss photonic crystal defect waveguides in InP", Appl. Phys. Lett. **84**, 3588–3590 (2004)
10. Fritze M., Knecht J., Bozler C., Keast C., Fijol J., Jacobson S., Keating P., LeBlanc J., Fike E., Kessler B., Frish M., and Manolatou C., "3D mode converters for SOI integrated optics", IEEE International SOI Conference, 165–166 (2002)
11. Taillaert D., Bogaerts W., Bienstman P., Zutter D.D., and Baets R., "An out-of-plane grating coupler for efficient butt-coupling from photonic crystal waveguides to single-mode fibers", IEEE J. Quant. Electron. **38**, 949–955 (2002)
12. Scheerlinck S., Van Laere F., Schrauven J., Taillaert D., Van Thourhout D., and Baets R., "Gold grating coupler for Silicon-on-insulator waveguides with 34% coupling efficiency", European Conference on Integrated Optics, ECIO 2007, Copenhagen, Denmark April 25–27 p. WF2 (2007)
13. Shani Y., Henry C.H., Kistler R.C., Orlowsky K.J., and Ackerman D.A., "Efficient coupling of a semiconductor laser to an optical fiber by means of a tapered waveguide on silicon", Appl. Phys. Lett. **55**, 2389–2391 (1989)
14. McNab S., Moll N., and Vlasov Y., "Ultra-low loss photonic integrated circuit with membrane-type photonic crystal waveguides", Opt. Express, vol. 11, no. 22, pp. 2927–2939, Nov. (2003)
15. Winn G., Fink Y., Fan S., and Joannopoulos J.D., "Omnidirectional reflection from a one-dimensional photonic crystal", Opt. Lett. **23**, 1573–1575 (1998)
16. Hosomi, K. and Katsuyama, T., "A dispersion compensator using coupled defects in a photonic crystal", IEEE J. Quantum Electron. **38**, 825–829 (2002)
17. Akahane Y., Asano T., Song B.S., and Noda S., "High-Q photonic nanocavity in a two-dimensional photonic crystal", Nature **425**, 944–947 (2003)
18. Fan S., Villeneuve P.R., Joannopoulos J.D., and Haus H.A., "Channel Drop Filters in Photonic Crystals", Opt. Express **3**, 4–11 (1998)
19. Dainese M., Swillo M., Thylen L., Qiu M., Wosinski L., Jaskorzynska B., and Shushunova V., "Narrow Band Coupler Based on One-Dimensional Bragg Reflection Waveguide", Proceedings of the 2003 Optical Fiber Communication Conference, March 23–28, 2003, Atlanta, Georgia, **1**, 44–46 (2003)
20. Fan S., Villeneuve P.R., Joannopoulos J.D., and Schubert E.F., "High Extraction Efficiency of Spontaneous Emission from Slabs of Photonic Crystals", Phys. Rev. Lett. **78:1717**, 3294–3297 (1997)
21. Notomi M., "Theory of light propagation in strongly modulated photonic crystals: Refractionlike behavior in the vicinity of the photonic band gap", Phys. Rev. **B 62**, 10696–10705 (2000)
22. Cubukcu E., Aydin K., Ozbay E., Foteinopoulou S., and Soukoulis C.M., "Negative refraction by photonic crystals", Nature **423**, 604–605 (05 Jun 2003)
23. Qiu M., Thylen L., Swillo M., and Jaskorzynska B., "Wave propagation through a photonic crystal in a negative phase refractive index region", IEEE J of Selected Topics in Quantum Electronics **9**, 106–110 (2003)
24. Berrier A., Mulot M., Swillo M., Qiu M., Thylen L., Talneau A., and Anand S., "Negative refraction at infrared wavelengths in a two-dimensional photonic crystal", Phys. Rev. Lett. **93**, 07390214 (2004)

25. Schonbrun E., Tinker M., Park W., and Lee J.-B., "Negative Refraction in a Si-Polymer Photonic Crystal Membrane", IEEE Photon. Technol. Lett. **17**, 1196–1198 (2005)
26. Liu L., Ao X., Wosinski L., and He S., "Compact polarization beam splitter employing positive/negative refraction based on photonic crystals of pillar type", Proc. SPIE **6352**, 635209–635211 (2006)
27. Luo C., Johnson S.G., Joannopoulos J.D., and Pendry J.B., "Subwavelength imaging in photonic crystals", Phys. Rev. **B 68**, 45115115 (2003)
28. Lin S.Y., Hietala V.M, Wang L., and Jones E.D., "Highly dispersive photonic band-gap prism", Opt. Lett. **21**, 1771–1773 (1996)
29. Momeni B., Huang J., Soltani M., Askari M., Mohammadi S., Rakhshandehroo M., and Adibi A., "Compact wavelength demultiplexing using focusing negative index photonic crystal superprisms", Opt. Express **14**, 2413–2422 (2006)
30. Altug H. and Vuckovic J., "Experimental demonstration of the slow group velocity of light in two-dimensional coupled photonic crystal microcavity arrays", Appl. Phys. Lett. **86**, 111102-1-3 (2005)
31. Part I Chap. 2 of this book
32. Povinelli M., Johnson S., and Joannopoulos J., "Slow-light, band-edge waveguides for tunable time delays", Opt. Express **13**, 7145–7159 (2005)
33. Md Zain A.R., Gnan M., Chong H.M.H., Sorel M., and De La Rue R.M., "Tapered Photonic Crystal Microcavities Embedded in Photonic Wire Waveguides With Large Resonance Quality-Factor and High Transmission", IEEE Photon. Technol. Lett. **20**, 6–8 (2008)
34. Xu Q., Almeida V.R., and Lipson M., "Experimental demonstration of guiding and confining light in nanometer-size low-refractive-index material", Opt. Lett. **29**, 1626–1628 (2004)
35. Almeida V.R., Qianfan Xu, and Lipson M., "Ultrafast integrated semiconductor optical modulator based on the plasma-dispersion effect", Opt. Lett. **30**, 2403–2405 (2005)
36. Hochberg M., Baehr-Jones T., Walker C., and Scherer A., "Integrated plasmon and dielectric waveguides", Opt. Express **12**, 5481–5486 (2004)
37. Genov D. A., Ambati M., and Zhang X., "Surface Plasmon Amplification in Planar Metal Films", IEEE J. of Quantum Electron. **43**, 1104–1108 (2007)
38. Soref R. and Lorenzo J., "All-silicon active and passive guided-wave components for $\lambda = 1.3$ and $1.6 \mu m$", IEEE J. Quantum Electron. **22**, 873–879 (1986)
39. Soref R., "The past, present and future of silicon photonics", IEEE J. Sel. Top. Quantum Electron. **12**, 1678–1687 (2006)
40. Lourdudoss S., Olsson F., Barrios C.A., Hakkarainen T., Berrier A., Anand S., Aubert A., Berggren J., Broeke R. G., Cao J., Chubun N., Seo S.-W., Baek J.-H., Chubun N., Aihara K., Pham A.-V., Ben Yoo S.J., Avella M., and Jiménez J., Heteroepitaxy and Selective Epitaxy for Discrete and Integrated Devices, Conference on Optoelectronic and Microelectronic Materials and Devices, COMMAD06, Perth, Western Australia (2006)
41. Fang A.W., Park H., Cohen O., Jones R., Paniccia M. J., and Bowers J.E., "Electrically pumped hybrid AlGaInAs-silicon evanescent laser", Opt. Express **14**, 9203–9210 (2006)
42. Rojo Romeo P., Van Campenhout J., Regreny P., Kazmierczak A., Seassal C., Letartre X., Hollinger G., Van Thourhout D., Baets R., Fedeli J.M., and Di Cioccio L., "Heterogeneous integration of electrically driven microdisk based laser sources for optical interconnects and photonic ICs", Optics Express **14**, 3864–3871 (2006)
43. Dai D., Liu L., Wosinski L., and He S., "Design and Fabrication of an Ultra-small Overlapped AWG Demultiplexers Based on α-Si Nanowire Waveguides", Electron. Lett. **42**, 400–402 (2006)
44. Qiu M. and Jaskorzynska B., "Design of a channel drop filter in a two-dimensional triangular photonic crystal", Appl. Phys. Lett. **83**, 1074–1076 (2003)
45. Jaskorzynska B., Zawistowski Z.J., Dainese M., Cardin J., and Thylén L., "Widely tunable directional coupler filters with 1D photonic crystal", Transparent Optical Networks, Proc. of 2005 7th International Conference **1**, 136–139 (2005)

46. Di Falco A., Assanto G., and Jaskorzynska B., "Analysis and design of a tunable wavelength-selective add-drop in liquid crystals on Silicon", Proc. of EOS Topical Meeting on Nanophotonics, Metamaterials and Optical Microcavities, 206–207, Paris, 16–19 October (2006)
47. Dainese M., Zhang Z., Swillo M., Wosinski L., Qiu M., and Thylén L., "Experimental demonstration of a vertically coupled photonic crystal filter", Proc. of 31st European Conference on Optical Communication, Glasgow, UK, Sept. 25–29, **2**, 185–186 (2005)

Part IV
Characterisation and Measurements of Nanostructures

9 Near Infrared Optical Characterization Techniques for Photonic Crystals

Romuald Houdré

Institut de Photonique et Electronique Quantique, Ecole Polytechnique Fédérale de Lausanne (EPFL), Station 3, Lausanne, 1015, Switzerland

Abstract. In this chapter we review the different techniques used for the optical characterization of photonic crystals (PhC). The chapter is divided in three parts, the first one on optical measurement techniques based on an external light source, the second one on techniques relying on the presence of a light source inside the PhC under investigation and the third one on more advanced techniques like scanning near field optical microscopy and Fourier imaging.

Key words: Optical characterization techniques; photoluminescence spectroscopy; cut-back method; end-fire technique; Fourier space imaging

9.1 Introduction

After the design and fabrication processes of a photonic crystal (PhC) structure have been completed, and the samples have been structurally checked against fabrication imperfection, it is now time to measure its optical properties and check whether the PhC structure behaves the way it was designed for and if not to search for the possible causes of the deviation from the nominal behaviour.

At this point, optical properties are a generic term that must be specified, because different experiments will obviously be used depending of the properties we are interested in. Several sub-domains, that will require completely different class of experiments, can already be distinguished:

1. Quantities related to the optical response; the transmission (T), the reflectivity (R), the diffraction efficiency (D), which can be separated in in-plane ($D_{\parallel}$) or out-of-plane (D_{∞}) in the case of planar PhC, the losses (L) and absorption (A). All these quantities usually are a function of the angular frequency (ω), the direction of propagation or the polarization.
2. Intrinsic quantities related to the band structure, dispersion curve or equi-frequency surfaces (EFS) and related quantities like group index dispersion.
3. Nonlinear properties, which are discussed in Part II of the present book.
4. Bulk properties of light propagating inside a PhC.
5. Properties of defect and localized states: resonance frequencies, quality factor.

6. Dynamic properties and time resolved investigations.
7. Emission properties, either spontaneous or stimulated emission and laser characteristics.

Furthermore, the selected experimental techniques will also depend on the source and detector available at the wavelength of consideration, visible, near infrared, far infrared or microwave. In this chapter we will mainly consider visible and near infrared experiments. Each technique will also have its advantages and inconvenient depending of the type of PhC, tri- or bi-dimensional. Within 2D-PhC we can further distinguish between low, high-index contrast type (i.e. membrane) or macroporous structures, which strongly differ by the nature of the electromagnetic mode that will couple in and out. Finally experimental set-up, requirements and quantities of interest will be different if the PhC was designed or is investigated in view of fundamental investigations or device development.

Before reviewing the details of each technique we should introduce a concept that is widely used: lithographic tuning. This comes from the consideration that most of the convenient light sources available have a reduced emission or tuneability bandwidth, typically 100 nm for a tuneable laser around 1.55 μm wavelength. Lithographic tuning makes profit of the scaling laws obeyed by PhC, which stand that any homothety of the real space dielectric map by a factor $s(\varepsilon(r) \rightarrow \varepsilon(r' = rs))$ leads to unchanged eigenstates (e.g. $H(r) \rightarrow H(r' = rs)$) with the wavevector and eigenfrequency scaled by a factor $1/s$ ($k \rightarrow k' = k/s$ and $\omega \rightarrow \omega' = \omega/s$). Immediate consequence of the scaling laws is the generalized used of reduced energy $u = a/\lambda$ and reduced wavevector $k = qa/2\pi$, where a is the lattice constant, λ the wavelength and q the wavevector. Lithographic tuning consists in extending the limited bandwidth of the light source by scanning the reduced energy u and varying 'a' while repeating the same homothetically scaled PhC structure and stacking the measured spectrum. The technique presents the additional advantage that, as measurements are performed around the same central wavelength, effect of material index or effective waveguide index dispersion is minimized and modelling with non-dispersive indices are most of the time sufficient.

9.2 External Probe Light Source

9.2.1 Reflectivity

Accurate and quantitative reflectivity or transmission measurements are delicate tasks. It usually consists in comparing the reflected or transmitted signal to a calibrated known mirror. Care has to be taken in order to have as much as possible identical path and detection for the reflected and reference signals. An example of such a set-up is shown in Fig. 9.1a and an example of reflectivity measurement obtained with such a set-up on a planar Fabry–Perot cavity is shown in Fig. 9.1b. Note that in a linear regime, the light source can be a wavelength tuneable source or a white light source and spectral resolution is performed afterwards. In practice the accuracy of the measurements can reach the fraction of a percent, however it

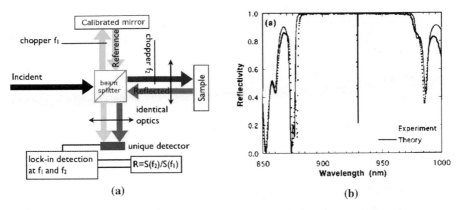

Fig. 9.1 (**a**) Example of a reflectivity measurement set-up. Both reference and sample arms are made as identical as possible, thanks to the use of two different chopping frequencies a unique detector can be used to simultaneously measure the signal and the reference. (**b**) Reflectivity spectrum of a GaAs/(Al,Ga)As planar microcavity. Reprinted from [1]

is not easy to measure directly reflectivity coefficients close to unity, which would imply being able to discriminate between e.g. $R = 0.999$ and $R = 0.997$. It is usually much more convenient to use the mirror to fabricate a high quality factor (Q) optical cavity and deduce R from the quality factor (Q) value.

To date first measurements on PhC like in [2] on opals of 0.11 μm polystyrene microspheres were very basic and used to show only a dip in the transmission spectrum without angular investigation for instance. Nowadays, direct R and T measurements are currently used to probe the photonic band-gap like in [3].

Much more advanced techniques based on an angular analysis of the reflectivity measurement was pioneered by R. Zengerle [4] and brought to a fully mature stage by M. Galli et al. [5]. Impinging light is reflected according to the grating diffraction law, which states the conservation of the in-plane component of the wavevector:

$$k_\parallel^{\text{ref}} = k_\parallel^{\text{inc}} + G = k_\parallel^{\text{inc}} + \sum_i m_i G_i \tag{9.1}$$

where G_i are basis vectors of the reciprocal lattice; $m_i = 0$ describes specular reflection while $m_i \neq 0$ correspond to diffraction processes. In principle such process is only governed be the reciprocal lattice vectors and does not depend on the band structure. However, for some specific set of incident k and wavelength the incident light can couple to a mode into the photonic crystal and gives rise to a dip in the reflected intensity spectrum and consequently provide information on the band structure $k(\omega)$ in a process similar to a Wood anomaly in a grating. An example of such experimental results obtained on macroporous silicon is shown in Fig. 9.2a and the corresponding analysis in Fig. 9.2b with the deduced dispersion. The method can also be applied to 3D PhC, in this case such a measurement probes mainly the density of states and its associated singularities [6]. As such the technique is limited to wavevector close to the zone centre of the Brillouin zone inside the air light cone.

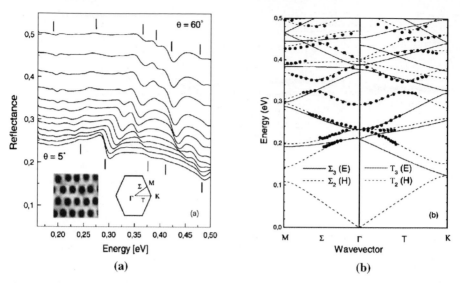

Fig. 9.2 (**a**) Reflectance of a macroporous silicon sample for TE-polarized light incident along the Γ-M orientation. The angle of incidence is varied from 5° to 60° with a step of 5°. *Vertical bars* mark the positions of 2D photonic modes for 5° and 60°. (**b**) *Points*: measured dispersion of the photonic bands, derived from the structures in reflectance curves; *solid* and *dashed lines*: photonic bands of the triangular lattice of air holes, separated according to parity with respect to the plane of incidence. Reprinted from [5]

This limitation can be overcome by the use of a prism coupler or a high index sphere, which allows an evanescent coupling between in-plane PhC wavevector and propagating field in the sphere as elegantly demonstrated in [7, 8] on a W1 waveguide on SOI with a ZnSe sphere.

9.2.2 End Fire

When more applied uses of PhC are considered, the type of characterization required is different and the whole PhC structure is more or less regarded as a black box whose parameter of interest are the input/output functionalities, even if at some point it will still be necessary to investigate the inside operation of the black box to understand the origin of an observed dysfunction. For this purpose the object that has to be measured should be placed as close as possible to a finalized real-world photonic device, including for example access and output ports. This is the purpose of the end-fire technique, which is schematically shown in Fig. 9.3. Going from the input light source to the output detection, an end fire set-up can be declined in many variants that consist in:

1. A light source, which can be a tuneable or a broad band source: laser, white light, light emitting diode (LED), super-luminescent LED, supercontinuum source.

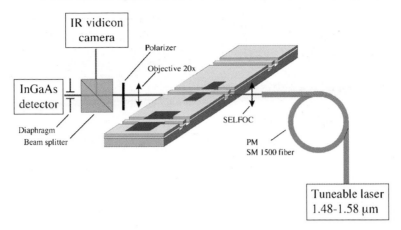

Fig. 9.3 Schematic of an end fire measurement set-up. Courtesy of Bozena Jaskorzynska

2. Transport to and from the device can be either fibred (preferably single mode) or in free space. Free space option allows the insertion of visualization system with an IR camera but tends to add sur-numerous degrees of freedom.
3. Light coupling in/out is done either with the help of a microscope objective or with tapered or microlensed fibre.
4. PhC being polarization sensitive, polarization control is required, ideally on both input and output side for polarisation analysis. This is usually achieved using polarisation maintaining fibre, polarisers, $\lambda/2$ and $\lambda/4$ retarding plates or multiple coiled fibres. Often this is limited to a TE/TM control and analysis.
5. The sample, which consists in different sections. Deep or shallow etched ridge access waveguide transporting light from the cleaved facet of the sample to the PhC structure, possibly a taper from the access waveguide to the PhC waveguide to adapt the very different mode shape and finally the PhC device itself.
6. The output arm is very similar to the input and terminates with a detector combined with a spectral analysis if a broadband source was used.

Two arbitrarily selected good examples of end fire set-up used for the characterization of advanced or complex devices can be found in [9] on a multi-port channel drop filter and [10] on PhC-based symmetric Mach–Zehnder interferometer. Ideally the measured spectra give a direct measurement of the optical response of the PhC device, in practice the signal is often degraded to a variable extend by interferences fringes (Fig. 9.4a) due to the presence of parasitic reflections in the sample (facets, tapers ...) depicted in Fig. 9.4b. Some of these reflections can be eliminated after optimisation of the design and processing or can also be exploited for characterization purposes.

In ideal cases, where parasitic reflections can be neglected, a simple method known as cut-back allows the measurement of insertion efficiencies and propagation losses (α) of a waveguide. It consists in plotting the total transmitted intensity on a semilog scale or in dB as a function of the waveguide length. The total trans-

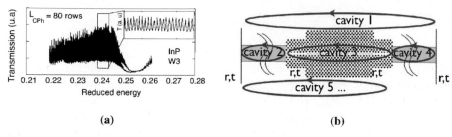

(a)   **(b)**

Fig. 9.4 (**a**) Transmission spectrum through a W3 PhC waveguide butt coupled with ridge waveguides. Reprinted from (Wild et al. 2005). (**b**) Possible multiple optical cavities in the presence of parasitic reflection at the cleaved facets and the injection tapers

mission follows the relation:

$$\left(\frac{S}{I}\right)_{dB} = T_{\text{insertion,dB}} - A_{\text{dB/cm}}L_{\text{cm}} = T_{\text{insertion,dB}} - 10\frac{\alpha}{\ln(10)}L. \qquad (9.2)$$

Intercept at the origin gives the insertion efficiency while the slope the propagation losses. An example of such procedure can be seen in [11] from which Fig. 9.5 is reprinted.

When back and forth reflections occur and an extra optical cavity giving rise to interferences fringes is present, an alternate method often referred as the Hakki–Paoli method can be used. It is based on a plot of a fringe contrast function vs. the cavity length. Simple derivation from Fabry–Perot transmission formula shows that:

$$f(u) = \ln\left(\frac{1-u}{1+u}\right) = \ln(R) - \alpha L \qquad (9.3)$$

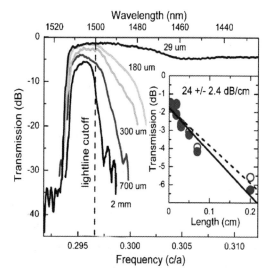

Fig. 9.5 Transmission spectrum and cut-back method used to measure record insertion and propagation losses on a SOI W1 waveguide. Reprinted from [11]

with:

$$u = \sqrt{\frac{T_{\min}}{T_{\max}}} = \sqrt{\frac{P_{\min}}{P_{\max}}}. \tag{9.4}$$

Note that, based on the fringe contrast, this method does not require quantitative transmission measurement; see e.g. [12]. Although exact theoretical model is missing, the method has also been applied in the case of several coupled cavities after Fourier filtering is performed to isolate the contribution of a single cavity.

An elegant and more accurate generalization of the method due to D. Hofstetter [13, 14] is based on the amplitude decay $A_{r,n}$ of the harmonics of the Fourier transform of the transmission spectrum which follows the exponential law:

$$A_{r,n}/A_0 = R^n e^{-n\alpha L}, \tag{9.5}$$

where n is the harmonic order and R the cavity facet reflectivity. Again plotting the harmonic decay rate as a function of the cavity length allows the determination of R and α [15]. Figure 9.6 shows the Fourier transform of the spectrum of Fig. 9.4a with such a harmonic decay law for the first four harmonics. Unfortunately, generalization to coupled cavities is not straightforward.

Interference fringes in Fabry–Perot cavities can also be used to measure dispersion curves. Careful examination of the transmission relation shows that the fringes are always equally spaced in k independently of the cavity medium dispersion, with a spacing $\Delta k = \pi/L$ proportional to the length L of the cavity:

$$T_{\text{FP}} = \frac{T^2}{1 + R^2 - 2R\cos\left(\frac{4\pi n L}{\lambda}\right)} = \frac{T^2}{1 + R^2 - 2R\cos(2kL)} \tag{9.6}$$

where T, R, L and n are the cavity mirror transmission, reflectivity, length and refractive index respectively. Behind the formulas the principle is that the mirrors at

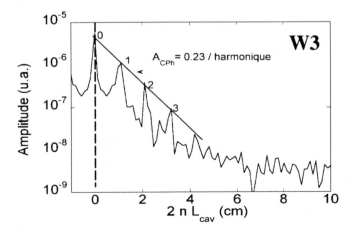

Fig. 9.6 Fourier transform of the spectrum of Fig. 9.4a showing an harmonic decay law for the first four harmonics. Reprinted from [15]

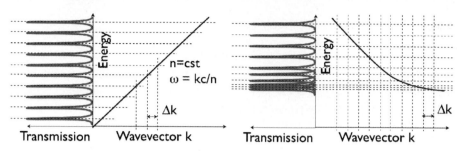

Fig. 9.7 (**a**) For a dispersion less medium, the k-space sampling, which manifests itself as resonances in the transmission spectrum lead to a constant spacing sampling in energy. (**b**) For a dispersive medium the k-space sampling allows the reconstruction of the material dispersion curve

both ends generate a new periodicity of characteristic length $2L$ in the structure, which leads to quantization effects periodic in the reciprocal space with a period $\Delta k = 2\pi/2L$. Note that if the period can be estimated from first order principle, exact k values depend of the phase of the reflectivity and are not known with the same accuracy [16]. The principle is schematized in Fig. 9.7. Example of application can be found in [17, 18] on the measurement of bulk 2D-PhC dispersion and [16] on W1 waveguide, an alternative technique based on Mach–Zehnder interferometry was used in [19].

Finally end fire techniques are often used for the measurement of high quality factor resonant cavities. Advanced techniques are detailed in Chap. 10 of this book. A generic configuration, depicted in Fig. 9.8, consists in measuring the transmission of a properly designed waveguide, which is laterally coupled to high-Q cavity [20] or coupling in and out into the cavity [21]. Once a mode is identified, the main difficulty comes from the fact that the measurement requires coupling to the probe waveguide which affects the quality factor according to:

$$\frac{1}{Q_{\text{meas}}} = \frac{1}{Q_{\text{int}}} + \frac{1}{Q_{\text{probe}}} \tag{9.7}$$

where Q_{int} is the intrinsic unloaded cavity Q, with only coupling to free space radiation, material losses and defects, Q_{probe} counts for the additional losses due to the coupling with the probe and Q_{meas} is the actual measured Q of the loaded cavity. It was shown [20] that for the design of Fig. 9.8 coupling to a cavity mode generates

Fig. 9.8 Schematic of PhC device including a cavity and a waveguide for cavity spectroscopy

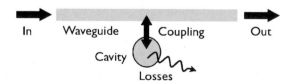

a dip in the transmission spectrum with:

$$Q_{int} = \frac{Q_{meas}}{\sqrt{T_{min}/T_{max}}} \qquad (9.8)$$

where T_{min} and T_{max} are the min and background transmission, note that the method does not require quantitative transmission measurement.

9.3 Internal Light Source Techniques

Measurements techniques where a light source is embedded inside or integrated in the close vicinity of the PhC structures have been developed for a large variety of purposes, versatility, quick characterization or investigation of the physics of light-matter interaction in cavity.

9.3.1 Internal Light Source

Internal light source (ILS) refers to a technique mainly developed in planar III–V PhC. The goal is to have a versatile technique that: does not require the full fabrication of device with access waveguides, allows the light source to be injected where needed and allows quantitative measurements. Historically, the first actually quantitative transmission measurements on 2D PhC were performed using ILS. The principle consists in the insertion of light emitters inside the planar waveguide. The emitters are either quantum wells (QW) [22] or quantum dots (QD) [17, 18], preferably inhomogeneous QD with a large emission band.

The general configuration for ILS experiments is illustrated in Fig. 9.9a. The photoluminescence (PL) of the light source embedded in planar waveguide is excited and propagates parallel to the surface as a guided mode and interacts with the PhC structure. The image of the guided beam emitted from the sample through a cleaved facet is coupled into spectrometer for spectral analysis. A polarizer allows one to select the TE or TM polarized component of the signal. In practice, some care has to be taken, due to refraction at the layer boundaries, three different beams come out from the cleaved facet after propagation through air, substrate and inside the planar waveguide, respectively. When the distance between the excitation spot and the edge is large enough the cross talk between the three signals is negligible and the selective analysis of the guided light can be performed by imaging the edge signal.

Figure 9.10 shows a typical image obtained when the collection optics is focused on the facet. The guided beam appears as a focused bright line along the edge while the black circle (diameter 4 µm) represents the conjugate image of the collection fibre. The guided contribution can be selected and spatially resolved aligning the circle with the bright line. Extensive descriptions can be found in [23] and [24].

The experimental procedure for PhC slab transmission measurements is illustrated in Fig. 9.9b. The reference signal $I_1(\lambda)$ and the PhC-related signal $I_2(\lambda)$ are measured in keeping constant the distance between the excitation points and

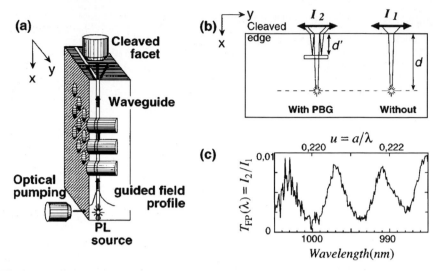

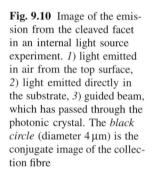

Fig. 9.9 a) Schematic of the internal light source experiment. b) Side view of a): configuration for measurements at normal incidence. The *rectangle* features the PhC slab. Multiple-beam interferences occur between the cleaved edge and the PhC pattern. c) Typical experimental transmission spectrum. Reprinted from [22]

Fig. 9.10 Image of the emission from the cleaved facet in an internal light source experiment. *1*) light emitted in air from the top surface, *2*) light emitted directly in the substrate, *3*) guided beam, which has passed through the photonic crystal. The *black circle* (diameter 4 μm) is the conjugate image of the collection fibre

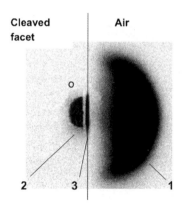

the cleaved facet. Provided that the PL emission remains homogenous, the ratio $I_1(\lambda)/I_2(\lambda)$ yields the absolute transmission spectrum $T(\lambda)$ of the PhC slab. Due to the limited width ($\Delta\lambda \approx 100$ nm) of the probed spectral range, "lithographic tuning" is used. PhC slabs with different lattice parameter (a) values and constant air filling factor (f) are measured and the whole photonic band-gap is explored as a function of the reduced frequency $u = a/\lambda$. An example of such measurement is shown Fig. 9.11.

The ILS measurement can also be extended to reflectivity measurement by making use of the Fabry–Perot cavity existing between the cleaved facet and the PhC (Fig. 9.9b), an example of a transmission spectrum exhibiting fringes allowing reflectivity measurement is shown Fig. 9.9c [22]. The freedom to position the light

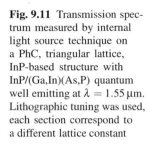

Fig. 9.11 Transmission spectrum measured by internal light source technique on a PhC, triangular lattice, InP-based structure with InP/(Ga,In)(As,P) quantum well emitting at $\lambda = 1.55\,\mu m$. Lithographic tuning was used, each section correspond to a different lattice constant

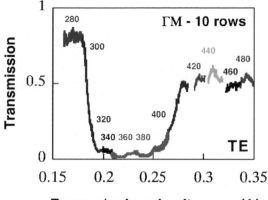

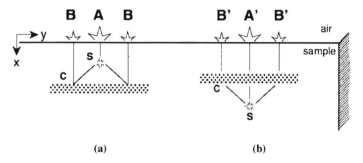

Fig. 9.12 Definition of measured diffraction efficiencies. (**a**) oblique incidence with reflection. (**b**) oblique incidence with transmission. Reprinted from [18]

source at wishes allows particular configurations that also enables the measurement of in-plane light diffraction from the PhC tile Fig. 9.12 and [17, 18]

The type of PhC structures that can be measured is not limited to crystal tile but, thank to its versatility, ILS has been used for the characterization of waveguides [25], bends [26] and Fabry–Perot cavities [24]. One of the limitation of the ILS technique is the residual absorption from light emitters layer that can hinder the propagation losses or quality factors for high performance structure. Some technique have been reported to overcome this issue [27]. Overall ILS is a very attractive technique because it allows a fast and simple quantitative measurement on structure without the adjunction of input–output access structures. This is a very efficient technique for processing validation and fast exploratory screening of PhC structures.

Concomitant with the ILS technique, but also applicable in principle to any other characterization techniques, we should mention the phenomenological model developed for the analysis of transmission spectrum including out of plane scattering losses, with the use of complex-valued refractive index in the air holes [28]. Such model proved to be very useful at least for processing validation and quantification of hole shape imperfections [29, 30].

9.3.2 Photoluminescence Spectroscopy

CW or time resolved PL experiments are also currently used to characterize PhC structure, most of the time for the characterization of optical cavities or investigation of alteration of the light-matter interaction. Although of fundamental interest from the physics point of view, there is little specific to report in term of optical characterization techniques, which consist in relatively standard and well documented PL and time resolved PL experiments. As examples we can mention the spectroscopy of optical resonators from out of plane emission [31,32], life time modification, Purcell effect, emission enhancement and inhibition experiments on 2D cavities [33], 1D photonic structures [34], bulk opals infiltrated with nanocrystals [35] and bulk 2D PhC [36].

9.4 Advanced Techniques

In the previous part we have reviewed the most frequently used methods of characterization for PhC. In the last part we will review more advanced tools that are less commonly used either because they are only of interest for very specific applications or because the complexity to be mastered to perform such measurements limits the number of experimental set-up to a few items worldwide.

9.4.1 Local Probe SNOM

Among the advanced techniques, scanning near-field optical microscope (SNOM) and atomic force microscope (AFM) are the most often used. In principle the experiment consists in probing the near field electromagnetic field with a local probe. The probe position is controlled in a similar fashion than in a scanning tunnelling microscope or AFM. The probe itself can be an AFM tip or an optical fibre tip, which will locally convert the evanescent mode into propagating modes either for collection or excitation. SNOM can be found in four different configurations:

1. Illumination: the fibre tip is used for excitation in connexion with a large area detector.
2. Collection: illumination is performed over a wide area and the fibre tip is used for collection.
3. Apertureless: illumination and collection are not spatially resolved and the tip is used for local modification of the optical response.
4. Evanescent: the probed field is the evanescent field in a total internal reflection (TIR) configuration.

SNOM will be described in details in the Chaps. 10 and 11. The reader can also refer to the review article by V. Sandogdhar [37]. We will only briefly review some examples. One of the first reports of SNOM on PhC images of planar PhC waveguides and bends in a modified end fire set-up [38] was used in the collection mode presents SNOM. Further K. Okamoto reports on optical mode characterization in

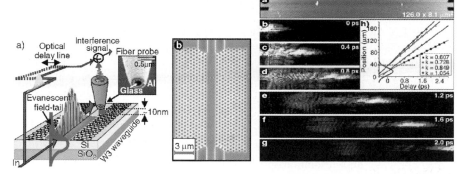

Fig. 9.13 *Left*: (*a*) Schematic representation of a pulse tracking experiment on a W3 PhCW. The evanescent field of a propagating pulse is picked up by a metal coated fibre probe with a subwavelength-sized aperture and interferometrically mixed with light from a reference branch. (*b*) Top view of the PhCW under study, although with a shorter device length. *Right*: A time-resolved NSOM measurement on a W3 PhCW (a 460 nm). (*a*) Topographic image of the structure rotated 90 with respect to Fig. 9.13(b). (*b–g*) The optical amplitude in the W3 PhCW for different reference times (400 ± 1 fs between frames; all frames have the same colour scale). Reprinted from [45]

planar PhC cavities in the collection or illumination mode [39]. SNOM is not limited to planar structure, example of transmission image in the illumination mode on opals are discussed in [40, 41]. 3D images of the field out-coupling from a PhC waveguide in micro-porous Si were also reported [42].

Special mention should be also made to time resolved and heterodyne SNOM. Usual SNOM delivers intensity images and phase information of the field is lost. Experimental knowledge of the full complex amplitude of the field is required for a complete and unequivocal reconstruction of the image source. Such phase information can be retrieved from the heterodyne SNOM techniques, which consist in the insertion of the SNOM set-up in one arm of a Mach–Zehnder interferometer [40, 41]. Only once amplitude is known, a Fourier transform gives access to the spectroscopy of the mode in the reciprocal space (*k*-space) and allows measuring the relative weight of Bloch harmonics in each Brillouin zone and their dispersion [43, 44]. It relies, however, on sophisticated modelling of the interaction between the near field and the tip to know the vectorial transfer function of the tip, which requires complex numerical post-processing. The same set-up, where now the phase shift is used as an optical delay line, can also be used for time resolved measurement [45]. Figure 9.13 shows such time resolved set-up and experimental results with the real-space observation of pulses propagating inside a PhC waveguide.

9.4.2 Fourier Imaging

The standard characterization techniques we have reviewed consist in the measurement of the optical response or light emission from the PhC structures. Although this gives access to a large number of important information, it is a type of a black box approach. It does not directly investigate light propagation inside the PhC and fundamental quantities such as dispersion curve or equi-frequency surfaces (EFS) are only inferred after data analysis. This has motivated the development of SNOM and heterodyne SNOM to achieve real space and Fourier space high-resolution imaging. However these methods can be cumbersome and time consuming. Imaging directly the Bloch mode in real and reciprocal space with old-fashioned classical optics is currently re-discovered in plasmonic [46, 47] and with nano-objects [48, 49], of which the following section will make extensive quotes.

The principle is to image with a high numerical aperture microscope objective out of plane scattered light from the PhC in e.g. a standard end fire experiment. This already provides high-resolution optical images, but moreover imaging the back focal plane of the collecting lens (a.k.a. Fourier plane) gives access to far field or k-space images (Fig. 9.14). In this plane every point is uniquely related to a direction of emission, thanks to geometrical optics, which is also uniquely related to a unique in-plane wave vector of the PhC Bloch mode thanks to in-plane wave vector conservation law. A set-up where the insertion or removing of a single lens allows switching from real and Fourier space is detailed in [48, 49].

Figure 9.15 shows an example of real space (Fig. 9.15a–d) and EFS imaging (Fig. 9.15e–h) of self collimating light propagation in a PhC tile. EFS closely matches theoretical calculations. Dispersion curves are directly obtained in plotting the wavevector values as a function of the frequency. An intensity modulation along the x direction whose associated spatial frequency increases for decreasing normalized frequency u can be seen in Fig. 9.15a–d. Fourier space imaging in Fig. 9.15e–h proves that this modulation arises from the standing wave produced by constructive interference between the forward (FW) and the backward (BW) propagating Bloch waves in the structure. This BW wave results from a reflection at the air-ridge waveguide interface. Note that that without any reflection, a constant intensity pattern would be measured. No interferences would be observed and the phase information and consequently the dispersion curve information would be lost in real space. In Fourier space, however, the spot at position $-k_x$ would still be present. Fourier space imaging is therefore a more universal and straightforward method to retrieve the dispersion relation than real-space imaging.

Fig. 9.14 Principle of Fourier imaging. In the back focal plane of the collecting lens every point is uniquely related to a direction of emission, thanks to geometrical optics, which is also uniquely related to a unique in-plane wave vector of the PhC Bloch mode thanks to in-plane wave vector conservation law

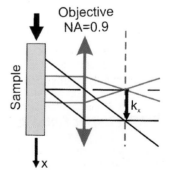

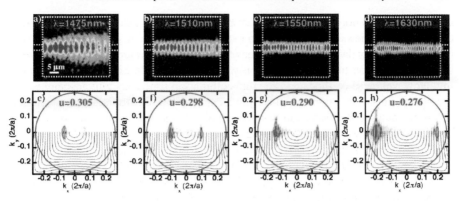

Fig. 9.15 (**a–d**) Near-field image of the Bloch wave mode in a square lattice PhC tile of 80 periods for different reduced energies u. Below each real space image is shown the corresponding far-field image from (**e**) to (**h**). The *thin black curves* in the lower half space of the far-field frame are the 2D PWE theoretical EFSs starting from $u = 0.315$ in the centre and drawn with a step of -0.005. The far-field image is limited by the maximum propagating wave vector $k_{max} = 0.2636 \cdot 2\pi/a$ (the *red circle* in (**e–h**) delimits the output pupil of the objective that defines k_{max}). The corresponding diffraction limited spatial resolution is $0.85\,\mu m$. Reprinted from [49]

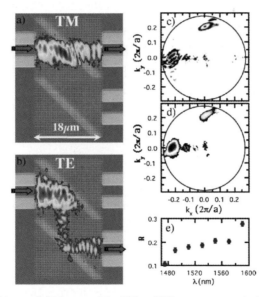

Fig. 9.16 (**a**) and (**b**) near-field image of the TM and TE waves, respectively, propagating in the double slab polarizer beam splitter (*colour coded*: increasing intensity from *blue* to *red*) superimposed with an optical microscope image of the structure (*black* and *white*), (**c**) emission diagram of the TE wave imaged in the Fourier space at $\lambda = 1550\,nm$, (**d**) same as **c**) with the emission of the second splitter filtered in the intermediate image plane, (**e**) reflection coefficient of the first slab deduced from the far-field pattern measured for different wavelengths as in **c**). Reprinted from [49]

PhC devices will eventually be integrated in more complex photonic circuits on a chip. Fourier imaging is a convenient tool for individual characterization in a complex environment. An example can be found in [48–50] with the demonstration of local investigation of a double PhC polarization beam splitter. The splitter consist in a PhC square tile operating in the self-collimation regime, in which is embedded a polarization sensitive 45° slab [49]. As explained by the authors, the transmission and reflection properties of each splitter cannot be extracted from the near-field images due to the overlap of incident and backscattered beams and the out-of-plane scattering signal at the different interfaces, which overlaps partly with the propagating beams at such small scales. The situation is more favourable in k-space. Indeed, k-space imaging provides a natural spatial selectivity of beams propagating along different directions, allowing the measurement of the relative intensity of each beam and the reflectivity coefficient of a single polarizer could be retrieved (Fig. 9.16).

This technique appears inherently limited to wavevector value inside the air light cone, while most of the present interesting structures are designed to operate below the light line in order to minimize propagation losses. Several ways have been implemented to break the light line limit:

1. The first one which is limited to real-space imaging relies on the use of defects or imperfections introduced deliberately or not. This method was in fact used long time ago by P.S. Russell to visualize light propagation in what was called at that time doubly periodic planar waveguide [51].
2. A second approach is inspired of techniques known as super-resolution in astronomy or medical imaging [48, 49] and references therein. The concept is the following: For a perfect infinitely long structure the Fourier space spectrum of the Bloch modes is a series of delta function. In most interesting structures these lines fall outside of the light cone so nothing can be measured. For an object of a finite size, the delta function is broadened and has finite value inside the light cone, this means that some information is leaking outside in the light cone on the form of a propagating wave. Such information can be collected in the far field and the whole spectrum can be recovered by analytical extension. With just the a priori knowledge of the existence of an internal periodicity N. Le Thomas and co-workers [48,49] could recover information in k-space values falling well away from the light cone and would be equivalent to a numerical aperture of 2.5.
3. A last approach consists in adding a vanishingly weak extra periodicity that folds back the Brillouin zone into the light cone. This is intrinsically the case when measuring coupled cavity waveguide as shown in [48,49] where the supercell periodicity folds back the entire dispersion curve inside the air light cone. The extra periodicity can also be deliberately included, as it is demonstrated on a measurement of slow light dispersion curve in a modified W1 PhC waveguide as it can be seen in Fig. 9.17 reprinted from [48, 49] or on SOI photonic wires [52].

Finally the high NA real and Fourier imaging is also a very powerful tool for sample screening, search for unexpected deviation or process validation. For example light scattering from very tiny residual gratings/in the range of 1–5 nm amplitude)

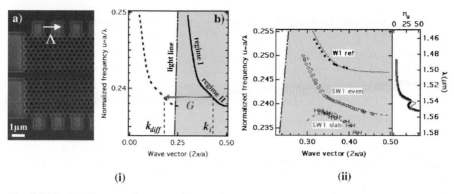

Fig. 9.17 i) (a) Electron microscopy image of a modified W1 waveguide with the linear probe gratings at the waveguide boundaries. **(b)** Illustration of the folding of the dispersion curve *dark line* into the *light cone dash line*. **ii)** *Left*: Experimental dispersion curves of the reference and modified W1. *Right*: Theoretical *grey* and experimental *black group*-index dispersion curves. Reprinted from [48]

located on the sidewalls of the waveguide could be observed [52]. This grating corresponds to a spatial modulation of $1.28\,\mu m$, which is exactly the dimension of the trapezoidal write subfield pattern used in the electron beam lithography Such grating induced by residual subfield stitching error had passed unnoticed through prior SEM characterizations.

9.5 Conclusion

In this chapter we have reviewed the different techniques used for the near infrared optical characterization of photonic crystals (PhC). One first section discussed optical measurement techniques based on an external light source: reflectivity and end fire measurement, the second section techniques relying on the presence of a light source inside the PhC under investigation: internal light source and photoluminescence and the third section more advanced techniques like scanning near field optical microscopy and Fourier imaging. A point that has not been exploited yet among all the already existing methods is their complementarily. Several techniques aim at measuring the same quantity, but with some subtle differences. A refined comparison of the different measurements should bring a very rich understanding of the underlying physical processes. For example investigation propagation of slow light modes in the same sample with reflectivity angular analysis, end fire transmission measurement, Fourier imaging and time resolved SNOM would certainly allow to unambiguously distinguish from true slow propagating modes, localized or resonant states, evanescent and interfaces states.

The advanced techniques presented will certainly a undergo a fast development in the following years for both quick device characterization (and possibly in-situ) and the advancement of the physical understanding of light propagation in PhC.

Acknowledgements The author would like to thank in particular N. Le Thomas for numerous, extremely useful and stimulating discussions and L. C. Andreani, H. Benisty, R. Ferrini, D. Hofstetter, J. Jágerská, T. F. Krauss, A. Talneau and C. Weisbuch and support from the COST P11 Action, the European projects ePIXnet (IST-004525), Funfox (IST-004582), and the Swiss NCCR-Quantum Photonics.

References

1. Stanley, R.P., Houdré, R., Oesterle, U., Gailhanou, M. and Ilegems, M.: Ultrahigh Finesse Microcavity with Distributed Bragg Reflectors. Appl. Phys. Lett. 65, 1883–1885 (1994).
2. Tarhan, I.I., Zinkin, M.P. and Watson, G.H.: Interferometric-technique for the measurement of photonic band-structure in colloidal crystals. Opt. Lett. 20, 1571–1573 (1995).
3. Vlasov, Y.A., Bo, X.Z., Sturm, J.C. and Norris, D.J.: On-chip natural assembly of silicon photonic bandgap crystals. Nature 414, 289–293 (2001).
4. Zengerle, R.: Light-Propagation in Singly and Doubly Periodic Planar Wave-Guides. J. Mod. Opt. 34, 1589–1617 (1987).
5. Galli, M., Agio, M., Andreani, L.C., Belotti, M., Guizzetti, G., Marabelli, F., Patrini, M., Bettotti, P., Dal Negro, L., Gaburro, Z., Pavesi, L., Liu, A. and Bellutti, P.: Spectroscopy of photonic bands in macroporous silicon photonic crystals. Phys. Rev. B 65, Artn 113111 (2002).
6. Pavarini, E., Andreani, L.C., Soci, C., Galli, M., Marabelli, F. and Comoretto, D.: Band structure and optical properties of opal photonic crystals. Phys. Rev. B 72, Artn 045102 (2005).
7. Galli, M., Belotti, M., Bajoni, D., Patrini, M., Guizzetti, G., Gerace, D., Agio, M., Andreani, L.C. and Chen, Y.: Excitation of radiative and evanescent defect modes in linear photonic crystal waveguides. Phys. Rev. B 70, Artn 081307 (2004).
8. Galli, M., Bajoni, D., Patrini, M., Guizzetti, G., Gerace, D., Andreani, L.C., Belotti, M. and Chen, Y.: Single-mode versus multimode behavior in silicon photonic crystal waveguides measured by attenuated total reflectance. Phys. Rev. B 72, Artn 125322 (2005).
9. Shinya, A., Mitsugi, S., Kuramochi, E. and Notomi, M.: Ultrasmall multi-port channel drop filter in two-dimensional photonic crystal on silicon-on-insulator substrate. Opt. Express 14, 12394–12400 (2006).
10. Sugimoto, Y., Tanaka, Y., Ikeda, N., Nakamura, H., Kanamoto, K., Ohkouchi, S., Watanabe, Y., Inoue, K. and Asakawa, K.: Fabrication and characterization of photonic crystal-based symmetric Mach–Zehnder (PC-SMZ) structures based on GaAs membrane slab waveguides. IEEE J. Sel. A. Comm. 23, 1308–1314 (2005).
11. McNab, S.J., Moll, N. and Vlasov, Y.A.: Ultra-low loss photonic integrated circuit with membrane-type photonic crystal waveguides. Opt. Express 11, 2927–2939 (2003).
12. Talneau, A., Mulot, M., Anand, S. and Lalanne, P.: Compound cavity measurement of transmission and reflection of a tapered single-line photonic-crystal waveguide. Appl. Phys. Lett. 82, 2577–2579 (2003).
13. Hofstetter, D. and Thornton, R.L.: Theory of loss measurements of Fabry–Perot resonators by Fourier analysis of the transmission spectra. Opt. Lett. 22, 1831–1833 (1997).
14. Hofstetter, D. and Thornton, R.L.: Measurement of optical cavity properties in semiconductor lasers by Fourier analysis of the emission spectrum. IEEE J. Quantum Electr. 34, 1914–1923 (1998).
15. Wild, B., Dunbar, L.A., Houdré, R., Duan, G.-H., Cuisin, C., Derouin, E., Drisse, O., Legouézigou, L., Legouézigou, O. and Pommereau, F.: Characterization and analysis of low loss two dimensional InP-based photonic crystal waveguides, Proceding of the 12th European Conference on Integrated Opics (ECIO), Grenoble, 2005 and Wild B. PhD thesis #3573 EPFL, Switzerland 2006 (http://library.cpfl.ch/theses/?nr=3573).
16. Notomi, M., Yamada, K., Shinya, A., Takahashi, J., Takahashi, C. and Yokohama, I.: Extremely large group-velocity dispersion of line-defect waveguides in photonic crystal slabs. Phys. Rev. Lett. 87, Artn 253902 (2001).

17. Labilloy, D., Benisty, H., Weisbuch, C., Smith, C.J.M., Krauss, T.F., Houdré, R. and Oesterle, U.: Finely resolved transmission spectra and band structure of two-dimensional photonic crystals using emission from InAs quantum dots. Phys. Rev. B 59, 1649–1652 (1999).

18. Labilloy, D., Benisty, H., Weisbuch, C., Krauss, T.F., Cassagne, D., Jouanin, C., Houdré, R., Oesterle, U. and Bardinal, V.: Diffraction efficiency and guided light control by two-dimensional photonic-bandgap lattices. IEEE J. Quantum Electr. 35, 1045–1052 (1999).

19. Vlasov, Y.A., O'Boyle, M., Hamann, H.F. and McNab, S.J.: Active control of slow light on a chip with photonic crystal waveguides. Nature 438, 65–69 (2005).

20. Akahane, Y., Asano, T., Song, B.S. and Noda, S.: Fine-tuned high-Q photonic-crystal nanocavity. Opt. Express 13, 1202–1214 (2005).

21. Kuramochi, E., Notomi, M., Mitsugi, S., Shinya, A., Tanabe, T. and Watanabe, T.: Ultrahigh-Q photonic crystal nanocavities realized by the local width modulation of a line defect. Appl. Phys. Lett. 88, Artn 041112 (2006).

22. Labilloy, D., Benisty, H., Weisbuch, C., Krauss, T.F., DeLaRue, R.M., Bardinal, V., Houdré, R., Oesterle, U., Cassagne, D. and Jouanin, C.: Quantitative measurement of transmission, reflection, and diffraction of two-dimensional photonic band gap structures at near-infrared wavelengths. Phys. Rev. Lett. 79, 4147–4150 (1997).

23. Benisty, H., Olivier, S., Weisbuch, C., Agio, M., Kafesaki, M., Soukoulis, C.M., Qiu, M., Swillo, M., Karlsson, A., Jaskorzynska, B., Talneau, A., Moosburger, J., Kamp, M., Forchel, A., Ferrini, R., Houdré, R. and Oesterle, U.: Models and measurements for the transmission of submicron-width waveguide bends defined in two-dimensional photonic crystals. IEEE J. Quantum Electr. 38, 770–785 (2002).

24. Ferrini, R., Leuenberger, D., Mulot, M., Qiu, M., Moosburger, J., Kamp, M., Forchel, A., Anand, S. and Houdré, R.: Optical study of two-dimensional InP-based photonic crystals by internal light source technique. IEEE J. Quantum Electr. 38, 786–799 (2002).

25. Olivier, S., Smith, C., Rattier, M., Benisty, H., Weisbuch, C., Krauss, T., Houdré, R. and Oesterle, U.: Miniband transmission in a photonic crystal coupled-resonator optical waveguide. Opt. Lett. 26, 1019–1021 (2001).

26. Olivier, S., Benisty, H., Weisbuch, C., Smith, C.J.M., Krauss, T.F., Houdré, R. and Oesterle, U.: Improved 60 degrees bend transmission of submicron-width waveguides defined in two-dimensional photonic crystals. J. Lightw. Technol. 20, 1198–1203 (2002).

27. Lombardet, B., Ferrini, R., Dunbar, L.A., Houdré, R., Cuisin, C., Drisse, O., Lelarge, F., Pommereau, F., Poingt, F. and Duan, G.H.: Internal light source technique free from reabsorption losses for optical characterization of planar photonic crystals. Appl. Phys. Lett. 85, 5131–5133 (2004).

28. Benisty, H., Labilloy, D., Weisbuch, C., Smith, C.J.M., Krauss, T.F., Cassagne, D., Beraud, A. and Jouanin, C.: Radiation losses of waveguide-based two-dimensional photonic crystals: Positive role of the substrate. Appl. Phys. Lett. 76, 532–534 (2000).

29. Ferrini, R., Houdré, R., Benisty, H., Qiu, M. and Moosburger, J.: Radiation losses in planar photonic crystals: two-dimensional representation of hole depth and shape by an imaginary dielectric constant. J. Opt. Soc. Amer. B 20, 469–478 (2003).

30. Ferrini, R., Leuenberger, D., Houdré, R., Benisty, H., Kamp, M. and Forchel, A.: Disorder-induced losses in planar photonic crystals. Opt. Lett. 31, 1426–1428 (2006).

31. Smith, C.J.M., Krauss, T.F., Benisty, H., Rattier, M., Weisbuch, C., Oesterle, U. and Houdré, R.: Directionally dependent confinement in photonic-crystal microcavities. J. Opt. Soc. Amer. B 17, 2043–2051 (2000).

32. Ochoa, D., Houdré, R., Ilegems, M., Benisty, H., Krauss, T.F. and Smith, C.J.M.: Diffraction of cylindrical Bragg reflectors surrounding an in-plane semiconductor microcavity. Phys. Rev. B 61, 4806–4812 (2000).

33. Chang, W.H., Chen, W.Y., Chang, H.S., Hsieh, T.P., Chyi, J.I. and Hsu, T.M.: Efficient single-photon sources based on low-density quantum dots in photonic-crystal nanocavities. Phys. Rev. Lett. 96, Artn 117401 (2006).

34. Viasnoff-Schwoob, E., Weisbuch, C., Benisty, H., Olivier, S., Varoutsis, S., Robert-Philip, I., Houdré, R. and Smith, C.J.M.: Spontaneous emission enhancement of quantum dots in a photonic crystal wire. Phys. Rev. Lett. 95, Artn 183901 (2005).

35. Lodahl, P., van Driel, A.F., Nikolaev, I.S., Irman, A., Overgaag, K., Vanmaekelbergh, D.L. and Vos, W.L.: Controlling the dynamics of spontaneous emission from quantum dots by photonic crystals. Nature 430, 654–657 (2004).
36. Fujita, M., Takahashi, S., Tanaka, Y., Asano, T. and Noda, S.: Simultaneous inhibition and redistribution of spontaneous light emission in photonic crystals. Science 308, 1296–1298 (2005).
37. Sandoghdar, V., Buchler, B., Kramper, P., Götzinger, S., Benson, O. and Kafesaki, M.: Scanning near-field optical studies of photonic devices. In: Photonic crystals, advances in design, fabrication and characterization. Busch, K., Lölkes, S., Wehrspohn, R.B. and Föll, H., Eds. pp. 215–237. Wiley-VCH, Weunheim (2004).
38. Bozhevolnyi, S.I., Volkov, V.S., Sondergaard, T., Boltasseva, A., Borel, P.I. and Kristensen, M.: Near-field imaging of light propagation in photonic crystal waveguides: Explicit role of Bloch harmonics. Phys. Rev. B 66, Artn 235204 (2002).
39. Okamoto, K., Loncar, M., Yoshie, T., Scherer, A., Qiu, Y. M. and Gogna, P.: Near-field scanning optical microscopy of photonic crystal nanocavities. Appl. Phys. Lett. 82, 1676–1678 (2003).
40. Flück, E., van Hulst, N.F., Vos, W.L. and Kuipers, L.: Near-field optical investigation of three-dimensional photonic crystals. Phys. Rev. E 68, Artn 015601 (2003).
41. Flück, E., Hammer, M., Otter, A.M., Korterik, J.P., Kuipers, L. and van Hulst, N.F.: Amplitude and phase evolution of optical fields inside periodic photonic structures. J. Lightw. Technol. 21, 1384–1393 (2003).
42. Kramper, P., Agio, M., Soukoulis, C.M., Birner, A., Muller, F., Wehrspohn, R.B., Gosele, U. and Sandoghdar, V.: Highly directional emission from photonic crystal waveguides of subwavelength width. Phys. Rev. Lett. 92, Artn 113903 (2004).
43. Engelen, R.J.P., Karle, T.J., Gersen, H., Korterik, J.P., Krauss, T.F., Kuipers, L. and van Hulst, N.F.: Local probing of Bloch mode dispersion in a photonic crystal waveguide. Opt. Express 13, 4457–4464 (2005).
44. Gersen, H., Karle, T.J., Engelen, R.J.P., Bogaerts, W., Korterik, J.P., van Hulst, N.F., Krauss, T.F. and Kuipers, L.: Direct observation of Bloch harmonics and negative phase velocity in photonic crystal waveguides. Phys. Rev. Lett. 94, Artn 123901 (2005).
45. Gersen, H., Karle, T.J., Engelen, R.J.P., Bogaerts, W., Korterik, J.P., van Hulst, N.F., Krauss, T.F. and Kuipers, L.: Real-space observation of ultraslow light in photonic crystal waveguides. Phys. Rev. Lett. 94, Artn 073903 (2005).
46. Giannattasio, A. and Barnes, W.L.: Direct observation of surface plasmon-polariton dispersion. Opt. Express 13, 428–434 (2005).
47. Massenot, S., Grandidier, J., Bouhelier, A., Colas des Francs, G., Markey, L., Renger, J., Gonzàles, U. and Quidant, R.: Polymer-metal waveguides characterization by Fourier plane leakage radiation microscopy. Appl. Phys. Lett. 91, Artn 243102 (2007).
48. Le Thomas, N., Houdré, R., Frandsen, L.H., Fage-Pedersen, J., Lavrinenko, A.V. and Borel, P.I.: Grating-assisted superresolution of slow waves in Fourier space. Phys. Rev. B 76, Artn 035103 (2007).
49. Le Thomas, N., Houdré, R., Kotlyar, M.V., O'Brien, D. and Krauss, T.F.: Exploring light propagating in photonic crystals with Fourier optics. J. Opt. Soc. Amer. B 24, 2964–2971 (2007).
50. Zabelin, V., Dunbar, L.A., Le Thomas, N., Houdré, R., Kotlyar, M.V., O'Faolain, L. and Krauss, T.F.: Self-collimating photonic crystal polarization beam splitter. Opt. Lett. 32, 530–532 (2007).
51. Russell, P.S.J.: Interference of Integrated Floquet-Bloch Waves. Phys. Rev. A 33, 3232–3242 (1986).
52. Jágerská, J., Le Thomas, N., Houdré, R., Bolten, J., Moormann, C., Wahlbrink, T., Ètyroký, J., Waldow, M. and Först, M.: Dispersion properties of silicon nanophotonic waveguides investigated with Fourier optics. Opt. Lett. 32, 2723–2725 (2007).

10 Characterization Techniques for Planar Optical Microresonators

René M. de Ridder, Wico C.L. Hopman*, and Edwin J. Klein

University of Twente, MESA+ Institute for Nanotechnology, Enschede, the Netherlands
*now at: Sensata Technologies, Almelo, the Netherlands

Abstract. Optical microresonators that are coupled to optical waveguides often behave quite similar to Fabry–Perot resonators. After summarising key properties of such resonators, three characterization methods will be discussed. The first involves the analysis of transmission and reflection spectra, from which important parameters like (waveguide) loss and coupling or reflection coefficients can be extracted. The second method is called transmission-based scanning near-field optical microscopy (T-SNOM) which allows to map out the intensity distribution inside a high-Q resonator with sub-wavelength resolution. For such resonators conventional SNOM suffers from inaccuracies introduced by the disturbing effect of the presence of the probe on the field distribution. T-SNOM avoids this problem by exploiting this disturbing effect. The third method is most applicable to large-size (many wavelengths across) resonators. It involves simultaneous analysis of scattered light and transmission/reflection spectra in order to relate spectral features to large-scale field distributions inside the resonator. Together, these techniques form a convenient toolbox for characterizing many different planar optical microresonators.

Key words: optical microresonators; optical measurements; SNOM; NSOM; AFM; scattered-light analysis

10.1 Introduction

Optical resonators have many applications, such as wavelength filtering, add-drop multiplexing of several wavelength channels, lasers, and field enhancement [1–4]. Recent developments in integrated optics towards high-refractive index contrast technology (e.g. silicon photonics [5]) and related fields such as photonic crystals have opened the road towards a strong miniaturisation of optical circuits. In particular microring resonators, e.g. [1, 2, 5, 6], and cavities in photonic crystals, e.g. [3–5, 7–10] have been intensively investigated. Integrated optical circuits with high functionality have been conceived [2, 11], based on arrays of microresonators. Conventionally, optical resonators are characterized by performing spectral transmission and/or reflection measurements. In these methods, the input-output relationships are analysed, and through device models, some data on internal parameters can be inferred. These methods will provide essential information on the functionality

and performance of the device in a system. For relatively simple resonator structures consisting of a few discrete elements, such as basic Fabry–Perot resonators,
these measurements are fully adequate. However, for more complicated structures
it will often be highly desirable to "look" inside the structure. This is especially the
case for nanostructured devices where small disturbances may strongly affect their
optical performance. Being able to map out the optical field distribution inside the
structure can be extremely helpful in diagnosing possible fabrication errors, and for
optimising the design, for example in order to maximise the quality factor Q.

In this chapter we will show how some internal parameters can be extracted
through analysis of spectral transfer functions. After that we will concentrate on two
different methods for looking inside the box. One method – that can provide sub-
wavelength resolution – is most applicable to rather small structures (up to about
$50 \times 50\,\mu m^2$) such as cavities in photonic crystals, whereas the second method finds
its main application in analysing extended structures such as gratings up to several
mm in length. We will start with an introductory section on Fabry–Perot-type resonators in order to briefly review their properties and key parameters that may also
apply to more complicated structures.

10.2 Fabry–Perot-Type Optical Resonators

One of the simplest types of resonator is the Fabry–Perot interferometer or etalon.
As shown in Fig. 10.1, it consists of two parallel semitransparent mirrors having
power reflectivities R_1 and R_2, separated by a given distance L. The space between
the mirrors is referred to as the cavity. We will restrict the treatment here to the case
of plane waves at normal incidence to the mirrors. Many practical resonators that
have a more complicated structure still behave quite similar to this simple Fabry–
Perot interferometer, e.g., an optical waveguide having two discontinuities that act
as mirrors (in the simplest Fabry–Perot solid state laser, these discontinuities are
the end facets showing Fresnel reflection, Fig. 10.2a), or a ring resonator that is
coupled to straight waveguides through the evanescent field (Fig. 10.2b). In that
case, assuming lossless devices, the power coupling coefficients κ_1 and κ_2 of the
directional couplers correspond to the power transmission coefficients $T_i = 1 - R_i$
of a Fabry–Perot. More generally, any cavity that is coupled to the outside world
through two ports can at some level of abstraction be described as a Fabry–Perot
resonator, for example a cavity in a photonic crystal (Fig. 10.2c), or – maybe less
obvious – a finite-length grating (Fig. 10.2d).

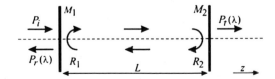

Fig. 10.1 Schematic of
Fabry–Perot resonator

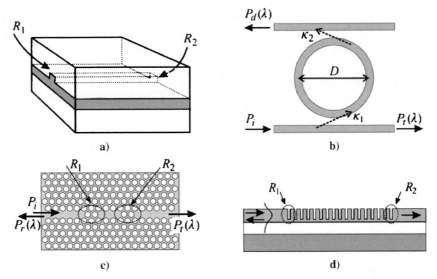

Fig. 10.2 Examples of Fabry–Perot-like resonators: **a)** waveguide with reflecting facets; **b)** ring resonator; **c)** cavity in a photonic crystal; **d)** finite-length waveguide grating

10.2.1 Resonance Condition; Free Spectral Range (FSR)

Resonance in a Fabry–Perot interferometer occurs if a wave after making a full roundtrip through the cavity (i.e. travelling twice the cavity length and reflecting once at each mirror) arrives in phase with itself. In a simple free-space Fabry–Perot, the phase shift at reflection is often neglected, and the roundtrip phase shift is determined by the optical path length nL, where n is the refractive index inside the cavity. The resonance condition is given by

$$\varphi = 2\beta L = m2\pi , \tag{10.1}$$

where φ is the roundtrip phase shift, the resonance order m is a positive integer, and β is the propagation constant of the wave inside the cavity:

$$\beta = \frac{\omega}{v_p} = \frac{\omega n}{c} = \frac{2\pi}{\lambda}n , \tag{10.2}$$

with angular frequency ω, phase velocity v_p, the speed of light in vacuum c, and the wavelength in vacuum λ. In waveguiding structures, the refractive index n should be replaced by the effective refractive index n_{eff}.

Conditions (10.1) and (10.2) imply the occurrence of multiple resonance frequencies ω_m:

$$\omega_m = m\pi c/(nL) . \tag{10.3}$$

The frequency (or wavelength) difference between two consecutive resonances is called the free spectral range (FSR). From (10.3) it seems to follow that the FSR is given by $\Delta\omega_{FSR} = \pi c/(nL)$. However, this is valid only for nondispersive media (n is constant); with dispersive media we should be a bit more careful. From (10.1) we find for consecutive resonances:

$$\Delta\varphi = 2\Delta\beta L = 2\pi . \tag{10.4}$$

If dispersion is not too strong, and $\Delta\omega_{FSR}$ is much smaller than the operating angular frequency, a linear approximation of dispersion is appropriate:

$$\Delta\beta = \frac{\partial\beta}{\partial\omega}\Delta\omega , \tag{10.5}$$

which, using (10.4) and the definitions of group velocity $v_g = \partial\omega/\partial\beta$ and group index $n_g = c/v_g$ leads to a more accurate expression for the FSR,

$$\Delta\omega_{FSR} = \pi c/(Ln_g) . \tag{10.6}$$

In terms of wavelength the FSR is expressed as

$$\Delta\lambda_{FSR} = \tfrac{1}{2}\lambda_0^2/(Ln_g) , \tag{10.7}$$

where λ_0 is the wavelength halfway between two considered resonances. It is clear that smaller cavities show larger FSR.

   In cases of strong dispersion and large FSR, the linear approximation of dispersion as used in (10.5) turns out to be not sufficiently accurate [12]. Analytic models are often lacking for the structures of interest, such as strong gratings and photonic crystal cavities. In such cases the dispersion curves $\beta(\omega)$ should be calculated numerically in order to verify the validity of (10.5) for the frequency range of interest. Another issue is the phase shift upon reflection. For simple cases such as Fresnel reflection at normal incidence at the interface between two dielectrics, the magnitude and phase of the reflection coefficient are well known. However, the situation is much less trivial if periodic structures are involved and evanescent waves occur that can be associated with a certain penetration depth into the reflecting medium. This leads to a frequency-dependent phase shift upon reflection, or, equivalently, a frequency-dependent effective cavity length. Such situations can be analysed using approximative coupled-mode models [13], or with "brute force" simulations, using e.g. finite-difference time-domain (FDTD) code.

10.2.2 Transfer Function

The transfer function of the simple Fabry–Perot can be calculated by considering the device as a feedback system for the electric field of the optical waves. The mirrors are characterized by their amplitude transmission and reflection coefficients t and r (transmitted field $E_t = tE_{in}$; reflected field $E_r = rE_{in}$); losses in the cavity are taken into account by assuming a complex propagation constant $\alpha + j\beta$ where α is the attenuation constant and j is the imaginary unit, so that a travelling wave in the cavity can be represented as a phasor $E = E_0 e^{-\alpha z} e^{-j\beta z}$. If we would not

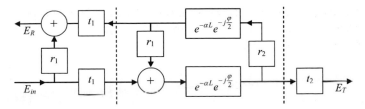

Fig. 10.3 Fabry–Perot cavity represented as a system with feedback. The block structure between the *vertical dashed lines* represents the process inside the cavity

consider any back-reflections, the transmitted field through a cavity with length L would be $E_T = E_{in}t_1t_2\,e^{-\alpha L}\,e^{-j\varphi/2}$ where we used (10.1). Taking into account the back-reflections results in a feedback scheme as illustrated by Fig. 10.3.

The field transmission of this system can be calculated to be

$$E_T = E_{in}\frac{t_1t_2\,e^{-\alpha L}\,e^{-j\frac{\varphi}{2}}}{1 - r_1r_2\,e^{-2\alpha L}\,e^{-j\varphi}}\;.\tag{10.8}$$

From this, the intensity transfer function is calculated as

$$T_{FP} \equiv I_T/I_{in} = |E_T/E_{in}|^2\;.\tag{10.9}$$

Substitution of (10.8) into (10.9) gives:

$$T_{FP} = \frac{T_0}{1 + B\sin^2(\varphi/2)}\;,\tag{10.10}$$

where T_0 and B are constants determined by mirror reflectivities and cavity losses,

$$T_0 \equiv \frac{e^{-2\alpha L}\,(1 - R_1)\,(1 - R_2)}{\left(1 - \sqrt{R_1R_2}\,e^{-2\alpha L}\right)^2}\;,\tag{10.11}$$

$$B \equiv \frac{4\sqrt{R_1R_2}\,e^{-2\alpha L}}{\left(1 - \sqrt{R_1R_2}\,e^{-2\alpha L}\right)^2}\;,\tag{10.12}$$

$$R_i = |r_i|^2\;.\tag{10.13}$$

Similarly the field reflection of this system can be calculated to be

$$E_R = E_{In}\frac{r_1 - r_2\,e^{-2\alpha L}\,e^{-j\varphi}}{1 - r_1r_2\,e^{-2\alpha L}\,e^{-j\varphi}}\;.\tag{10.14}$$

From this, the intensity reflection function is calculated as

$$R_{FP} \equiv I_R/I_{In} = |E_R/E_{In}|^2\;.\tag{10.15}$$

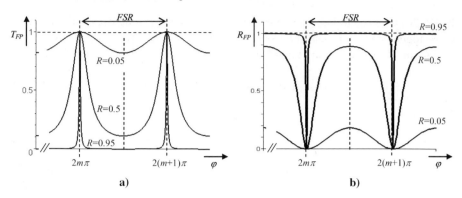

Fig. 10.4 a) Transmission and **b)** reflection spectra of lossless Fabry–Perot resonator with equal mirrors for 3 different reflectance values

Substitution of (10.14) into (10.15) gives:

$$R_{\text{FP}} = 1 - \frac{C}{1 + B\sin^2(\varphi/2)}, \tag{10.16}$$

where C is given by:

$$C = \frac{(1 - R_1)(1 - R_2 e^{-4\alpha L})}{(1 - \sqrt{R_1 R_2}\, e^{-2\alpha L})^2}. \tag{10.17}$$

For the case of a lossless cavity ($\alpha = 0$), and identical mirrors ($R_1 = R_2 = R$), Fig. 10.4 shows the transfer and reflection spectra $T_{\text{FP}}(\varphi)$ and $R_{\text{FP}}(\varphi)$ with different values of R as parameter. It shows that high reflectivity is needed for obtaining narrow transmission peaks with deep minima in between.

10.2.2.1 Width of a Transmission Peak (FWHM)

The width of a resonance peak is usually given as the full width at half maximum (FWHM), i.e. the difference $\Delta\omega_{\text{FWHM}}$ between two frequencies at both sides of the peak maximum, where the transmission is half that at the resonance frequency ω_m. At frequencies $\omega_m \pm 1/2\Delta\omega_{\text{FWHM}}$, (10.10) should evaluate to $T_0/2$, or

$$\sin^2(\varphi/2) = 1/B. \tag{10.18}$$

Combining (10.1), (10.2), and (10.3) we rewrite φ as:

$$\varphi = m2\pi \frac{\omega}{\omega_m} \frac{n(\omega)}{n(\omega_m)}, \tag{10.19}$$

where we wrote the possibly dispersive refractive index explicitly as a function of frequency. Using (10.19) in (10.18) and substituting $\omega_m \pm 1/2\Delta\omega_{\text{FWHM}}$ for ω, we get:

$$\sin^2\left\{ m\pi \frac{\omega_m \pm \frac{1}{2}\Delta\omega_{\text{FWHM}}}{\omega_m} \frac{n(\omega)}{n(\omega_m)} \right\} = \frac{1}{B}. \tag{10.20}$$

In (10.20) we need to know the refractive index at three frequencies that are rather close together in high-quality resonators. Therefore, the error will be small (and evaluation of (10.20) much more convenient) if we neglect dispersion and set

$$n\left(\omega_m - \tfrac{1}{2}\Delta\omega_{\text{FWHM}}\right) = n\left(\omega_m + \tfrac{1}{2}\Delta\omega_{\text{FWHM}}\right) = n(\omega_m),\qquad(10.21)$$

so that (10.20) simplifies to

$$\sin^2\left\{m\pi \pm \tfrac{1}{2}m\pi\left(\Delta\omega_{\text{FWHM}}/\omega_m\right)\right\} = 1/B.\qquad(10.22)$$

If the cavity losses are sufficiently low, then $1/B$ will be a small number, and only a small error will be made if we replace the sine function by its argument (e.g., for $R > 0.9$ and $\alpha L < 0.05$, $B \geq 90$, $\sin(\ldots) \leq 0.01$, and the error made in the value of the argument will be below 0.2%). Due to the symmetry and periodicity of the $\sin^2()$ function, the $m\pi$ term and the $\pm$ sign can be safely omitted, leading to:

$$\Delta\omega_{\text{FWHM}} = \omega_m \frac{2}{m\pi\sqrt{B}}.\qquad(10.23)$$

10.2.2.2 Quality Factor Q

A quantity of great practical importance is the quality factor Q, which can be defined in a number of equivalent ways. An operational definition that allows direct measurement of Q is the relative peak width, which can also be directly derived from (10.23):

$$Q \equiv \omega_m/\Delta\omega_{\text{FWHM}} = \tfrac{1}{2}m\pi\sqrt{B}.\qquad(10.24)$$

For frequencies near resonance, (10.10) can be written as a Lorentzian function, in terms of the resonance frequency ω_m and Q only. Near resonance, we write:

$$\omega = \omega_m + \Delta\omega,\qquad(10.25)$$

with $\Delta\omega \ll \omega_m$. This allows us to again neglect dispersion, so that the argument in the sine function in (10.10) becomes similar to that in (10.22) with $\Delta\omega_{\text{FWHM}}$ replaced by $\omega_m - \omega$:

$$\sin^2\left(\varphi/2\right) = \sin^2\left(m\pi + m\pi\left(\omega_m - \omega\right)/\omega_m\right) \cong \left(m\pi\left(\omega_m - \omega\right)/\omega_m\right)^2.\qquad(10.26)$$

Inserting (10.26) into (10.10), and using (10.24), results in an expression, valid only for a narrow frequency range centred on a resonance peak, showing the Lorentzian shape of such peaks:

$$T\left(\omega\right) \cong \frac{T_0}{1 + B\left(\frac{m\pi}{\omega_m}\right)^2\left(\omega_m - \omega\right)^2} = T_0\frac{1}{1 + \left(\frac{2Q}{\omega_m}\right)^2\left(\omega_m - \omega\right)^2}.\qquad(10.27)$$

With (10.23), an equivalent expression can be found in terms of ω_m and $\Delta\omega_{\text{FWHM}}$.

10.2.3 Field Distribution

The operation of a Fabry–Perot resonator was introduced with Fig. 10.1, showing forward and backward propagating waves in the cavity. These travelling waves interfere to form standing waves, as shown schematically in Fig. 10.5. The order of such resonant modes is given by the integer m that occurs in (10.1).

Low-order resonances are usually easily identified by their frequencies, which are relatively wide apart. For higher-order modes, the relative frequency difference $(\omega_{m+1} - \omega_m)/\omega_m$ becomes smaller, making it more difficult to identify a particular resonant mode.

At resonance, the field strength inside the cavity may become much larger than the input field. Looking at the schematic of Fig. 10.3, it can be seen that the amplitude E_t of the output wave is a factor t smaller than E_{right} of the right-travelling wave inside the cavity. On the other hand, we know that for a lossless cavity at resonance $E_{\text{out}} = E_{\text{in}}$. The amplitude E_{left} of the left-travelling wave inside the cavity is a factor r smaller than E_{right}.

Intensity maxima in the cavity arise where right- and left-travelling waves are in phase so that they add up to $E_{\text{max}} = E_{\text{left}} + E_{\text{right}} = E_{\text{right}}(1 + r) = E_{\text{in}}(1 + r)/t$. Using $t^2 = T = 1 - R$, and $r^2 = R$, the intensity enhancement is calculated as

$$I_{\text{max,cavity}}/I_{\text{in}} = \left(E_{\text{max,cavity}}/E_{\text{in}}\right)^2 = \left(1 + \sqrt{R}\right)^2 / (1 - R) , \qquad (10.28)$$

which becomes quite large if R approaches 1. This phenomenon is often used for local enhancement of optical effects, in particular nonlinear optics. The enhancement is limited by optical loss. In the case of planar optics not only propagation loss due to absorption or scattering should be considered but also out-of-plane diffraction, especially with periodic structures such as gratings and photonic crystals.

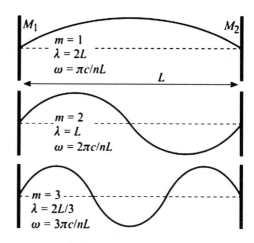

Fig. 10.5 Schematic picture of the lowest 3 longitudinal resonant modes in a Fabry–Perot resonator

10.2.4 Higher-Order Transversal Modes; Degeneracy

In guided-wave optics the light is confined in transversal directions (perpendicular to the direction of propagation) by dielectric waveguides. If the waveguide inside the Fabry–Perot cavity supports only a single guided mode (Fig. 10.6a), the analysis given above is valid if the refractive index is replaced by the effective index of the waveguide mode.

$$n_{\text{eff}} = \beta \lambda / (2\pi) \,, \tag{10.29}$$

where β is the modal propagation constant and λ is the wavelength in vacuum.

If the waveguide supports multiple modes, things get a bit more complicated. In the example of Fig. 10.6b, two modes are considered, having slightly different propagation constants β_0 and β_1, hence slightly different effective indices $n_{\text{eff},0}$ and $n_{\text{eff},1}$. The resonance conditions (10.1) for these modes will therefore be somewhat different, so that in general these modes will be in resonance at different fre-

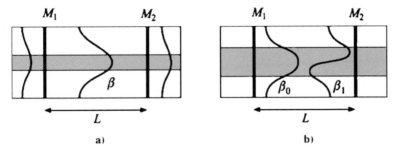

Fig. 10.6 Waveguide-based Fabry–Perot resonator; the mirrors are represented symbolically as *thick vertical lines.* **a)** single-mode waveguide; **b)** bimodal waveguide

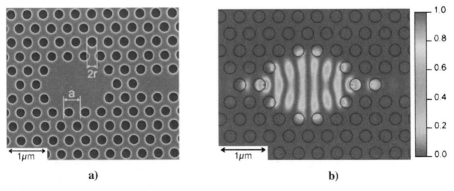

Fig. 10.7 Fabry–Perot-like resonator in a photonic crystal slab. **a)** Scanning electron microscope (SEM) image of the device that was fabricated (at IMEC, Leuven, Belgium) in a silicon on insulator (SOI) technology [5]; silicon ($n = 3.45$) top layer thickness 220 nm, lattice period $a = 440$ nm, air hole radius $r = 270$ nm. **b)** Intensity distribution of resonant mode at $\lambda = 1550$ nm, calculated with a 2D-finite-difference time-domain (FDTD) method, using an effective index method for determining the silicon slab effective index $n_{\text{eff}} = 2.9$ [14]

quencies. However, at a given frequency one transversal mode may resonate at order m_0, whereas the other mode resonates at a different order m_1. This occurs if $n_{\text{eff},0}/m_0 = n_{\text{eff},1}/m_1$.

In that case the standing wave pattern can be an arbitrary superposition of the two resonant modes, which are then said to be degenerate. In a more general case of nonrectangular two-dimensional resonators, the transversal (waveguide) and longitudinal (Fabry–Perot) modes cannot be determined independently, and it is a nontrivial task to calculate the composite modes. An example of such a resonator, in this case a cavity in a photonic crystal, schematically shown in Fig. 10.2c, can be seen in Fig. 10.7 together with a typical modal intensity distribution.

10.3 Spectral Analysis

Transmission and reflection spectra (or "drop" and "through" spectra in the case of ring resonators) represent the important basic filtering or add-drop multiplexing functionality of these devices. These spectral filter characteristics can usually be measured conveniently using either a white-light source in combination with a spectrometer, or a tuneable laser with a suitable photodetector.

10.3.1 Determining Cavity Loss and Mirror Reflectivities

An important parameter that is not always easy to measure directly is the loss of a mode in a waveguide. Often other origins of wave attenuation, such as fibre-chip coupling loss, and unknown cavity mirror reflectivities or coupling coefficients make propagation loss measurements difficult. Knowledge of propagation loss is important for the operation of resonators (Q), and for almost all other applications.

The spectral response curves of a Fabry–Perot-type resonator provide sufficient information to calculate propagation loss α as well as the reflectivities R_i (or coupling coefficients κ_i in the case of ring resonators). The equations required to obtain the values of α and R_i can be found from the expressions that give the on- and off-resonance power ratios of the transmitted and reflected spectra.

The on/off resonance power ratio M_T of the signal that is transmitted by the Fabry–Perot is defined as the ratio of the transferred power at resonance ($\varphi = m2\pi$) to the transferred power at anti-resonance ($\varphi = m2\pi + \pi$):

$$M_T = \frac{T_{\text{OnRes}}}{T_{\text{OffRes}}} = \frac{\left.\frac{T_0}{1+B\sin^2(\varphi/2)}\right|^{\varphi=2\pi}}{\left.\frac{T_0}{1+B\sin^2(\varphi/2)}\right|^{\varphi=\pi}} = 1+B = \left(\frac{1+\sqrt{R_1 R_2}\,e^{-2\alpha L}}{1-\sqrt{R_1 R_2}\,e^{-2\alpha L}}\right)^2 , \quad (10.30)$$

which can be rewritten as:

$$\sqrt{R_1 R_2}\,e^{-2\alpha L} = \frac{\sqrt{M_T}-1}{\sqrt{M_T}+1} . \quad (10.31)$$

Similarly the on/off resonance reflected power ratio M_R is defined as:

$$M_R = \frac{R_{\text{OffRes}}}{R_{\text{OnRes}}} = \frac{1 - \frac{C}{1+B\sin^2(\phi/2)}\Big|^{\phi=\pi}}{1 - \frac{C}{1+B\sin^2(\phi/2)}\Big|^{\phi=2\pi}} = \frac{1 - \frac{C}{1+B}}{1 - C} = \left(\frac{\frac{\sqrt{R_1}+\sqrt{R_2}\,e^{-2\alpha L}}{1+\sqrt{R_1 R_2}\,e^{-2\alpha L}}}{\frac{\sqrt{R_1}-\sqrt{R_2}\,e^{-2\alpha L}}{1-\sqrt{R_1 R_2}\,e^{-2\alpha L}}}\right)^2.$$

$$(10.32)$$

In order to obtain an explicit expression for α, we first take the square root of the product of (10.30) and (10.32):

$$\sqrt{M_R M_T} = \frac{\sqrt{R_1}+\sqrt{R_2}\,e^{-2\alpha L}}{\sqrt{R_1}-\sqrt{R_2}\,e^{-2\alpha L}}, \qquad (10.33)$$

which can be rearranged to give:

$$\frac{\sqrt{R_2}}{\sqrt{R_1}}e^{-2\alpha L} = \frac{\sqrt{M_R M_T}-1}{\sqrt{M_R M_T}+1}. \qquad (10.34)$$

For Fabry–Perot cavities with $R_1 = R_2 = R$ this reduces to:

$$e^{-2\alpha L} = \frac{\sqrt{M_R M_T}-1}{\sqrt{M_R M_T}+1}, \qquad (10.35)$$

which, combined with (10.31) can be used to find the values of α and R.

10.4 Transmission-Based SNOM (T-SNOM)

10.4.1 Origin of the Method

Scanning near-field optical microscopy (SNOM, NSOM) is a well-known method for probing optical field distributions with sub-wavelength resolution [15–17]. The basic principle is shown in Fig. 10.8. An optical fibre which is modified to have an extremely sharp tip (diameter in the order of 100 nm) is brought in close proximity to the optical field to be probed. For obtaining the highest resolution, the tip is often clad with metal. At the very end of the tip a tiny hole is made with a diameter ranging from a few tens to a few hundreds of nm. It would typically be positioned in the evanescent field tail of guided waveguide modes. A small amount of light is scattered by the presence of the tip and collected by the fibre that guides the photons to a sensitive optical detector.

By scanning the tip across the surface of the device under analysis, and recording the amount of light captured by the fibre at each position, an image can be constructed that may represent the local optical intensity distribution. A problem with this method is that the probe may affect the optical field distribution. This is especially problematic when analysing microcavities, since the original structure will be significantly modified by the mere presence of the dielectric fibre material and its metal cladding. Also the light scattering process that is essential for SNOM

Fig. 10.8 Fibre tip of scanning near-field optical microscope (SNOM) interacting with the evanescent field of a resonant mode of a microcavity in a photonic crystal slab (cross-sectional view)

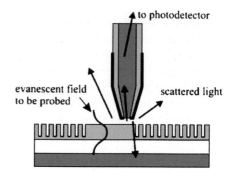

operation may be an important loss mechanism which can strongly affect Q of a resonant structure. Only a small fraction of the scattered light will be captured by the fibre and guided towards the detector, so that sensitivity and noise issues may easily arise.

In our work we originally intended to use SNOM measurements for characterizing the microresonators that we designed and fabricated. Because of the considerations mentioned above, we wanted to test the loading effect of a SNOM-like probe on a microresonator.

In order to check the effects of a probe, we measured the transmission through the Fabry–Perot-like resonator while scanning the probe over the device. For convenience we used an atomic force microscope (AFM) with a Si_3N_4 or a Si tip with a radius in the order of 10 nm, which is significantly smaller than a typical SNOM tip. Still, we observed a strong effect of the probe on the light transmission through the resonator. We decided to turn this drawback into an advantage, as shown below.

10.4.2 Set-Up

It should be expected that the strongest interaction of the nanoprobe with the optical field will occur at locations with highest field intensity. The effect of the probe may

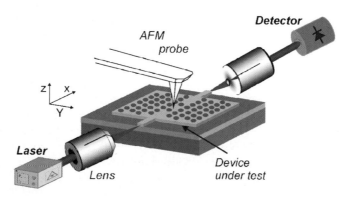

Fig. 10.9 Principle of transmission-based scanning near-field optical microscopy (T-SNOM) [14]

be both loss (due to scattering or waveguiding inside the probe), which mainly affects Q, and detuning of the cavity (the probe forms a "dielectric load" or becomes part of the cavity, thus changing its shape and effective index), bringing it into or out of resonance for a given frequency. If we measure the transmitted light through the cavity for each position of the probe, we can map out the locations of strong interaction on the surface of the resonator. The setup is shown schematically in Fig. 10.9. Since the image is obtained by scanning the optical near field with a probe while the signal is derived from the optical transmission through the device under test, we named this method transmission-based scanning near-field optical microscopy (T-SNOM).

10.4.3 Image Construction – Contrast

The method has two main advantages compared to conventional SNOM. The measurement is in fact a comparison between the undisturbed situation (no probe interacting with the optical field) and the disturbed situation, whereas in SNOM only photons are detected in the disturbed situation. A low optical SNOM signal may therefore be caused either by a small intensity in the undisturbed case, or it can be due to the disturbance by the probe itself that may cause the resonator to go out of resonance, which can only happen when the probe is at a location of high undisturbed intensity. With T-SNOM we actually exploit this disturbance, giving rise to less ambiguity in the result: a strong change of the optical signal will always be caused at a probe position with high undisturbed intensity. The second advantage is that the T-SNOM optical signal can be much stronger than that of SNOM. A conventional SNOM probe only detects the photons that are captured by the fibre. For obtaining high spatial resolution the opening in the tip should be small and the tunnelling efficiency of photons through this aperture becomes low. This may give rise to noise problems and, hence, long signal integration times. By contrast, a T-SNOM image is typically obtained with the undisturbed device in resonance, so that a large optical signal is transmitted to the detector. The probe interaction causes sharp dips in this transmission. This typically results in a better signal to noise ratio. A drawback of T-SNOM is that it works well only for resonant structures or other interferometric devices. The loss induced by the tiny probe on a large waveguide will be rather small, resulting in low contrast images. On the other hand, this is a situation where SNOM is generally reliable, so the two methods are mutually complementary. It should be mentioned that SNOM is a more versatile instrument since it does not depend on waveguiding structures and it can also be used in a mode that the fibre tip is used for locally illuminating a sample.

Using an AFM scanning head makes it possible to generate topographic data simultaneously with the optical data. Traditional SNOM provides similar facilities. The result of a T-SNOM measurement of the photonic crystal resonator shown in Fig. 10.7a is given in Fig. 10.10. The AFM (with a 10 nm radius Si_3N_4 tip, shown in Fig. 10.11a) was operating in contact mode with closed-loop control.

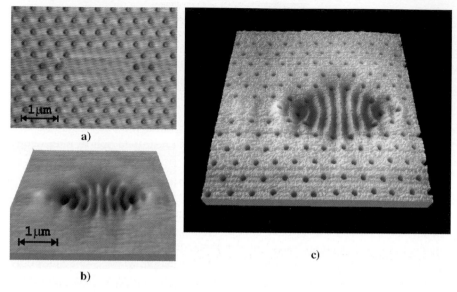

a)

b)

c)

Fig. 10.10 T-SNOM imaging of the photonic crystal resonator shown in Fig. 10.7a. **a)** Topography measured by AFM. **b)** Optical transmission through the resonator at the undisturbed resonance wavelength $\lambda_r = 1539.25$ nm, as a function of AFM tip position; *dark regions* correspond to low light transmission, hence strong interaction of the AFM tip with the optical field. **c)** Composite image, combining data of **a)** and **b)**, showing the location of the optical field distribution with respect to the resonator geometry [14]

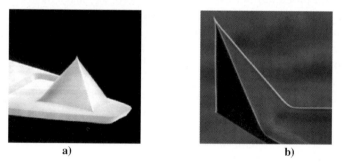

a)   b)

Fig. 10.11 SEM images of AFM probes used. **a)** Silicon nitride with top angle 23° and tip radius 10 nm. **b)** Silicon with 17° and tip radius 7.5 nm

In our experiments the scanning speed was limited by the response time of the photodetector. The image of Fig. 10.10, having a resolution of 256×256 pixels on a 20 nm grid, was taken in approximately 45 s.

Using this small dielectric tip with a refractive index ($n = 2.0$) smaller than the silicon slab effective index ($n_{\mathrm{eff}} \cong 2.9$), the dominant effect of the probe on the resonator was detuning (shifting the resonant wavelength to larger values) rather than inducing scattering loss. This is illustrated by Fig. 10.12 showing a T-SNOM image of the same resonator, taken at a slightly larger wavelength of 1541.5 nm,

Fig. 10.12 T-SNOM image similar to Fig. 10.10, but wavelength, now 1541.5 nm (off resonance). Detuning by the AFM tip could bring it into resonance again, resulting in inverted image contrast [14]

where the device is off resonance. The tuning action of the tip brings it on resonance again, now producing increased light transmission at locations of strong mechano-optical interaction. This resulted in an image with inverted contrast.

10.4.4 Measurements of Nanomechano-Optical Interaction

We measured the effect of small horizontal and vertical tip displacements on the tuning and on the quality factor of the resonator. Most of these experiments were done with a Si tip, see Fig. 10.11b, which has a stronger optical effect than a Si_3N_4 one, because of its higher refractive index ($n = 3.45$).

Transmission spectra for different tip positions in contact with the resonator surface are shown in Fig. 10.13a. Positioning the tip right at the hot spot (labelled A in the inset of the figure), almost completely kills the resonance. The dependence of the transmission on the distance of the tip from the hot spot is shown in Fig. 10.13b. For this structure the vertical dependence is especially strong: less than 100 nm displacement of the tiny tip results in more than 50% modulation of the transmission: this is a truly nanomechano-optical effect.

The tuning effects of the Si and Si_3N_4 AFM tips have been compared in Fig. 10.14. As expected, the Si tip has a much stronger effect than the Si_3N_4 one, especially on the cavity loss. For tuning purposes the Si_3N_4 tip is preferred.

10.4.5 Modelling T-SNOM

T-SNOM has been modelled in order to verify our intuitive understanding, and its operation has been simulated [18]. The purpose was to check whether the T-SNOM image does represent the standing wave pattern in a resonator with sufficient accuracy. A full and truly accurate model of T-SNOM should be three-dimensional. Because this is not practical with the currently available computing resources, the models were restricted to two dimensions. Two models have been used, a cross-sectional view in a plane perpendicular to the photonic crystal slab ("side view"), and an effective-index-based "top view", see Fig. 10.15.

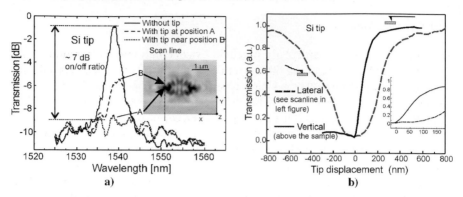

Fig. 10.13 Nanomechano-optical interaction. **a)** Transmission spectra for the situation without probe; with a Si probe positioned at the antinode labelled *A*; and at a location *B*, as indicated in the *inset*. **b)** Transmission (at $\lambda = 1539.25$ nm) versus Si tip displacement in the z and y direction (see the *dotted* "scan line" in the *inset* of the *left* figure) [14]

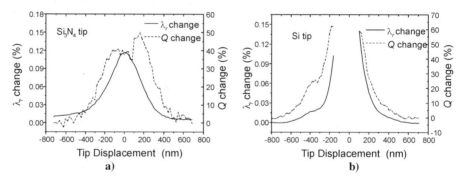

Fig. 10.14 Comparison of different tips. Nanomechano-optical tuning of the resonance wavelength and Q by lateral displacement with respect to hot spot A, in contact along the scanline shown in the inset of Fig. 10.13a. **a)** Si_3N_4 probe; **b)** Si probe. Close to the hot spot no data could be obtained because the transmission had dropped below the noise level [14]

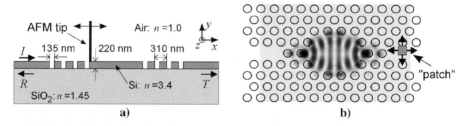

Fig. 10.15 Two-dimensional T-SNOM models. **a)** Side view. The resonator is modelled as a slab waveguide cavity enclosed by two short Bragg gratings; the AFM-probe is modelled as a semi-infinite plane perpendicular to and in contact with the slab. **b)** Top view. The slab waveguide outside the photonic crystal holes is modelled by its effective index; the probe is represented by a small "patch" of material having an increased refractive index

10.4.5.1 Side-View Model

The side-view model is particularly suited for evaluation using an eigenmode propagation method, because of the limited number of different cross-sections, all with rectangular-shaped elements, making it possible to subdivide the region in a small number of horizontal layers and vertical slices. In this case the so-called quadrilateral eigenmode propagation (QUEP) [19] has been selected.

The model was first used to calculate the transmission, reflection and scattering spectra for several probe positions. The results for the undisturbed situation and the maximally disturbed case are shown in Fig. 10.16. As the next step, the model was recalculated at the undisturbed resonance wavelength, for many positions of the probe along the length of the resonator, thus simulating the scanning action of the tip. The resulting transmission versus tip position graph is shown in Fig. 10.17a. The

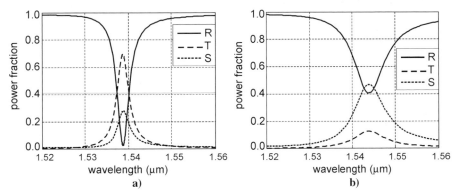

Fig. 10.16 Calculated reflection (R), transmission (T) and scattered power (S) spectra; $R + T + S = 1$. **a)** Unperturbed resonator. **b)** Silicon probe (50 nm wide) at "hot spot" location [18]

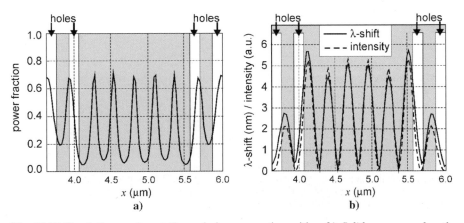

Fig. 10.17 Simulation results. **a)** Transmission versus tip position. **b)** *Solid curve*: wavelength shift as calculated by combining the transmission graph (this figure, part **a**.) with the transmission spectrum of 10.16a. *Dashed curve*: calculated intensity distribution in the resonator

changes in transmission in this graph have been translated into equivalent shifts of the resonance wavelength, using the previously measured spectra of Fig. 10.16. This equivalent wavelength shift has been plotted together with the calculated intensity distribution in Fig. 10.17b. The two graphs match quite well, indicating that indeed T-SNOM data can be used for determining field distributions inside this type of resonators.

Another calculation, illustrated in Fig. 10.18, simulates the effect of probe height variation. In contact ($y = 0$), the transmission is low (strong detuning), and as the tip is raised the transmission increases. It is interesting to see that at a distance of approximately $0.5\,\mu m$, there is a maximum transmission which is higher than in absence of the probe (large y). This is somewhat unexpected because the wavelength was selected for resonance without the probe. In fact, this simulation reproduces an effect that can also be seen in the measurement data of Fig. 10.10. At both the left-hand and the right-hand sides of the cavity there is a small region with a somewhat brighter colour than the background, signalling an increased transmission as well. The calculated field distributions without the tip and with the tip at the "critical height" of $0.5\,\mu m$ (Fig. 10.19), show that the presence of the tip at that location

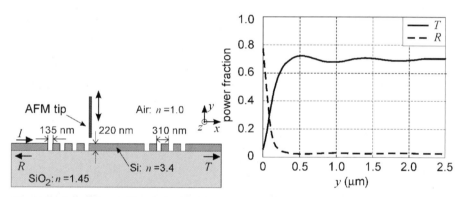

Fig. 10.18 Simulation of probe height variation. **a)** Configuration: the z-position of the probe is at the hot spot; y is varied. **b)** Calculated transmission and reflection versus probe height [18]

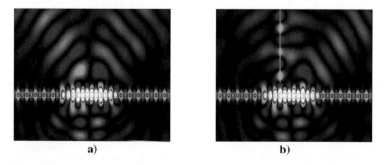

Fig. 10.19 Calculated field distributions at the undisturbed resonance wavelength. **a)** Without probe. **b)** With probe at $0.5\,\mu m$ height above hot spot [18]

somewhat suppresses the radiation loss due to out-of-plane diffraction. The space between the resonator surface and the probe tip acts as an additional cavity, which is in antiresonance at the critical height.

A similar model has also been used for modelling the dynamic behaviour of T-SNOM when it is operated in so-called tapping mode (introducing a vertical vibration of the tip) [20]. The strongly nonlinear relationship between transmission and tip height requires a suitable weighing function in order to obtain the correct time-averaged transmitted power. Tapping mode operation is beneficial for tip durability and for reducing the risk of producing artefacts by dragging along dust particles with the tip, but it reduces the lateral resolution.

10.4.5.2 Top-View Model

The top-view model, which is illustrated in Fig. 10.15b, is numerically somewhat more challenging than the side-view model because of the lack of straight interfaces. This makes it impractical to use an eigenmode propagation method, because the number of slices should be chosen very large in order to obtain a good approximation to the geometry of the problem. Because eigenmode calculations for all these slices are expensive, another method – the two-dimensional finite-difference time-domain (FDTD) method – usually considered as "brute force" comes out more favourably.

In this model, the vertical structure is accounted for by a local effective index. The silicon slab making up the "backbone" of the structure is represented by the effective index that it would have for a guided optical slab mode; the photonic crystal holes are approximated by an effective index of 1 (air), and the AFM probe is accounted for by simulating its dielectric loading effect through a locally increased effective index (the "patch" in Fig. 10.15b). The results, shown in Fig. 10.20, confirm that T-SNOM measurements in Fabry–Perot-like resonators provide a good approximation of the actual intensity distribution of the standing wave pattern inside the resonator.

In conclusion, T-SNOM is a useful extension of the available techniques for characterizing resonant optical structures. It provides a way to map out the intensity distribution inside a resonator, thus resolving the mode structure. It helps to identify

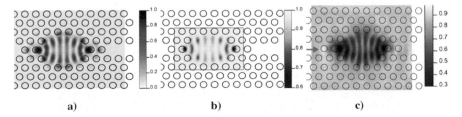

Fig. 10.20 Top-view model, comparison to measurement. **a)** Calculated intensity distribution. **b)** Simulated T-SNOM response by recalculating the model many times for different "patch" locations in order to represent the T-SNOM probe scanning. **c)** Actual T-SNOM measurement data overlaid with structure data [18]

"hot spots" that are particularly useful for nanomechano-optical interaction that can be exploited for sensors and extremely compact and low power optical switching devices. By exploiting an effect that is harmful to other SNOM-type measurements, T-SNOM avoids some of the known uncertainties, while operating with relatively high speed and with good signal-to-noise-ratio. Extensive modelling has confirmed the validity of T-SNOM measurements on Fabry–Perot-like microresonators.

10.5 Far-Field Scattering Microscopy (FSM)

Although near-field optical measurements, like those described in the preceding section, currently provide the highest available spatial resolution, there are situations where such methods are impractical. This is especially the case if large areas need to be imaged, and/or if the time required for collecting the image data by serial scanning becomes prohibitive, or if the optical structure has a thick cladding layer which makes the evanescent field inaccessible.

An alternative method for looking inside an optical device is using the light that is scattered from the waveguiding regions. A condition for this to work is that small sub-wavelength-size scatter centres should be randomly and homogeneously distributed in the waveguiding and/or cladding materials (Rayleigh scattering regime). In that case, which occurs for many practical optical waveguide technologies, the scattered light intensity is proportional to the guided light intensity. Even in the case of materials showing negligible Rayleigh scattering, the method can sometimes be used if discrete scatterers such as waveguide discontinuities are present at locations of interest in the structure. Then, again the scattered light intensities at such intended discrete scatterers is a measure for the local guided light intensities, so that scattered-light images can provide information about power ratios at selected locations inside the structure.

Because the scattered light is to be collected by a lens at a distance much larger than the wavelength, the maximum attainable resolution is restricted to the order of a wavelength by the diffraction limit. Although this precludes the imaging of the field structure inside microcavities, such as shown in Fig. 10.10, other useful applications exist, such as waveguide loss measurements [21] and mapping of intracavity power of a microring resonator [22]. In this section we will discuss the application of the method to measure intensity distributions inside a waveguide grating [12].

10.5.1 Set-Up

The experimental set-up is schematically shown in Fig. 10.21. Our specific FSM implementation was designed to facilitate the association of scattered light images, that are captured by a CCD camera, with simultaneously measured transmission spectra using a tuneable laser and a photodetector.

Requirements for the camera are linearity and high sensitivity in the wavelength region of interest. In our case the camera could resolve 320×240 pixels with 12 bits per pixel. In combination with a microscope lens, at the highest magnification each

Fig. 10.21 Principle of far-field scattering microscopy (FSM), allowing to associate scattered-light images with specific features in device transmission spectra [12]

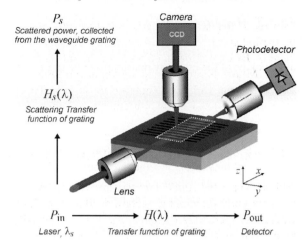

pixel maps to a 800×800 nm area on the object under test. The laser could be tuned in the range of 1470–1600 nm with 1 pm wavelength resolution.

The device under investigation is a waveguide grating such as drawn schematically in Fig. 10.22. This structure can be considered as a Fabry–Perot-like resonator. In the non-grated waveguide sections the usual guided modes can propagate. In the grating section the eigenmodes are the Floquet–Bloch modes of the periodic structure. The transition between grated and ungrated sections are optical discontinuities which partially reflect the guided modes. Hence, a finite-length grating acts as a Fabry–Perot resonator for the Floquet–Bloch modes.

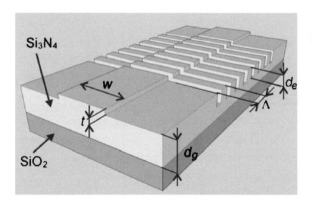

Fig. 10.22 Schematic drawing of shallow ridge waveguide into which a grating has been etched. Materials: Si_3N_4 guiding layer, $n = 1.981$; SiO_2 bottom cladding, $n = 1.445$; top cladding either air ($n = 1$) or polymer ($n = 1.5$). Dimensions: $w = 7\,\mu m$; $d_g = 275$ nm; $t = 5$ nm; $\Lambda = 470$ nm; $d_e = 50$ nm; device lengths of 500, 1000, and 2000 periods have been used

10.5.2 Measurements

A typical set of measurement data is shown in Fig. 10.23. The camera images show an envelope of the intensity distribution; the high spatial frequencies of the Floquet–Bloch modes cannot be resolved with this method.

10.5.3 Interpretation

The large scale envelope variations (orders of magnitude larger than the wavelength of the light) may be somewhat surprising. This phenomenon can be explained from long-range phase relationships between the Fourier components of the (propagating) Floquet–Bloch mode on the one hand and the overall propagation phase of the field, using a coupled-mode model. These features are reproduced by model calculations using a two-dimensional bidirectional eigenmode propagation method, shown in Fig. 10.24 [12, 13].

As described in [12], the amplitude enhancement in the resonance peak can be used for calculating the group index of slow light that occurs in the regime close to the photonic band edge. For the structure considered here, a maximum group index of 14 was found using this method. Among the several slow-light characterization methods discussed in [12], this method was found to be among the more reliable ones. Related to this also Q of such resonances could be conveniently determined.

In conclusion, the facility to simultaneously obtain both spectral transmission data and spatial scattered light distributions from a finite-length periodic structure

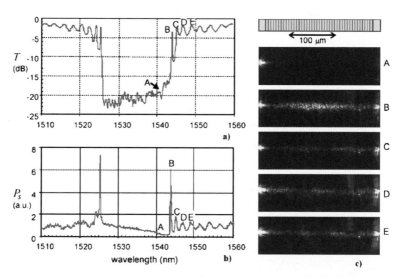

Fig. 10.23 Combined spectral and far-field scattering measurements. **a)** Normalised transmission spectrum of the waveguide grating, showing a stop-band. **b)** Spatially averaged scatter spectrum, derived from camera images. **c)** Typical scattered-light images at 5 different wavelengths, indicated by letters *A–E*, also shown in graphs **a)** and **b)** *A*: inside the stop-band, *B–E* Fabry–Perot like grating resonances of order 1-4. Top figure symbolically shows the grating location and a length scale

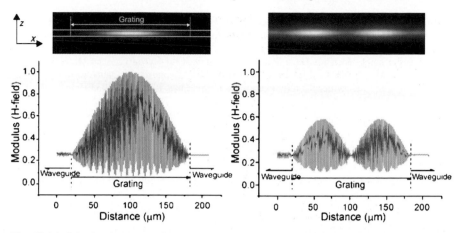

Fig. 10.24 Calculated magnetic field amplitude distribution along the length of the grating. *Top*: field distribution in the grated waveguide cross-section (see coordinate axes in Fig. 10.21). *Bottom*: field distribution along the centre line of the waveguiding layer. *Left*: lowest order mode (corresponding to label *B* in Fig. 10.23); *right*: second order mode (label *C*). It can be seen that the large-scale distribution is actually an envelope of the Floquet–Bloch mode amplitude

was found helpful for understanding the phenomena near the band edge in such structures. The long-range periodic intensity variation of Fabry–Perot like resonances of the Floquet–Bloch modes that were theoretically predicted could be observed with the far-field scattering microscopy method.

10.6 Summary

Three classes of characterization techniques for optical microresonators have been described, viz. analysis of input-output characteristics, transmission-based SNOM, and far-field scattering microscopy. Their typical application fields and specific advantages and disadvantages have been discussed.

Acknowledgements The contributions of Anton Hollink, Kees van der Werf, Hugo Hoekstra, Ronald Dekker, Didit Yudistira, Remco Stoffer, Henk van Wolferen, Marco Gnan, Chris Roeloffzen, Wim Bogaerts, Vinod Subramaniam, Frans Segerink are gratefully acknowledged. This work was supported by NanoNed, a national nanotechnology programme coordinated by the Dutch ministry of Economic Affairs, and was also supported by the European Network of Excellence ePIXnet.

References

1. Morand A., Zhang Y., Martin B., Phan Huy K., Amans D., Benech P., Verbert J., Hadji E., Fédéli J.-M.: Ultra-compact microdisk resonator filters on SOI substrate. Opt. Express **14**, 12814–12821 (2006)

2. Klein E.J., Geuzebroek D.H., Kelderman H., Sengo G., Baker N., Driessen A.: Reconfigurable optical add-drop multiplexer using microring resonators. IEEE Photon. Technol. Lett. **17**, 2358–2360 (2005)

3. Hendrickson J., Richards B.C., Sweet J., Mosor S., Christenson C., Lam D., Khitrova G., Gibbs H.M., Yoshie T., Scherer A., Shchekin O.B., Deppe D.G.: Quantum dot photonic-crystal-slab nanocavities: Quality factors and lasing. Phys. Rev. B. **72** 193303(1-4) (2005)

4. Notomi M., Shinya A., Mitsugi S., Kira G., Kuramochi E.,Tanabe T.: Optical bistable switching action of Si high-Q photonic crystal nanocavities. Opt. Express **13**, 2678–2687 (2005)

5. Bogaerts W., Baets R. Dumon P., Wiaux V., Beckx S., Taillaert D., Luyssaert B., Van Campenhout J., Bienstman P., Van Thourhout D.: Nanophotonic waveguides in silicon-on-insulator fabricated with CMOS technology. J. Lightwave Technol. **23**, 401–412 (2005)

6. Niehusmann J., Vörckel A., Bolivar P.H., Wahlbrink T., Henschel W., Kurz H.: Ultrahigh-quality-factor silicon-on-insulator microring resonator. Opt. Lett. **29**, 2861–2863 (2004)

7. Foresi J.S., Villeneuve P.R., Ferrera J., Thoen E.R., Steinmeyer G., Fan S., Joannopoulos J.D., Kimerling L.C., Smith H.I., Ippen E.P.: Photonic-bandgap microcavities in optical waveguides. Nature **390**, 143–145 (1997)

8. Notomi M., Shinya A., Mitsugi S., Kuramochi E., Ryu H.Y.: Waveguides, resonators and their coupled elements in photonic crystal slabs: Opt. Express **12**, 1551–1561 (2004)

9. Akahane Y., Asano T., Song B.S., Noda S.: Fine-tuned high-Q photonic-crystal nanocavity. Opt. Express **13**, 1202–1214 (2005)

10. Sauvan C., Lecamp G., Lalanne P., Hugonin J.P.: Modal-reflectivity enhancement by geometry tuning in photonic crystal microcavities. Opt. Express **13**, 245–255 (2005)

11. Little B.E., Chu S.T., Pan W., Kokubun Y.: Microring resonator arrays for VLSI photonics. IEEE Photon. Technol. Lett. **12**, 323–325 (2000)

12. Hopman W.C.L., Hoekstra H.J.W.M., Dekker R., Zhuang L., de Ridder R.M.: Far-field scattering microscopy applied to analysis of slow light, power enhancement, and delay times in uniform Bragg waveguide gratings. Opt. Express **15**, 1851–1870 (2007)

13. Hoekstra H.J.W.M., Hopman W.C.L., Kautz J., Dekker R., de Ridder R.M.: A simple coupled mode model for near band-edge phenomena in grated waveguides. Opt. Quantum Electron. **38**, 799–813 (2006)

14. Hopman W.C.L., Van Der Werf K.O., Hollink A.J.F., Bogaerts W., Subramaniam V., De Ridder R.M.: Nano-mechanical tuning and imaging of a photonic crystal micro-cavity resonance. Opt. Express **14**, 8745–8752 (2006)

15. Betzig E., Trautman J.K., Harris T.D., Weiner J.S., Kostelak R.L.: Breaking the diffraction barrier: optical microscopy on a nanometric scale. Science **251**, 1468–1470 (1991)

16. Hsu J.W.P.: Near-field scanning optical microscopy studies of electronic and photonic materials and devices. Mater. Sci. Eng. Rep. **33**, 1–50 (2001)

17. Gersen H., Karle T.J., Engelen R.J.P., Bogaerts W., Korterik J.P., van Hulst N.F., Krauss T.F., Kuipers L.: Real-space observation of ultraslow light in photonic crystal waveguides. Phys. Rev. Lett. **94**, 073903(1-4) (2005)

18. Hopman W.C.L., Stoffer R., de Ridder R.M.: High-resolution measurement of resonant wave patterns by perturbing the evanescent field using a nanosized probe in a transmission scanning near-field optical microscopy configuration. J. Lightwave Technol. **25**, 1811–1818 (2007)

19. Hammer M.: Quadridirectional eigenmode expansion scheme for 2D modeling of wave propagation in integrated optics. Opt. Commun. **235**, 285–303 (2004)

20. Hopman W.C.L., Van Der Werf K.O., Hollink A.J.F., Bogaerts W., Subramaniam V., De Ridder R.M.: Experimental verification of a simple transmission model for predicting the interaction of an AFM-probe with a photonic crystal micro-cavity in tapping mode operation. IEEE Photon. Technol. Lett. **20**, 57–59 (2008)

21. McNab S.J., Moll N., Vlasov Y.A.: Ultra-low loss photonic integrated circuit with membrane-type photonic crystal waveguides. Opt. Express **11**, 2927–2939 (2003)

22. Klunder D.J.W., Tan F.S., Van Der Veen T., Bulthuis H.F., Sengo G., Docter B., Hoekstra H.J.W.M., Driessen A: Experimental and numerical study of SiON microresonators with air and polymer cladding. J. Lightwave Technol. **21**, 1099–1110 (2003)

11 On SNOM Resolution Improvement

Tomasz J. Antosiewicz[1], Marian Marciniak[2], and Tomasz Szoplik[1]

[1] University of Warsaw, Faculty of Physics, Pasteura 7, 02-093 Warsaw, Poland
[2] National Institute of Telecommunications, Department of Transmission and Optical Technologies, Szachowa 1, 04-894 Warsaw, Poland

Abstract. Spatial transversal resolution of scanning near-field optical microscopes depends on the distance between a sample and the aperture of a tip and aperture diameter. When the tip aperture – sample distance is kept constant due to the shear-force tuning fork technique then the detectable amount of light passing through the probe decides on the smallest aperture diameter. Energy throughput is limited by the aperture diameter which is smaller than the cut-off diameter $D = 0.6\lambda/n$ of modes guided in a tapered fibre. Beyond the cut-off the propagation vector becomes imaginary and only evanescent waves reach the orifice. In simulations using the finite-difference time-domain method we develop a concept of enhanced light transmission through tapered-fibre metal-coated corrugated tips. Corrugation of the interface between the fibre core and metal coating, which is structured into parallel grooves of different profiles curved inward the core, introduces efficient photon–plasmon coupling. Corrugated tips may lead to better than 20 nm SNOM resolution.

Key words: scanning near-field optical microscope – SNOM; SNOM resolution; SNOM probes; photon–plasmon coupling; tapered-fibre metal-coated corrugated SNOM probes

11.1 Introduction

In the classical theory of diffraction the size of the apertures was greater than the wavelength, the screen was not conductive and the fields were treated as scalar values. These limiting assumptions were the reasons behind Bethe's work [1], in which he analysed diffraction of electromagnetic waves on small holes in terms of vector theory. He was motivated by interest in microwave radiation of a small hole in a cavity. A sub-wavelength pinhole in a perfectly conducting thin plane screen radiates into the near-field with intensity proportional to $(a/\lambda)^4$, where $a < \lambda$ is the hole diameter. The result came from calculation of fields from bogus magnetic charges and currents in the plane of the hole. For the case of field distribution in and near the pinhole Bouwkamp [2] adjusted this result by ensuring the continuity of the electric field. The next step in the understanding of diffraction on tiny holes was made Leviatan [3], who calculated the diffracted field assumxing quasi-static surface magnetic currents and quasi-static surface electric charges distributed in the pinhole plane with $a = \lambda/10$. He showed, that at distances bigger than $\lambda/2$ radiation of a small aperture can be approximated as a field of a dipole. These considerations on near-field diffracted light on small holes became important in connection with a new idea in microscopy.

For hole-object distances a few times smaller than a, an object may deform the aperture radiation field. Interaction of illuminating and scattered light remains an issue in the characterization of photonic crystals. Ways to alleviate this problem are discussed in the previous Chaps. 9 and 10.

A breakthrough in microscopy based on light collecting apertures was introduced by Synge [4], who proposed imaging by the use of a sub-wavelength aperture scanning a surface at a small distance. After years of oblivion his idea has led to the development of scanning near-field microscopy.

Microwave imaging with resolution beyond the diffraction limit came into reality in the seventies [5]. The first demonstration of optical resolution $\lambda/20$ [6] resulted in the development of the scanning near-field optical microscope (SNOM) [7]. In the last 20 years near-field microscopy techniques were developed to meet specific needs in the fields of photonic crystals, biology, plasmonics and nanooptics [8, 9].

Near-field techniques are based on imaging through sub-wavelength apertures and make use of both propagating and evanescent waves and light–plasmon coupling [10–12]. Near-field illumination and far-field detection is employed in aperture SNOMs working in scattering mode and field-enhanced SNOMs based on Raman scattering. In turn, far-field illumination and near-field detection is employed in scanning tunnelling optical microscopes (STMs) and in SNOMs working in collection mode. In Chap. 10 an efficient combination of SNOM and atomic force microscope (AFM) measurements called transmission-based SNOM is reported which leads to single nanometres resolution [13].

In the following sections we address the most important issues which helped to increase SNOM resolution. We start with a description of light transmission through sub-wavelength size holes in surface plasmon (SP) supporting metal layers [14–16]. In fact, this enhanced transmission was discovered later than a metal coated tapered fibre SNOM probe [6]. Recent progress in plasmonics assures that improvement of SNOM resolution is possible. Then we present a new description of emission of light by metal-coated probes that is consistent with experimental results. In the following sections we summarize the state-of-the-art in the techniques of SNOM apertureless tapered metal tips and tapered-fibre metal-coated aperture probes [17]. In the last section we present a tapered-fibre metal-coated aperture probe which due to corrugations of the core-coating interface increases efficiency of photon–plasmon coupling. The idea promises more accurate optical characterization techniques for research in photonic crystals, plasmonics, and metamaterials [18, 19].

11.2 Transmission of Light Through Sub-wavelength Apertures and Aperture Arrays in Metal Films

Transmission through arrays of holes with a constant holes-to-screen area ratio and varying hole diameters (that is hole number) was studied in the microwave spectral range with the application to frequency selective surfaces. In these experiments 100% transmission was observed, however, the holes in question had relatively big radii in comparison to the wavelength and transmission for individual holes was large. Recently, extraordinarily large light transmission through arrays of holes of sub-wavelength radii normalized to their area was discovered [14]. In the experiment the intensity of observed spectral maxima of transmitted light was more than twice larger than those impinging on the apertures. The increased transmission is attributed to the excitation of surface plasmons-polaritons (SPPs) on the metal film. SPPs are efficiently excited, when their momenta are equal to the sum of the momenta of light and the grating:

$$
k_{sp}(\omega) = \frac{\omega}{c} \sqrt{\frac{\varepsilon_m(\omega)\varepsilon_d}{\varepsilon_m(\omega)+\varepsilon_d}} = nk_0(\omega)\sin(\theta) \pm pg ,
\tag{11.1}
$$

where k_{sp} is the SP wavevector; the metal dielectric function is complex $\varepsilon_m(\omega) = \varepsilon'_m(\omega) + i\varepsilon''_m(\omega)$, where $|\varepsilon'_m| \gg |\varepsilon''_m|$; ε_d and n are the dielectric constant and refractive index of a lossless material surrounding the metal layer; $k_0 = 2\pi/\lambda$ is the wavevector of incident light in vacuum; θ is the angle of incidence; p is an integer and $g = 2\pi/b$ is the wavevector of a 1-D array of period b. For perpendicular illumination the generated SPPs depend only on that period, as light contributes no momentum. When the angle of incidence changes, surface plasmons of varying resonance frequencies satisfying Eq. (11.1) are excited.

Plasmons generated on the incident side of the film couple to plasmons on the back side, where they radiate according to Eq. (11.1). The observed transmittance spectra are the result of constructive coupling of light to plasmons and then to radiation for those frequencies, for which plasmon momentum is equal to the grating momentum.

Similar experiments were conducted for screens with nanoholes arranged in a Penrose pattern. The difference in the observed transmission spectra results from the fact that a Penrose pattern has two main vectors defining the lattice. Thus the resonance condition set in Eq. (11.1) is fulfilled for two wavelengths for perpendicular incidence and two distinct maximums appear. However, the normalised intensity peaks are lower for the Penrose pattern lattice than for the regular lattice, which has only one main maximum [20].

11.3 A New Model of Charge Density Distribution on SNOM Probe Rims

To understand the physical limit and furthest practical boundary of resolution achievable with aperture probe SNOM we need a valid model of near- and far-field radiation emitted from a tapered-fibre metal-coated probe. Below we present an analytical calculation of forward emission modes of an aperture tapered optical fibre tip with metal coating [18].

Several SEM images of tip apexes show random nonuniformities of tip metal coatings that arise from both etching optical fibres and metal sputtering [9,17]. Even for linearly polarized light surface irregularities result in an induced azimuthal periodicity of charge density distribution on the rim of an aperture of a tapered optical fibre tip. Thus charge density distributions can be considered as quasi-dipoles and multi-quasi-dipoles located solely on an edge. Assumed smear of induced dipoles differs from previously used models where far-field radiation from mathematical dipoles and multipoles was considered [21–23]. We admit that the assumed location of charges exclusively on a rim does not fulfil the continuity condition that should hold. The model neglects the radial decay of charge density what allows to write a simple formula with azimuthal dependence

$$\rho(r,\varphi) \propto \cos(N\varphi)\cos(\omega t)\,\delta(r-R)\,,\qquad(11.2)$$

where N is the number of quasi-dipoles induced on the aperture circumference and R is a hole radius. The rim charge density distributions of Eq. (11.2) are illustrated in Fig. 11.1. In all cases the charge density amplitudes are the same, what results from a comfortable assumption that the absolute charge increases linearly with the value of N. In reality, however, distribution of charge between quasi-dipoles of different order N depends on an individual shape of probe random irregularities.

To calculate the near- and far-field forward emission modes we need to know rim current values which we calculate using the law of charge and current conservation

$$\frac{\partial}{\partial t}\rho + \nabla j = 0\,,\qquad(11.3)$$

where j is the current density. From the charge and current densities we calculate in 3D the relevant field distributions. Time-retarded equations are employed because

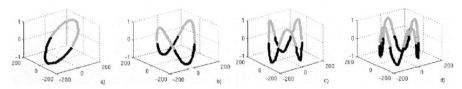

Fig. 11.1 Rim charge density distributions in arbitrary units for: **a)** $N = 1$; **b)** $N = 2$; **c)** $N = 3$; **d)** $N = 4$. Values above and below zero represent positive and negative charges, respectively. Reprinted from [18]

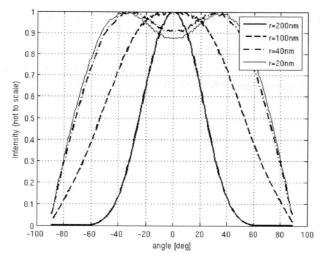

Fig. 11.2 Far-field angular intensities radiated by diluted dipoles of various apertures. Reprinted from [18]

the relation $r \leq R \ll \lambda$ is not valid for quasi-static approximations previously used to describe SNOM probes radiation. When the beam coupled to the fibre probe is linearly polarized, the quasi-dipole contributes the most to emitted radiation. Far-field angular intensities radiated by a quasi-dipole on rims of various apertures are shown in Fig. 11.2. In the plots of the angular intensity a minimum appears when the wavelength-to-aperture radius ratio $\lambda/r > 3.75$.

Modelled far-field angular intensities are in agreement with the experimental result of Obermüller and Karrai shown in Fig. 11.3 [24]. They measured the intensity values on a semicircular surface parallel to the polarization plane. The observed far-field on-axis minimum was never explained by previously presented theoretical models. However, an angular periodicity of rim charge density assumed before has led to qualitative confirmation of experimentally measured intensity distributions in the near-field of a tip [25].

We admit that the proposed model does not take into account tunnelling of light through the metal coating as well as aperture clearance radiation what leads to an over optimistic estimation of achievable resolution. Therefore, theoretical plots in Fig. 11.2 are narrower than the experimental plots of Fig. 11.3. FDTD simulations of Gaussian beam propagation in a tapered metal-coated optical fibre reported in [18] prove that the model neglects also the non-zero value of the field within the area of the aperture.

FDTD simulations of [18] confirm previous, e.g. [23, 25], observations, that the diameter of emitted beam and consequently SNOM resolution do not depend on the wavelength used, but on the rim aperture and the distance between sample and tip. When the tip–sample distance is kept small with the shear-force technique, higher order multi-quasi-dipoles do not decrease the resolution. However, when the tip–sample distance increases above the usual working distance in shear-force mi-

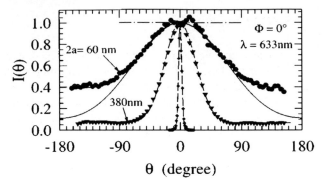

Fig. 11.3 Normalized angular dependence of the transmitted intensity $I(\theta)$ measured for aperture diameters of $2a = 60$ nm ($ka = 0.30$), $2a = 380$ nm ($ka = 1.89$) and $2a = 3.2$ μm ($ka = 15.9$). The detector is scanned in the plane of the polarization ($\theta = 0$). The *dash-dotted line* spanning $-90°$ to $90°$ corresponds to Bethe's theory. The *dashed lines* corresponds to the best Gaussian fit. The *full line* corresponds to the case of magnetic and electric radiating dipoles of strength 2 and 1, respectively, and placed perpendicular to each other in the plane of the aperture. Figure and caption reprinted from [24]

croscopy, the beams radiated by high order multi-quasi-dipoles widen and resolution decreases.

The model stresses the role of random nonuniformities of tip metal coating on the polarization of radiated beams in spite of linear polarization of light coupled to the tapered fibre. A single quasi-dipole radiates a beam that is almost linearly polarized, as shown in Fig. 11.4a. Higher order multi-quasi-dipoles emit beams with complex polarization patterns of $2\pi/N$ rotational symmetry [Fig. 11.4b and c]. When such beams have high intensity, then observation of polarization dependant features of a sample and the use of polarization dependant imaging enhancement reported in [26, 27] become impossible.

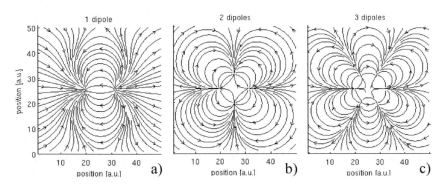

Fig. 11.4 Polarization of the electric field for wavelength $\lambda = 500$ nm at a distance of 25 nm from the radiating plane for an arbitrary time frame for **a)** 1 quasi-dipole, **b)** 2 quasi-dipoles, **c)** 3 quasi-dipoles. Reprinted from [18]

11.4 Types of SNOM Probes

The crucial rule of near-field microscopy is that the narrower the confinement of a scanning light beam the better is the resolution. According to Novotny and Hecht [9] a practical limit of the resolution is given by the sum of the diameter of the aperture of a tapered-fibre metal-coated probe and twice the skin depth d of the metal coating, that is the distance at which the incident electric field intensity decreases to $1/e$ of its value at the surface, where

$$d = \frac{\lambda}{4\pi\sqrt{\varepsilon_m}} \, . \tag{11.4}$$

The value $2d$ corresponds to light penetration into metal layers on both sides of the aperture. At wavelengths higher than $\lambda = 500\,\text{nm}$ on a flat silver surface the skin depth d is about $10\,\text{nm}$, however, it decreases with the thickness of the metal layer. For wavelengths below $500\,\text{nm}$ the skin depth sharply increases with frequency and after a maximum at about $300\,\text{nm}$ decreases [28].

Component $2d$ does not set the lowest resolution limit at $20\,\text{nm}$. According to the model presented in the previous section, radiation of the rim charges is in accordance with the experimental results [24]. We rely on simulations based on time-retarded equations, where the near- and far-field angular intensities radiated by diluted dipoles of various apertures were calculated with space discretization of $1\,\text{nm}$ and $100\,\text{nm}$, respectively [18]. We are not aware of any experimental data for cylindrical surfaces but expect that decay of electromagnetic fields depends on surface shape and on axially symmetric surfaces should be faster than exponential.

11.4.1 Tapered-Fibre Metal-Coated Aperture SNOM Probes

Transmission properties of sub-wavelength apertures of SNOM probes can be described as propagation of light in cylindrical metal waveguides with decreasing diameter. Light is coupled into fibres and passes through a narrowing metal channel which attenuates the signal and reduces the number of sustained modes. Guiding properties of such waveguides are critical for imaging quality and resolution. Thorough work by Novotny and Hafner shows that metal-coated cylinders guide a number modes and their type is restricted by the diameter of the dielectric core [29]. As its size decreases the waveguide supports less and less modes until the last remaining are the HE11 propagating mode and the HE1 surface one, if excited. However, even the HE11 mode is stopped at the cut-off diameter equal to $0.6\lambda/n$ and only an evanescent field remains in the absence of the surface mode. The amount of energy reaching the apex of the tip depends on the distance z for which evanescent solutions are present and vanishes proportionally to $\exp(-kz)$.

An example of such a probe formed by laser-heating and pulling from a fibre-optic cable is shown in Fig. 11.5a. These tips are characterized by small taper angles and have typically low transmission on the order of 10^{-8}–10^{-4} for apertures from 30 to $100\,\text{nm}$ [30]. To circumvent this drawback a couple of ways have been proposed.

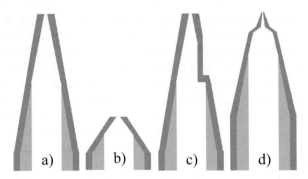

Fig. 11.5 SNOM probes (core, cladding and metal coating thicknesses are not to scale) **a)** typical tapered fibre probe, **b)** large taper angle probe, **c)** asymmetric probe, **d)** triple tapered probe

One approach, through modification of the shape of probes, is to decrease the distance over which the evanescent wave decays. This is done by increasing the taper angle. Various papers have documented experiments in which large taper angles of 80° and more were fabricated by etching techniques, usually in hydrofluoric acid (Fig. 11.5b). Another solution is the use of an asymmetric probe (Fig. 11.5c) suggested in [31] for which the excitation efficiency of the HE11 mode is high and leads to a tenfold increase in emitted light intensity. A similar concept of increasing the HE11 mode strength was used in the design of a triple tapered probe (Fig. 11.5d). Such a tip also has the advantage, that part of the light can be focused on the entrance to the last, narrow part of the probe. In addition, the distance for which an evanescent solution exists is reduced. The overall reported intensity increase reached three orders of magnitude [32].

The second approach involves the use of a dielectric with a refractive index n higher than that of silica, which is frequently used for the visible range probes. Because the cut-off diameter D depends on the wavelength, which in turn depends on the medium and is smaller for media with bigger n, D will be smaller [33]. This approach has been successfully used in silicon-based cantilever probes in the wavelength regime for which Si is transparent. The refractive index of silicon at $\lambda = 1\,\mu m$ is about $n = 3.57$ and gives $\lambda/n = 280\,nm$, which is smaller than most optical wavelengths achievable in silica probes.

In the third approach, the excitation of surface plasmons in the probe can increase light throughput. Careful design of the probe aperture also influences the spatial distribution of the field in the vicinity of the probe, thereby offering control over the profiles of the radiated beams. The creation of strongly localized optical spots using SP assisted sharp ridge nanoaperture [34] and I-shaped aperture in a pyramidal cantilever probe [35] has been recently reported.

11.4.2 Apertureless Tapered Metal SNOM Probes

Solid metal tapered wires with sharp tips generate better confined high intensity near-fields than the above described aperture tapered-fibre metal-coated probes [9, 17]. Localized surface plasmons at the tip apex are efficiently excited with a far-field collimated polarized laser beam illuminating the tapered wire from the direction perpendicular to wire axis. Excitation of SPs is most efficient when linearly polarized light has its electric field parallel to the tip axis. A sample is illuminated simultaneously by the far-field of the source, which excites plasmons on the tip, and near-field radiation of the tip. Intensity of the second one usually exceeds that of the first one by up to three orders of magnitude. Even with a high contrast of irradiations, separation of a signal generated by the near-field from noise introduced by the far-field creates a problem difficult to solve [36,37]. To avoid troublesome signal enhancement and noise removal procedures new technological solutions are sought.

In a recent approach [38], nanofocusing is achieved in a laterally tapered gold stripe waveguide on a sapphire substrate. The crux of the efficient photon-plasmon coupling is that SPP modes propagating on the substrate side of the interface (the high-index side) are excited when an illuminating beam hits a sub-wavelength hole array etched at the basis of the 60 μm long tip. The ascendant high-index side SPP modes are asymmetric, that is have opposite signs of the transverse electric field on both sides of metal layer. Symmetric SPP modes propagating predominantly on the low-index side of the interface, that is gold-air side of the waveguide, focus much worse that the asymmetric ones. Better focusing of predominantly asymmetric modes on the gold-sapphire interface was measured with upconversion luminescence from erbium dopant. Verhagen et al. [38] reported collimation of a beam propagating at the gold-sapphire interface to diameters below 100 nm. This approach to SNOM probe formation may appear simpler than pulling and etching necessary to fabricate tapered-fibre probes.

In the other approach the questions of detrimental far-field illumination and technically difficult coupling of light to tips are theoretically settled with the idea of tapered-fibre metal-coated probes without an aperture at the apex [39–41]. Optical fibre core modes couple to plasmon waves on the outer surface of the metal coating with efficiency reaching 10% [42]. It is worth testing if in apertureless tapered-fibre metal-coated probes this efficiency may increase when the interface between the fibre core and metal layer is corrugated as suggested recently for aperture probes [19] and if symmetric SPPs on the outer surface (the low-index side) of the metal coating are excited. A near-field beam collimation to tens of nanometres in a computer experiment was recently reported [42].

Focusing electromagnetic waves to nearly $\lambda/100$ spots is possible in other than visible spectral ranges. A solid metal tapered tip with periodic corrugations is used to focus THz radiation [43]. A corrugated metal cone tapered from a 0.2 to 0.02 mm diameter focuses 0.5 mm wavelength radiation to about a 20 μm size spot. At terahertz frequencies, metals behave as perfect conductors, that is electromagnetic waves virtually do not exist inside the metal and extend a considerable distance away. Sub-wavelength size structuring of the metal surface facilitates excitation of highly confined, so called spoof SPPs similar to those in optical range. This gives new perspectives for high resolution THz inspection, especially when we keep in mind that in this case a near-field means several millimetres. Important applications of THz imaging are expected both in medicine and security.

11.5 Corrugated Metal-coated Tapered Fibre SNOM Probes

In aperture SNOM probes, as mentioned earlier, aperture diameter reduction leads to resolution enhancement at the expense of the loss of signal intensity. We propose a novel method based on our analysis of rim charge radiation [18] to improve probe energy throughput. The idea is based on the excitation of propagating SPPs and localized plasmons which are not limited by the cut-off diameter. The role of proposed corrugations of the dielectric core-metal coating interface inside a tapered fibre tip is to enhance the generation of surface plasmons. The corrugations enable coupling of the wavevectors of incident light to those of travelling plasmon waves via the spatial frequency components of Fourier spectra of grooves. At the rim of the tip SPPs increase current intensity, what results in enhanced quasi-dipolar radiation.

The idea of corrugations of the interface between the dielectric core and metal coating [19] is coherent with previous studies on light transmission through tiny holes surrounded by surface corrugations and plasmon guiding by means of structured surfaces [15, 43–45].

Below we do not dwell on fabrication techniques of corrugated SNOM probes. Our aim is to optimise their energy throughput keeping in mind that the number of grooves should be reduced to a minimum and grooves should be shallow and as wide as possible, to ease the etching procedure, e.g. [46, 47].

We consider a fibre probe tapered from a 2 µm diameter to 50 nm at the aperture. The fibre core is made of dispersionless silica and coated with a 70 nm thick layer of silver. The taper angle chosen is 20 degrees to reduce the simulation volume and keep space for grooves. Moreover, at this small taper angle corrugations distinctly manifest their impact on energy throughput. We consider 40 and 30 nm shallow grooves with circular and oval profiles of respectively 80 and 180 nm widths shown in Fig. 11.6a,b. We compare their performance with that of a smooth tip (Fig. 11.6c). For the wavelength range 450–600 nm the following parameters of probe structure are scanned: a lattice period Λ of corrugations; an offset, that is the relative shift s of the lattice with respect to the apex; and a width w of oval grooves. The FDTD simulations in 2D are made in with 0.5 nm and 3D ones with 4 nm space discretization. We admit that 2D FDTD simulations give results more optimistic than those made in 3D.

Fibres are illuminated with a CW Gaussian beam and a broad-band Gaussian impulse with such dispersion that 6σ is equal to the core diameter. Both signals are

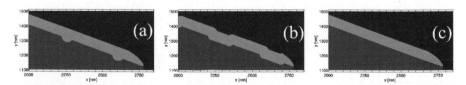

Fig. 11.6 Modelled tip structures (**a**) with circular, (**b**) and oval grooves, (**c**) and the smooth one. Colours indicate: glass core – *dark grey*, metal coating – *light grey*, vacuum – *black*. The pictures show, because of clarity, only the symmetrically cut, narrow end of the tips [19]

linearly polarized with the electric field in the plane of the structure for 2D simulations.

To illustrate the influence of introduced corrugations on the energy distribution inside the probe we present the time-averaged Poynting vector lengths in Fig. 11.7. From left to right in columns, the energy distributions are shown for increasing wavelengths 450, 510 and 600 nm. Cut-off diameters increase correspondingly. In a smooth probe, for all wavelengths the incident wave does not propagate farther than the cut-off diameter (Fig. 11.7a–c). In this case a propagating SPP on the SiO_2–Ag interface cannot be excited, because wave vectors of SPP and incident light are mismatched. For 510 nm wavelength photon–plasmon coupling is the most efficient for both groove profiles (Fig. 11.7e,h). In the narrow part of the tips beyond the cut-off we observe energy localised at the core-coating interface, which is a characteristic feature of SPs. Moreover, the excited plasmons are localized and their appearance depends on the relative positions of scattering centres (grooves) and the propagating wave cut-off.

Figure 11.8 shows dependence of energy transmission on light wavelength calculated for circular and oval grooves with different lattice constants. For six circular

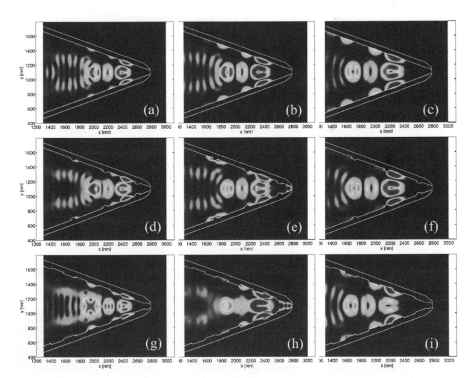

Fig. 11.7 Time-averaged energy distributions in three analysed structures: smooth probe (**a**)–(**c**), and probes with six circular grooves of $\Lambda = 445$ nm and $s = 173$ nm (**d**)–(**f**), and six oval corrugations of $\Lambda = 325$ nm and $s = 87$ nm (**g**)–(**i**). Wavelengths are 450 nm, 510 nm and 600 nm from left to *right columns*, respectively

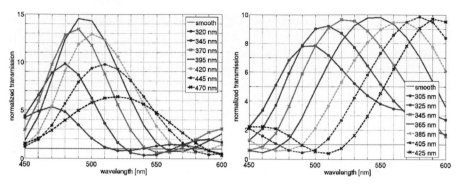

Fig. 11.8 Transmission for SNOM probes with six corrugations and different lattice constants calculated for the spectral range 450–600 nm: (*left*) circular and (*right*) oval. *Inlets* give lattice constant Λ values. For each wavelength values are normalized to transmission of a smooth tip

corrugations of 80 nm width the maximum transmission exceeds that of a smooth probe by a factor of 14 for the optimum lattice-to-width ratio $\Lambda/w = 5$. For oval grooves, which are more attractive from etching feasibility point of view, there is no such sharp dependence. Enhancement factors are close to 10 for oval groove width 180 nm and a range of lattice constants from 325 to 425 nm.

Synthesized results of simulations are presented in Fig. 11.9a–d. For both groove profiles, the maximum transmissions normalized with respect to that of a smooth tip are given for different ranges of lattice constants and offsets. For six circular grooves of 80 nm width enhancement factor close to 15 is calculated for $\Lambda = 395$ nm and $s = 180$ nm at 495 nm wavelength. Spectral positions of maximum enhancements move toward longer waves with increase of both lattice constants and offsets. For 180 nm wide six oval grooves enhancement factors close to 10 appear in a range of lattice constants from 325 to 425 nm what is a positive feature. Namely, in spite of technological inaccuracies the same enhancement factor is possible, although wavelength tuning may be required. The enhancement factor of probes with oval corrugations is linearly dependent on the offset values, bigger the offset, smaller the gain. This is disadvantageous since the last groove cannot be arbitrarily close to the apex because of tip fragility. For oval grooves spectral positions of maximum enhancements show a similar tendency as for the circular ones. Etching of six grooves in tapered silica fibre cannot be an easy task and calculated enhancement factors may serve only as a reference for more realistic corrugation structures with smaller number of grooves.

In simulations we observe that the corrugations do not improve the width of the emitted signal which still is defined by the aperture diameter and double skin depth. This is shown in Fig. 11.10. Thus, improvement of resolution should come from the decrease of the apex diameter while keeping the output energy at detectable levels.

Achievable widths of grooves depend on available fabrication technology [48–50]. To our knowledge nanogroove etching in tapered SiO$_2$ fibres was never considered. It is reasonable to expect that wider grooves are easier to implement. We assess the influence of groove width on the transmission properties of corrugated

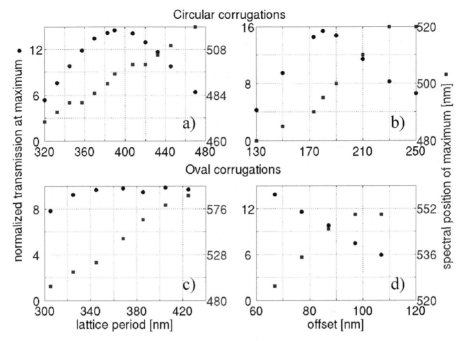

Fig. 11.9 Normalized transmissions at maximum (*circles*, *left* y-axis) and the spectral position of the maximum (*squares*, *right* y-axis) calculated for six circular (**a**), (**b**) and six oval (**c**), (**d**) corrugations as a function of lattice constant (**a**), (**c**) and offset (**b**), (**d**). In plots (**a**) and (**c**) the offsets are constant and equal 173 nm and 200 nm respectively; in plots (**b**) and (**d**) the lattice periods are constant and equal 395 nm and 365 nm respectively. These plots have additional data points which were omitted in Fig. 11.8 for clarity

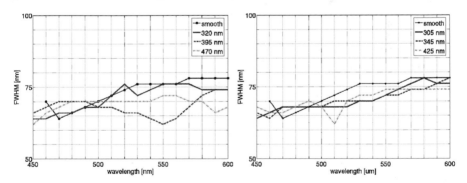

Fig. 11.10 Comparison of FWHM for a smooth probe and probes with circular (*left*) and oval (*right*) corrugations of different lattice constants

SNOM probes. Grooves width can be changed by shifting the position of either end. We change the width of the corrugations by 10 nm increments, separately for the near and far end of the groove with respect to the probe aperture. In Fig. 11.11 we observe, that in the first case for wider corrugations (Fig. 11.11(left)) the intensity of

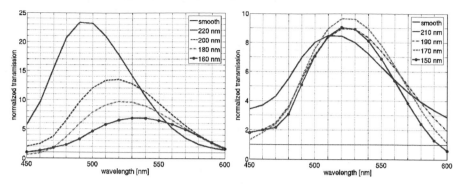

Fig. 11.11 Normalized transmission for SNOM probes with oval corrugations of varied widths. With respect to Fig. 11.6b: (*left*) the position of the left side of the corrugations is kept constant and the right is varied, i.e. the grooves are longer and approach the probe end; (*right*) the position of the right side of the corrugations is kept constant and the left is varied, i.e. the grooves are longer or shorter, but begin at the same place with respect to the probe aperture. Plots are labelled by the groove width

radiated energy reaches a gain factor of 23 and the spectral location of the maximum shifts towards shorter wavelengths.

This large enhancement results from SPs concentrating in increasingly smaller volumes between the first corrugation and the tip apex. Their energy density is higher and thus the radiation flux is greater than for plasmons bound in larger volumes. A similar tendency of increasing transmission with a decreasing offset is also observed when we only change the distance between the lattice and the probe apex but not the groove width. This gives the means of increasing energy transmission by shifting the corrugations closer to the apex of the probe. However, grooves cannot be arbitrarily close to the end of the tip because its durability would be limited.

The second way of increasing the width of the corrugations does not affect the groove-apex distance. Figure 11.11(right) shows that an increasing width with a constant groove-apex distance has a negligible impact on the performance of SNOM probes, as both the maximum transmission value and spectral location remain virtually unchanged. This is advantageous for fabrication purposes, as etching relatively wide grooves presents less of a challenge than narrow ones.

A positive result, from the point of view of practical realisation, is shown in Fig. 11.12. When the number of oval grooves is reduced from six to two enhancement factors do not change considerably. For a wide range of separations between the grooves we observe the relative enhancement by one order of magnitude within a range of wavelengths 500–600 nm. When the separation-to-width ratio increases by 50%, enhancement factors drop to about 3.

Finally, we calculate the energy throughput of tapered fibre metal-coated corrugated probes of different apex diameters. As baseline we use radiation properties of a classical, smooth 50 nm aperture probe shown in Fig. 11.13a as a horizontal line with unity transmission. We observe that corrugated probes with large diameters have specific spectral properties – distinct maxima at 530 nm – which are connected

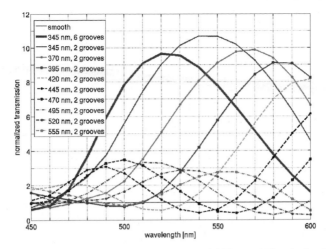

Fig. 11.12 Transmission of tips with two oval grooves of 180 nm width calculated with respect to the transmission of a smooth metal-coated probe. Distances between grooves are shown in the *inlet*. Plots are compared with the *thick line* corresponding to a probe with six grooves

with position of the wavelength dependent cut-off diameter relative to the corrugations. In the spectral range 450–500 nm decrease of the apex diameters results in small changes of relative transmission, however, corrugated probes with diameters 30 nm and 10 nm have better throughput performance than the smooth one of 50 nm apex. In the spectral range 500–600 nm decrease of the apex diameter to 10 nm drastically lowers energy throughput of corrugated probes, which falls below levels emitted by a classical 50 nm probe.

The above results are supported by calculation of FWHM values for the same spectral range and apex diameters from 50 to 10 nm. Figure 11.13b shows that within 490–600 nm range the FWHM of beam emitted from the smallest diameter tip is about 3 times smaller than that of 50 nm probe. Moreover, only for this wavelength range all considered probes emit a well collimated beam via the apex and a low level of noise resulting from tunnelling through the metal-coating. For smaller wavelengths the width of the beams increases, because light leakage through the silver coating increases with respect to the intensity of light emitted via the apex.

Detailed modelling of the tapered-fibre metal-coated SNOM probes should be done in cylindrical coordinates, e.g. [51, 52]. In the performed calculations the reduction of the number of dimensions to two influences the obtained results. Nevertheless, it is demonstrated that propagating surface plasmons on the dielectric-core metal-coating interface have a cut-off, while the localised SPs do not. On the basis of early simulations we expected, that this result will be confirmed in 3D FDTD simulations.

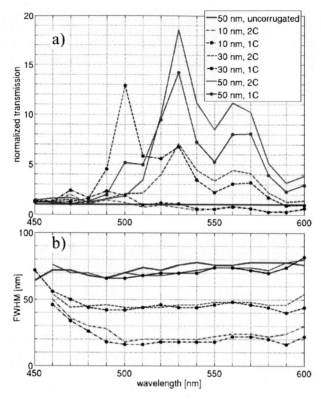

Fig. 11.13 a) Transmission of SNOM probes with one (1C) and two (2C) oval corrugations and different aperture diameter values calculated for the spectral range 450–600 nm. For each wavelength transmission values are normalized to transmission of a smooth tip with a 50 nm diameter. **b)** FWHM of SNOM probes with one (1C) and two (2C) oval corrugations and different diameter values calculated for the spectral range 450–600 nm

11.6 Summary

Improvement of resolution of SNOM has remained a challenge since the first optical imaging in 1984. Several solutions were proposed and commercially available level of 50 nm resolution has been achieved. The future progress in plasmonic and nanophotonic devices demands resolutions even better than an eighth part of the plasmon wavelength which on a Ag/SiO$_2$ interface is equal to about 250 nm and corresponds to $\lambda = 440$ nm in vacuum. The idea of tapered-fibre metal-coated aperture probes with corrugations of the core – metal coating interface promises such an improvement under the condition that a technique of etching narrow and shallow grooves in glass is developed. A resolution limit given by the sum of the apex diameter and two skin depths of a metal used for coating can be improved due to reduction of the first component.

Acknowledgements This research was sponsored by the Polish grants 37/COS/2006/03 and N N507 445534. We also acknowledge the support from the COST Action P11 and 6FP Network of Excellence Metamorphose. Numerical computations were performed in the Interdisciplinary Centre for Mathematical and Computational Modelling (ICM), University of Warsaw, grant number G26-9.

References

1. Bethe, H.A.: Theory of diffraction by small holes. Phys. Rev. **66**, 163–182 (1944).
2. Bouwkamp, C.J.: On Bethe's theory of diffraction by small holes. Philips Res. Rep. **5**, 321–332 (1950).
3. Leviatan, Y.: Study of near-zone fields of a small aperture. J. Appl. Phys. **60**, 1577–1583 (1986).
4. Synge, E.H.: A suggested method for extending the microscopic resolution into the ultramicroscopic region. Phil. Mag. **6**, 356 (1928).
5. Ash, E.A. and Nichols, G.: Super-resolution aperture scanning microscope. Nature **237**, 510–512 (1972).
6. Pohl, D.W., Denk, W. and Lanz, M.: Optical stethoscopy: Image recording with resolution $\lambda/20$. Appl. Phys. Lett. **44**, 651–653 (1984).
7. Dürig, U., Pohl, D.W. and Rohner, F.: Near-field optical-scanning microscopy. J. Appl. Phys. **59**, 3318–3327 (1986).
8. Sakoda, K.: Optical properties of photonic crystals. Springer, Berlin (2005).
9. Novotny, L. and Hecht, B.: Principles of nano-optics. Cambridge University Press, Cambridge (2006).
10. Zayats, A.V. and Smolyaninov, I.I.: Near-field photonics: surface plasmon polaritons and localized surface plasmons. J. Opt. A: Pure Appl. Opt. **5**, S16–S50 (2003).
11. Zayats, A.V., Smolyaninov, I.I. and Maradudin, A.A.: Nano-optics of surface plasmon polaritons. Phys. Rep. **408**, 131–314 (2005).
12. Maier, S.A.: Plasmonics: Fundamentals and Applications. Springer, New York (2007).
13. Hopman, W.C.L., Stoffer, R. and de Ridder, R.M.: High-resolution measurement of resonant wave patterns by perturbing the evanescent field using a nanosized probe in a transmission scanning near-field optical microscopy configuration. J. Lightwave Technol. **25**, 1811–1818 (2007).
14. Ebbesen, T.W., Lezec, H.J., Ghaemi, H.F., Thio, T. and Wolff, P.A.: Extraordinary optical transmission through sub-wavelength hole arrays. Nature **391**, 667–669 (1998).
15. Lezec, H.J., Degiron, A., Devaux, E., Linke, R.A., Martin-Moreno, L., Garcia-Vidal, F.J. and Ebbesen, T.W.: Beaming light from a sub-wavelength aperture. Science **297**, 820 (2002).
16. Genet, C. and Ebbesen, T.W.: Light in tiny holes. Nature **445**, 39–46 (2007).
17. Kim, J.H. and Song, K.B.: Recent progress of nano-technology with NSOM. Micron **38**, 409–426 (2007).
18. Antosiewicz, T.J. and Szoplik, T.: Description of near- and far-field light emitted from a metal-coated tapered fibre tip. Opt. Express **15**, 7845–7852 (2007).
19. Antosiewicz, T.J. and Szoplik, T.: Corrugated metal-coated tapered tip for scanning near-field optical microscope. Opt. Express **15**, 10920-10928 (2007).
20. Bravo-Abad, J., Fernandez-Dominguez, A. I., Garcia-Vidal, F.J. and Martin-Moreno, L.: Theory of extraordinary transmission of light through quasiperiodic arrays of sub-wavelength holes. Phys. Rev. Lett. **99**, 203905 (2007).
21. Shin, D.J., Chavez-Pirson, A. and Lee, Y.H.: Multipole analysis of the radiation from near-field optical probes. Opt. Lett. **25**, 171–173 (2000).
22. Drezet, A., Woehl, J.C. and Huant, S.: Diffraction by a small aperture in conical geometry: Application to metal-coated tips used in near-field optical microscopy. Phys. Rev. E **65**, 046611 (2002).

23. Arnoldus, H.F. and Foley, J.T.: Highly directed transmission of multipole radiation by an interface. Opt. Commun. **246**, 45–56 (2005).
24. Obermüller, C. and Karrai, K.: Far field characterization of diffracting circular aperture. Appl. Phys. Lett. **67**, 3408–3410 (1995).
25. Drezet, A., Huant, S. and Woehl, J.C.: In situ characterization of optical tips using single fluorescent nanobeads. J. Lumin. **107**, 176–181 (2004).
26. Durkan, C. and Shvets, I.V.: Polarization effects in reflection-mode scanning near-field optical microscopy. J. Appl. Phys. **83**, 1837–1843 (1998).
27. Gademann, A., Shvets, I.V. and Durkan, C.: Study of polarization-dependant energy coupling between nearfield optical probe and mesoscopic metal structure. J. Appl. Phys. **95**, 3988–3993 (2004).
28. Wang, Z., Cai, X., Chen, Q. and Li L.: Optical properties of metal-dielectric multilayers in the near UV region. Vacuum **80**, 438–443 (2006).
29. Novotny, L. and Hafner, C.: Light propagation in a cylindrical waveguide with a complex, metallic, dielectric function. Phys. Rev. E. **50**, 4094–4106 (1994).
30. Valaskovic, G.A., Holton, M. and Morrison, G.H.: Parameter control, characterization, and optimization in the fabrication of optical fiber near-field probes. Appl. Opt. **34**, 1215 (1995).
31. Yatsui, T., Kourogi, M. and Ohtsu, M.: Highly efficient excitation of optical near-field on an apertured fiber probe with an asymmetric structure. Appl. Phys. Lett. **71**, 1756–1758 (1997).
32. Yatsui, T., Kourogi, M. and Ohtsu, M.: Increasing throughput of a near-field optical fiber probe over 1000 times by the use of a triple-tapered structure. Appl. Phys. Lett. **73**, 2090–2092 (1998).
33. Kuznetsova, T.I. and Lebedev, V.S.: Transmission of visible and near-infrared radiation through a near-field silicon probe. Phys. Rev. B **70**, 035107 (2004).
34. Jin, E.X. and Xu, X.: Obtaining super resolution light spot using surface plasmon assisted sharp ridge nanoaperture. Appl. Phys. Lett. **86**, 111106 (2005).
35. Tanaka, K., Tanaka, M. and Sugiyama, T.: Creation of strongly localized and strongly enhanced optical nearfield on metallic probe-tip with surface plasmon polaritons. Opt. Express **14**, 832–846 (2006).
36. Adam, P.M., Royer, P., Laddada, R. and Bijeon, J.L.: Apertureless near-field optical microscopy: Influence of the illumination conditions on the image contrast. Appl. Opt. **37**, 1814–1819 (1998).
37. Hecht, B., Bielefeld, H., Pohl, D.W., Novotny, L. and Heinzelmann, H.: Influence of detection conditions on near-field optical imaging. J. Appl. Phys. **84**, 5873–5882 (1998).
38. Verhagen, E., Polman, A. and Kuipers, L.: Nanofusing in laterally tapered palsmonic waveguides. Opt. Express **16**, 45–57 (2008).
39. Novotny, L., Pohl, D. and Hecht, B.: Scanning near-field optical probe with ultrasmall spot size. Opt. Lett. **20**, 970–972 (1995).
40. Ashino, M. and Ohtsu, M.: Fabrication and evaluation of a localized plasmon resonance probe for near-field optical microscopy/spectroscopy. Appl. Phys. Lett. **72**, 1299–1301 (1998).
41. Janunts, N.A., Baghdasaryan, K.S., Nerkararyan, K.V. and Hecht, B.: Excitation and superfocusing of surface plasmon polaritons on a silver-coated optical fibre tip. Opt. Commun. **253**, 118–124 (2005).
42. Ding, W., Andrews, S.R. and Maier, S.A.: Internal excitation and superfocusing of surface plasmon polaritons on a silver-coated optical fibre tip. Phys. Rev. A **75**, 063822 (2007).
43. Maier, S.A., Andrews, S.R., Martin-Moreno, L. and Garcia-Vidal, F.J.: Terahertz surface plasmon-polariton propagation and focusing on periodically corrugated metal wires. Phys. Rev. Lett. **97**, 176805 (2006).
44. Baida, F.I., Van Labeke, D. and Guizal, B.: Enhanced confined light transmission by single sub-wavelength apertures in metallic films. Appl. Opt. **42**, 6811–6815 (2003).
45. Martín-Moreno, L., García-Vidal, F.J., Lezec, H.J., Degiron, A. and Ebbesen, T.W.: Theory of highly directional emission from a single sub-wavelength aperture surrounded by surface corrugations. Phys. Rev. Lett. **90**, 167401 (2003).

46. Lazarev, A., Fang, N., Luo, Q. and Zhang, X.: Formation of fine near-field scanning optical microscopy tips. Part I. By static and dynamic chemical etching. Rev. Sci. Instrum. **74**, 3679–3683 (2003).
47. Yang, J., Zhang, J., Li, Z. and Gong, Q.: Fabrication of high-quality SNOM probes by pretreating the fibres before chemical etching. J. Microsc. **228**, 40–44 (2007).
48. Matsumoto T., Ichimura T., Yatsui T., Kourogi M., Saiki T. and Ohtsu M.: Fabrication of a near-field optical fiber probe with a nanometric metallized protrusion. Opt. Rev. **5**, 369–373 (1998).
49. Mononobe, S., Saiki, T., Suzuki, T., Koshihara, S. and Ohtsu, M.: Fabrication of a triple tapered probe for near-field optical spectroscopy in UV region based on selective etching of a multistep index fiber. Opt. Commun. **146**, 45–48 (1998).
50. Haber, L.H., Schaller, R.D., Johnson, J.C. and Saykally, R.J.: Shape control of near-field probes using dynamic meniscus etching. J. Microsc. **214**, 27–35 (2004).
51. Baida, F.I., Van Labeke, D. and Pagani, Y.: Body-of-revolution FDTD simulations of improved tip performance for scanning near-field optical microscopes. Opt. Commun. **225**, 241–252 (2003).
52. Baida, F.I., Belkhir, A., Van Labeke, D. and Lamrous, O.: Sub-wavelength metallic coaxial waveguides in the optical range: Role of the plasmonic modes. Phys. Rev. B **74**, 205419 (2006).

Part V
Simulation Techniques

12 Photonic Crystals: Simulation Successes and some Remaining Challenges

Trevor M. Benson[1] and Peter Bienstman[2]

[1] George Green Institute for Electromagnetics Research, University of Nottingham,
   Nottingham NG7 2RD, UK
[2] Department of Information Technology, Ghent University-IMEC
   St. Pietersnieuwstraat 41, 9000 Ghent, Belgium

Abstract. A wide range of software now exists for the design and simulation of photonic crystals. The sophistication and high performance specifications of many photonic crystal devices mean that techniques must typically not only provide accurate vector results, or at least a reliable error estimate, but also be able to deal with multi-scale problems, intricate materials properties including non-linearity, arbitrary geometries and multi-physics effects. The design process also demands consideration of process variation and performance optimisation issues.

   In this chapter we review some of the modelling and simulation activities that have formed the activities of Working Group 2 of the COST P11 Action on 'The physics of linear, non-linear and active photonic crystals' and place these achievements within some more general trends in electromagnetics modelling. It will be seen that although time-domain numerical techniques such as the finite difference time-domain (FDTD) and transmission line modelling (TLM) methods have come to the forefront in recent years, principally driven by their flexibility, other techniques still have significant roles to play in the design process and in efficient, accurate and thorough simulation investigations.

Key words: modelling methods; simulation; slow wave structures; photonic wire; second harmonic generation; membrane waveguides

12.1 Introduction

Significant attention and investment has been made in providing computer simulation packages for photonics devices and systems. The emergence of photonic crystals, photonic crystal fibres, all-optical signal processing, plasmonics, nano- and micro- resonators and photonic molecules, optical memory and device synthesis continue to further the requirements on simulation techniques. Techniques must typically not only provide accurate vector results, or at least a reliable error estimate, but also be able to deal with multi-scale problems, intricate materials properties and arbitrary geometries. The questions posed by the designers are largely common across the broad spectrum of computational electromagnetics, including: Will the model solve my problem? Are there any constraints? Can you deal with time-varying and non-linear problems? What size of problem is realistic? How accurate is the solution?

The driving force for many photonic crystal devices is improved performance within a compact footprint. This in turn demands the use of materials with high refractive index contrast, giving rise to strong reflections and radiation losses as well as technological-associated problems such as the influence of sidewall roughness and other imperfections on device performance. The modelling of non-linear effects in these structures is also necessary because for many useful functions, such as versatile wavelength conversion and second-harmonic generation, advantage is taken of the strong confinement in photonic-crystal based waveguides and cavities to reduce switching energies to practical levels. It almost goes without saying that 2D models are often inadequate for the advanced design work required for such components and the cutting-edge need is for robust 3D vector tools.

The period of the COST P11 project (2003–2007) has seen significant improvements in the performance of photonic crystal-based waveguides and switching devices. In this chapter we review some of the techniques that have been used within Working Group 2 of the COST P11 Action to help meet new challenges, present an overview of some of the comparative studies of techniques that have been made and then draw attention to some emerging themes. We start with a brief overview of methods available, categorized for convenience into time- and frequency-domain methods. Most of these techniques were reviewed during the COST P11 Training School on 'Modelling and simulation techniques for linear, nonlinear and active photonic crystals' held at The University of Nottingham in June 2006 (www.nottingham.ac.uk/ggieemr) when demonstrations were given using software developed in various research laboratories and commercial tools from Photon Design (http://www.photond.com/) and RSoft (http://www.rsoftdesign.com/).

12.2 Time-Domain Techniques

Time-domain numerical techniques such as the finite-difference time-domain (FDTD) [1,2] and transmission line modelling TLM [3] methods are flexible, adaptable and compatible with parallel computing techniques. They seem particularly well suited to the simulation of some of the linear and non-linear properties of compact photonic crystal structures although residual problems remain with problem discretization (stairstep error), the long run-times needed to obtain precise frequency responses via Fourier transformation, and the difficulty of directly studying statistical variations because of meshing resolution. TLM schemes based on unstructured 2D triangular [4] or 3D tetrahedral [5] meshes have recently been proposed to ameliorate the effect of stair-step error. The use of unstructured meshes incurs it own computational overheads. However unstructured meshes generally require fewer mesh elements for a given quality of boundary description and this typically means that the net computational requirements of unstructured algorithms are far less than those of structured algorithms. Other features of these models that are important for meeting the emerging requirements of photonics modelling are (i) effective FDTD [6] and TLM methods [7, 8] for calculating the broadband electromagnetic responses for problems that involve complex material behaviour and (ii) the development of a general TLM node, derived from rigorous field analysis, that allows small objects of almost arbitrary shape to be embedded within large scale TLM cells [9].

The time-domain beam-propagation method (TD-BPM) allows the simulation of transients, dynamics and bi-directional propagation in optical structures without the potentially large memory and computational overheads of FDTD and TLM [10–13]. The technique assumes a slowly varying envelope approximation and, by using an implicit scheme that is unconditionally stable, a large time step may be used. General material properties can be efficiently described by using a Z-transform representation of the frequency-dependent complex permittivity [14, 15] and the bandwidth restrictions imposed by the slowly varying envelope approximation can be relaxed using Padé approximants in the time axis [16].

Modelling the physics of time varying electromagnetic phenomena by means of a Time-Domain Volterra Integral Equation (TDIE) [17] has also led to the development of numerical algorithms with significant computational advantages over FDTD and TLM. In the TDIE algorithms fields are directly discretized only in those regions where the structure under consideration differs from a simple background medium. This yields significant scope for memory reduction, albeit requiring sufficient time history. Further advantages offered by the TDIE scheme are that the kernels of the Volterra equations intrinsically satisfy the radiation conditions at infinity, negating the need to explicitly construct absorbing or perfectly matched boundaries, and the fact that they can give intuitive information on observable behaviour. However, the complexity of devices often prohibits full 3D and vectorial Volterra Integral Equation treatment formulations of a given problem. Recently a hybrid TLM-Volterra Integral Equation technique was applied to the modelling of electro-absorption travelling wave modulators [18].

12.3 Frequency-Domain Techniques

A frequency-domain technique is one which calculates the steady state behaviour at a particular frequency. Many frequency-domain methods are available to the designer, including the finite difference method, finite element method and method of lines which were reviewed in [2]. The eigenmode expansion method (discussed below), source model technique [19, 20] and Integral Equation methods [21–25] are also worthy of note. Comparisons of some of these methods have been made as part of the activities of Working Group 2 of the COST P11 Action, Physics of linear, non-linear and active photonic crystals as described on the COST P11 web pages, http://w3.uniroma1.it/energetica/, in [26, 27], and in Sect. 12.4.

Numerical techniques such as the finite difference or finite element methods explicitly sample fields throughout the entire problem space. Although relatively simple and flexible numerical algorithms result, computational resources (memory and run-time) scale with the *volume* of the full problem space. Furthermore the computational workspace must be truncated with a suitable boundary condition. They can be easily outperformed by other approaches, for example Integral Equation formulations or eigenmode expansion (mode matching) methods, for certain classes of geometry. One of the recurrent challenges for numerical methods is the need to embed small-scale features that significantly affect overall system behaviour within a larger workspace and the trade-off between accuracy and computational efficiency that can result. Within the remit of COST P11 examples of where this challenge occurs include Photonic Band-gap (PBG) structures and quantum wells, but the same generic problem occurs when modelling thin wires in Electromagnetic Compatibility problems [9].

The eigenmode expansion method (EEM) is a tool that has been widely used in the COST P11 modelling exercises. The EEM divides the structure to be studied into piecewise constant sections in the longitudinal direction. The field in each section is described as a superposition of the eigenmodes of the section, calculated from the transverse refractive index profile. Radiation loss can be incorporated by using perfectly matched layers as an outer terminating layer in each transverse section. At the junctions between the longitudinal sections established mode-matching is applied to determine reflection and transmission matrices between the various modes. These interface matrices are combined to obtain a scattering matrix for the entire structure. For a given input excitation, the mode amplitudes and by superposition the total field can be determined throughout the structure. Details of this method can be found in [28] with a subsequent expansion to non-linear materials described in [29]. An implementation of the algorithm in the CAMFR simulation software [28] can be freely downloaded from http://camfr.sourceforge.net/.

Much present day simulation and design work still involves approximations. Relying solely on rapidly increasing computer power is not a sustainable strategy for overcoming the limitations of present day simulation packages. For example a time-domain method, such as the TLM or FDTD method, can result in extremely large computational problems especially when three spatial dimensions plus time are considered. Beam propagation methods (BPMs) [2, 10] have proved an invaluable tool for the designers of integrated photonics for several decades. Many of the assumptions underlying the original BPM formulations have been relaxed using new formulations and algorithms (see [10] for example). The computational efficiency of BPMs means they still have much to offer, but some residual issues can be highlighted.

The BPM method makes simplifying assumptions to reduce problem complexity, typically that propagation is predominately along one particular direction and that single frequency operation can be considered. Important improvements to the method include its extension to the wide-angular propagation via the Padé approximants [30, 31], the extension to the vector case [32, 33] and implementation of bi-directional propagation schemes (see for example [34]). It is this idea of using Padé approximants which was later extended in the development of the wideband finite difference time-domain beam propagation method described in Sect. 12.2. The main computational hold-up for many wide-angle 3-D BPM algorithms is the need to solve a very large, though sparse, matrix problem at each propagation step. Du Fort Frankel [35] and novel iterative Alternating Direction Implicit (ADI) schemes [36] address this difficulty.

A serious challenge facing every numerical modelling technique is the lack of universality when being applied to a scope of geometrically different problems. The concept of structure-related, non-orthogonal FD-BPM algorithms can overcome both stair-stepping and wide-angled propagation limitations, [37]. These algorithms adopt structure-related (SR) coordinate systems that naturally follow the local geometry of the structures, offering the designer an extra degree of freedom in approach and method. The use of a non-uniform triangular grid for the numerical simulation of waveguides of arbitrary cross-section has been proposed in [38], with the primary aim of overcoming the numerical uncertainty associated with the stair-stepping approximation to the refractive index profile that is inevitable when modelling such waveguides using a rectangular grid and which can prove particularly problematic for structures with high refractive index contrast.

An alternative approach to the development of simulators for large optical circuits is to use a circuit model [39]. Such a model is usually based on a transmission or scattering matrix where the coefficients can be based on analytical, numerical, experimental or even empirical data. It differs from a numerical model in that it is an equivalent representation of the behaviour of a particular object without reference to its physical layout or technology. The advantage of such an approach is that it provides a very fast route for the study of large circuits in both frequency- and time-domains. Its speed also makes it particularly well suited to design optimisation and sensitivity (tolerance) analysis.

Hybrid simulation techniques, where different simulation tools are used to model those parts of a particular problem to which they are best suited, can also offer clear computational advantage for certain configurations. Examples include the embedding of a numerical algorithm within a global region described by an Integral Equation method [40], and the analysis of three-dimensional fibre-to-chip coupling using the finite difference beam propagation method to model the fields in the planar chip guide and the semi-analytical free space radiation mode method to model the fields in the fibre guide [41].

Frequency-domain methods are less general than time-domain approaches, and dynamic and non-linear behaviour is less obvious, but this is countered by a trivial treatment of material dispersion (i.e. $\varepsilon(\omega)$ can be specified at each frequency), substantially shorter simulation time [15,42] and results that are often more straightforward to interpret. Frequency-domain techniques can also offer problem-specific advantages. When analysing resonators, for example, closely spaced resonances can be easily resolved by successive frequency-domain analyses, and eigenfrequencies and eigenstates are accessed directly rather than from the Fourier transform of a transient response.

12.4 COST P11 Modelling Exercises

Having overviewed a number of general issues in computational electromagnetic design and some of the available tools we now describe some of the modelling exercises proposed under WG2 of the COST P11 Action. Some of the other modelling tasks proposed included an exercise on plasmonic structures aimed at designing a system of silver nanorods to enhance electromagnetic field, an exercise on polarization rotation in bent waveguides concerning the coupling or TE- and TM-like modes induced by the curvature of a dielectric waveguides, a slit-groove diffraction problem and a task on the modelling of spontaneous emission in organic light emission diodes with photonic crystals to improve the out-coupling. Details of these exercises can be found on the WG2 section of the COST P11 web pages, http://w3.uniroma1.it/energetica/WG2.htm. A summary of a further simulation exercise on the rigorous modelling of a high fill factor photonic crystal fibre (PCF) with an elliptically deformed core and noncircular air holes, that was carried out within Working Group 3 of the Cost P11 Action, can be found in [26]. The paper describes the used of different numerical methods to compare several practically important fibre characteristics, such as the spectral dependence of the phase and the group effective indices, the birefringence, the group velocity dispersion and the confinement.

12.4.1 Slow Wave Structures (SWS)

Slow wave structures are important photonic crystal based devices for enhancement of non-linear effects. Inside the slow wave structure the group velocity of an optical is can be significantly reduced, and this leads to enhancement of self-phase modulation and frequency mixing effects. The aim of this task was to model non-linear slow wave structures with an increasing level of detail. The basic building block is a one-dimensional Fabry–Perot cavity, initially consisting of a Kerr medium sandwiched between two identical reflectors of fixed reflectivity (transmission). The model was then extended to the case of practical distributed Bragg reflectors.

Finally transmission through multiple repetitions of the basic block was considered, where the exit reflector of the first block is connected to the incident reflector of the second block by a quarter wave layer, Fig. 12.1.

A comparative analysis of results obtained using three numerical methods operating in the time-domain and two numerical frequency-domain models was published in [27] and showed good agreement between various models used. It was shown

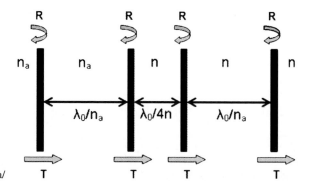

Fig. 12.1 Schematic diagram of a slow wave structure investigated as a COST P11 modelling exercise. Reproduced from http://w3.uniroma1.it/energetica/

that the frequency-domain methods are typically much more efficient than the time-domain methods, by a factor approaching two orders of magnitude in the examples reported in the paper. A feature of this structure is the presence of optical bistability. The time-domain methods were less able to accurately reconstruct the spectrum where sharp transitions with frequency occur, as occurs in the presence of bistability. At the boundary of the bistable regions strong peaks occur in the group delay. Consequently the time-domain techniques require large simulation times to match the accuracy of the frequency-domain methods. On the other hand, the frequency-domain methods could not easily reveal regions where stable solutions do not exist and only the time-domain techniques can predict the self-pulsation that is found with a sufficiently high number of cavities and non-linearity. It was concluded that "it appears that a single numerical method is not sufficient to fully and effectively characterize a SWS in the nonlinear regime, but a comparative analysis including both time-domain and frequency-domain methods is necessary for an accurate investigation."

12.4.2 Photonic Wire

Another popular exercise within WG2 of the COST P11 Action was the simulation of the properties of a silicon-on-insulator (SOI) photonic wire, Fig. 12.2. Silicon photonic wires are single mode structures comprised of a small silicon core of high refractive index (say 3.5) clad by air and silicon dioxide regions of much lower refractive indices (1 and 1.45 respectively). In the structure studied the silicon waveguide core has a width of 500 nm and a thickness of 220 nm, these small dimensions being chosen to maintain single mode behaviour. This configuration is a difficult one to model because of the leakage of the fundamental mode into the high refractive index silicon substrate. This typically leads to a modal index of order $2.4 - j3 \times 10^{-8}$ and the challenge is to model the small imaginary part to high accuracy.

The results of the silicon photonic wire comparison were presented in [43] and the analysis techniques compared included the effective index, eigenmode expansion, perturbation, plane wave expansion and finite elements methods. Interestingly propagation losses measured at the Photonics Research Group at the University of Ghent prior to the commencement of the simulation exercise suggested experimental losses as low as $2.4 \pm 1.6\,\text{dB cm}^{-1}$. The corresponding imaginary part of the modal index is actually of the order of 10^{-6} to 10^{-5} and this illustrates just how

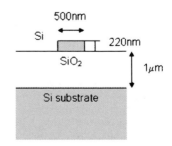

Fig. 12.2 Silicon photonic wire structure investigated as a COST P11 modelling exercise. Reproduced from http://w3.uniroma1.it/energetica/

demanding the fabrication constraints are when dealing with high-index contrast structures with sub-wavelength feature size.

12.4.3 Second Harmonic Generation

There is a growing interest in the non-linear quadratic interactions of the second order non-linear coefficient $\chi^{(2)}$ due to its ability to provide versatile and efficient frequency conversion, such as second harmonic generation (SHG), frequency up and down conversion and parametric amplification [44]. The FDTD method can provide time-domain modelling of a pulsed beam in $\chi^{(2)}$ material over a wide bandwidth. However, the method requires tremendous computational effort even when analysing a 2D waveguide, [45]. The alternative time-domain beam-propagation method (TD-BPM) allows the simulation of optical structures in the time-domain without the potentially large memory and computational overheads of FDTD and TLM. In order to model second harmonic generation in non-linear optical devices two parabolic coupled non-linear wave equations, for the fundamental field and the second harmonic field are solved simultaneously in the time-domain [15]. The method described in [15] is wide-angled and bi-directional and so is able to account for all wave-medium interactions such as scattering and reflection. As part of the COST P11 exercise this model was applied to non-linear optics problems, including second-harmonic generation in slow wave structures, where its results were compared against results obtained using the non-linear eigenmode expansion method. The structure studied is shown in the schematic diagram of Fig. 12.3 and the modelling challenges include the need to include radiation losses, up to 20 periods and to achieve phase matching.

The second harmonic conversion efficiency is defined as the ratio between the final total second harmonic intensity and the total initial fundamental field intensity. The spectra of the second harmonic field conversion efficiency for a twenty period grating, calculated using the EEM [29] and the TD-BPM methods, are compared in Fig. 12.4. The two sets of results show remarkable agreement, especially consider-

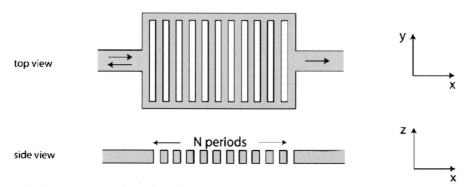

Fig. 12.3 Schematic view of the 2D air/Al$_{0.3}$Ga$_{0.7}$As PBG waveguide grating structure studied. Adapted from http://w3.uniroma1.it/energetica/

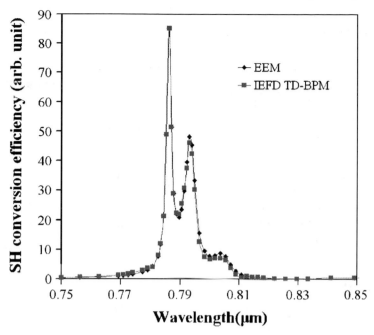

Fig. 12.4 A comparison of the SH conversion efficiency for the 2D nonlinear 20 period PBG waveguide grating obtained from the IEFD TD-BPM with that of the Eigenmode Expansion method (EEM). Reproduced from [15] with the permission of the author

ing that the EEM is a frequency-domain technique so its spectrum is obtained from a set of calculations at each required frequency whereas the time-domain data is obtained from coupled time-domain runs, assuming centre wavelengths of 1.55 and 0.775 μm for the fundamental and second harmonic fields. A more detailed comparison of the two approaches [46] reveals the relative merits of the two analysis approaches. For instance the EEM method is very efficient, and the accuracy of the TD-BPM deteriorates away from the assumed centre wavelength(s) but on the other hand a time-domain method is needed to study the dynamics of pulse propagation in such structures.

12.4.4 Photonic Crystal Membrane Waveguides

Photonic crystal membrane waveguides combine a thin guiding layer, typically of high refractive index semiconductor material, suspended in air or on a lower refractive index (e.g. oxide) film. A regular pattern of holes, of order 100 nm in diameter, is then patterned into the thin semiconductor layer to produce forbidden bands. The introduction of single or multiple line defects can produce strongly confined waveguides whereas other geometries, such as point defects, work as compact, high-Q resonators. In order to understand the low-loss limits of photonic crystal membrane waveguides 3D modelling is essential. A number of methods have been applied to this task, in both frequency- and time-domains. In the frequency-domain the loss is accessed via the imaginary part of the complex propagation. This can be calculated using, for example, a 3D numerical technique such as the finite difference method [47], or a modal expansion technique [48,49]. Two techniques have been used to quantify the loss using time-domain numerical methods. In the first [50], whole waveguide structure is discretized in 3D space and the total power, derived from the Poynting vector, is calculated at two closely spaced planes in the guide from which a loss per unit length is evaluated. In the second [51], simulations are performed for two different photonic crystal membrane waveguide lengths, including input and output waveguide sections. Whilst this approach is computationally intensive it is general and more representative of the experimental situation. Results from the various methods show good agreement [51]. Once again the comparisons show the flexibility of the time-domain methods (for example the ability to model the effect of surface roughness or disorder in a physically realistic way), being balanced against the computational efficiency of the frequency-domain methods.

12.5 A Flavour of Some Emerging Problems

Frequency-domain or circuit-based methods are the tools of choice when optimising device performance against some systems-imposed requirement. As we have seen, defects in membrane photonic crystals have led to the development of compact optical components. Many groups have applied optimisation techniques to refine some of these structures, for example bends [52] and splitters [53]. The provision of various functionalities through arrays of scattering centres whose position is not constrained by the layout of a formal lattice is also being considered, a concept referred to as an Optical Application Specific Integrated Circuit or OASIC [54]. In the context of a Genetic Algorithm (GA) optimisation tool a very large number of possible trial designs is considered, the performance of each of which must be assessed using electromagnetic simulation. Calculation errors must be traded-off against runtime as, with the simulation parameters dynamically changed in response to the accuracy needed at different stages in the GA cycle. Furthermore, methods of different accuracies can be used at different stages of optimisation by ensuring adequate control of solver accuracy.

Recent progress in the fabrication of photonic nano-materials and sub-wavelength optics provides the enabling science and technology for nano-scale optics. However, the understanding of the interplay between light and nanostructures, notwithstanding the dynamic and all-optical control of such structures, is still incomplete. The electromagnetic description of small-scale metallic structures, for instance 1D and 2D gratings and local field enhancement due to the generation of plasmons in thin films and metal nano-particles also provide ongoing challenges with many practical applications. Existing general tools such as mode matching and FDTD codes are widely used to address these problems, which are also well suited to treatments using specialist nodal models within TLM or accurate sub-sampling techniques within a finite-difference based method. Much attention has been paid to accurate finite difference discretizations for both propagation and mode solving work, in particular the correct treatment of fields at material boundaries [55–57] and also the ability to describe fine features such as quantum wells or metal-dielectric plasmonic waveguides [15, 58, 59] using a coarser mesh than that dictated by the smallest feature size. Generally speaking, however, much work still remains to be done in order to capitalise on new characteristics obtained by combining materials properties and geometrically patterning, especially when seeking all-optical operation at the highest speeds.

12.6 Conclusions

Techniques in Computational Electromagnetics have developed considerably, to the point where a wide range of powerful modelling techniques is available for photonic simulation and design. In this paper we have reviewed some of these techniques and their application to the modelling of linear, nonlinear and active photonic crystals. Numerical simulation has a major innovative role to play in understanding and exploiting the basic mechanisms of light-matter interaction in micro- and nanostructured materials, including metals (plasmonics), especially the trade-offs between strong localization, propagation losses and non-linear effects. This said, the modeller's point of view was expressed very clearly by Freude (COST P11 Training School on 'Modelling and simulation techniques for linear, nonlinear and active photonic crystals', June 2006) "Blind trust in numerical data is dangerous; some basic understanding of the algorithm is necessary; one often operates at the limits of resource; so when are the data reliable?".

Time-domain numerical techniques are extremely flexible and have the advantage of compatibility with parallel computing techniques. However, they are generally computationally demanding, especially when three space dimensions are considered and depending on the particular problem to be solved other approaches can offer significant computational and accuracy advantages, and/or additional physical insight to device behaviour. The computational efficiency of long-established techniques such as the beam propagation method means that they still have a considerable role to play in photonics design. One remaining challenge within the domain of photonic crystals is to integrate such (time-domain) electromagnetic codes with models of other physical processes, such as thermal and electronic effects, that can not only impact on device performance but also occur on different characteristic timescales

Acknowledgements We acknowledge, with thanks, the valued contributions from the research groups and individuals who have contributed to the COST P11 Working Group 2 exercises.

References

1. Taflove A., Hagness S.C.: Computational electrodynamics: The finite difference time-domain method, Artech House, (1995).
2. Scarmozzino R., Gopinath A., Pregla R., Helfert S.: Numerical techniques for modeling guided-wave photonic devices, IEEE Jor. of Selected Topics in Quantum Electronics, Vol. 6(1), 150–162, (2000).
3. Christopoulos, C.: The Transmission-Line Modeling Method: TLM. Piscataway, NJ: IEEE Press, (1995).
4. Sewell P., Wykes J.D., Benson T.M., Christopoulos C., Thomas D.W.P., Vukovic A.: Transmission-line modeling using unstructured triangular meshes, IEEE Trans. Microwave Theory and Techniques, Vol. 52(5), 1490–1497, (2004).
5. Sewell P., Benson T.M., Christopoulos C., Thomas D.W.P., Vukovic A., Wykes J.D.: Transmission line modeling (TLM) based upon unstructured tetrahedral meshes, IEEE Trans Microwave Theory and Techniques, Vol. 53, 1919–1928, (2005).
6. Luebbers R.J., Hunsberger F.: FDTD for Nth-order dispersive media, IEEE Trans. Antennas Propag., Vol. 40(11), 1297–1301, (1992).
7. Paul J., Christopoulos C., Thomas D.W.P.: Generalized material models in TLM-Part I: Materials with frequency-dependent properties, IEEE Trans. Antennas Propag., Vol. 47(10), 1528–1534, (1999).
8. Janyani, V.: Modelling of dispersive and nonlinear materials for optoelectronics using TLM, PhD thesis University of Nottingham, (2005).
9. Liu Y., Sewell P., Biwojno K., Christopoulos C.: A generalized node for embedding subwavelength objects into 3-D Transmission-Line Models, IEEE Trans. Electromagnetic Compatibility, Vol. 47(4), 749–955, (2005).
10. Yamauchi, J.: Propagating Beam Analysis of Optical Waveguides, Research Studies Press Limited, (2003), ISBN 0 86380 265 6.
11. Shibayama J., Yamahira A., Mugita T., Yamauchi J., Nakano H.: A finite-difference time-domain beam-propagation method for TE- and TM-wave analyses, J. Lightwave Technol., Vol. 21(7), 1709–1715, (2003).
12. Shibayama J., Muraki M., Yamauchi J., Nakano H.: Comparative study of several time-domain methods for optical waveguide analyses, J. Lightwave Technol., Vol. 23(7), 2285–2293, (2005).
13. Koshiba M., Tsuji Y., Hikari, M.: Time-domain beam propagation method and its application to photonic crystal circuits, J. Lightwave Technol., Vol. 18(1), 102–110, (2000).
14. Hu B., Sewell P., Vukovic A., Paul J., Benson T.M.: General approach for Nth order dispersive material, IEE Proceedings: Optoelectronics, 153(1), 13–20, (2006).
15. Hu B.: Advanced Beam Propagation Methods for the Analysis of Integrated Photonic Devices, PhD Thesis, University of Nottingham, (2006).
16. Lim J.J., Benson T.M., Larkins E.C., Sewell P.: "Wideband finite difference time-domain propagation method" Microwave and Optical Technology Letters, Vol. 34(4), 243–247, (2002).
17. Nerukh, A.G., Scherbatko I.V., Marciniak M.: Electromagnetics of modulated media with applications to photonics. Warsaw, Poland: Nat. Inst. Telecommun., (2001).
18. Vukovic A., Bekker E.V., Sewell P., Benson T.M., Paul J., Sakhnenko N.K., Nerukh A.G.: Proc. 23rd Annual Review of Progress in Applied Computational Electromagnetics, Verona, 532–537, March (2007).
19. Hochman A., Leviatan Y.: Analysis of strictly bound modes in photonic crystal fibers by use of a source-model technique, J. Opt. Sco. Am. A21, 1073–1081, (2004).
20. Hochman A., Leviatan Y.: Calculation of confinement losses in photonic crystal fibers by use of a source-model technique, J. Opt. Soc. Am. B22, 474–480, (2005).
21. Boriskina S.V., Nosich A.I.: Radiation and absorption losses of the whispering-gallery-mode dielectric resonators excited by a dielectric waveguide, IEEE Trans. Microwave Theory Tech., Vol. 47, 224–231, (1999).

22. Boriskina S.V., Benson T.M., Sewell P., Nosich A.I.: Highly efficient full-vectorial Integral Equation solution for the bound, leaky and complex modes of dielectric waveguides. IEEE Journal of Selected Topics in Quantum Electronics, Vol. 8(6), 1225–1232, (2002).
23. Boriskina S.V., Benson T.M., Sewell P., Nosich A.I.: Q-factor and emission pattern control of the whispering-gallery modes in notched microdisk resonators, IEEE J. Selected Topics in Quantum Electronics, Vol. 12(1), 52–58, (2006).
24. Smotrova E.I., Nosich A.I., Benson T., Sewell P.: Cold-cavity thresholds of microdisks with uniform and non-uniform gain: quasi-3D modeling with accurate 2D analysis, IEEE J. Selected Topics in Quantum Electronics, Vol. 11(5), 1135–1142, (2005).
25. Smotrova E.I., Nosich A.I., Benson T., Sewell P.: Optical coupling of whispering-gallery modes in two identical microdisks and its effect on photonic molecule lasing, IEEE J. Selected Topics in Quantum Electronics, Vol. 12(1), 78–85, (2006).
26. Szpulak M., Urbanczyk W., Serebryannikov E., Zheltikov A., Hochman A., Leviatan Y., Kotynski R., Panajotov K.: Comparison of different methods for rigorous modeling of photonic crystal fibers, Optics Express, Vol. 4(2), 5699–5714, (2006).
27. Morichetti F., Melloni A., Petracek J., Bienstman P., Priem G., Maes B., Lauritano M., Bellanca G.: Self-phase modulation in slow-wave structures: A comparative numerical analysis, Optical and Quantum Electronics, Vol. 38, 761–780, (2006).
28. Bienstman P., Baets R.: Optical modelling of photonic crystals and VCSELs using Eigenmode Expansion and Perfectly Matched Layers, Optical and Quantum Electronics, Vol. 33, 327–341, (2001).
29. Maes B., Bienstman P., Baets R.: Modeling second-harmonic generation by use of mode expansion, J. Opt. Soc. Am. B 22, 1378–1383, (2005).
30. Hadley G.R.: Wide-angle beam propagation using Pade approximant operators, Opt. Lett., Vol. 17(20), 1426–1428, (1992).
31. Yevick, D.: IEEE Photon. Technol. Lett., Vol. 12, 1636–1638, (2000).
32. Clauberg R., Von Allmen P.: Vectorial beam propagation method for integrated optics, Electron. Lett., Vol. 27, 654, (1991).
33. Huang W.P., Xu C.L.: Simulation of three-dimensional optical waveguides by a full-vector beam propagation method, IEEE J. Quantum Electron., Vol. 29, 2639, (1993).
34. Lu Y.Y.: Some techniques for computing wave propagation in optical waveguides, Commun. Comput. Phys., Vol. 1(6), 1056–1075, (2006).
35. Sewell P., Benson T.M., Vukovic A.: A stable Du Fort Frankel beam propagation method for lossy structures and those with Perfectly Matched Layers, Journal of Lightwave Technology, Vol. 23(1), 374–381, (2005).
36. Bekker E.V., Sewell P., Benson T.M., Vukovic A.: Wide-angle alternating-direction implicit finite-difference BPM, Proceedings 9th International Conference on Transparent Optical Networks, Vol. 1, 250–253, (2007), IEEE Catalog Number 07EX1796C, ISBN 1-4244-1249-8.
37. Benson T.M., Sewell P., Sujecki S., Kendall P.C.: Structure related beam propagation, Optical and Quantum Electronics, Vol. 31, 689–703, (1999).
38. Hadley G.R.: Numerical simulation of waveguides of arbitrary cross-section, Int. J. Electron. Commun., Vol. 58, 1–7, (2004).
39. Melloni A., Floridi M., Morichetti F., Martinelli M.: Equivalent circuit of Bragg gratings and its application to Fabry–Pérot cavities, Journ. Opt. Society Am. A, Vol. 20(2), 273–281, (2003).
40. Benson T.M., Sewell P., Vukovic A., Christopoulos C., Thomas D.W.P., Nosich A.: Hybrid methods for efficient electromagnetic simulation, Proceedings Progress in Electromagnetics Research Symposium March (2006), Cambridge, USA, p. 182.
41. Hu B., Sewell P., Vukovic A., Lim J.J., Benson T.: Numerical techniques for multimode interference couplers, Proc SPIE 5728, 174–183, (2005), Y Sidorin and CA Waechter eds.
42. Vukovic A., Benson T.M., Sewell P., Bozeat R.J.: "Novel hybrid method for efficient 3D fibre to chip coupling analysis" IEEE J. Selected Topics in Quantum Electronics, Vol. 8, 1285–1293, (2002).

43. Bienstman P., Selleri S., Rosa L., Uranus H.P., Hopman W.C.L., Costa R., Melloni A., Andreani L.C., Hugonin J.P., Lalanne P., Pinto D., Obayya S.S.A., Dems M., Panajotov K.: Modelling leaky photonic wires: A mode solver comparison, Optical and Quantum Electronics, Vol. 38(9-11), 731–759, (2006)
44. Agrawal G.P.: Contemporary nonlinear optics, New York: Academic Press, (1992)
45. Dumeige Y., Raineri F., Levenson A.: Second-harmonic generation in one-dimensional photonic edge waveguides, Phys. Rev. E, Vol. 68, 066617-1–066617-7, (2003).
46. Maes B., Bienstman P., Baets R., Hu B., Sewell P., Benson T.: Modeling second-harmonic generation in high-index-contrast devices, Optical and Quantum Electronics, published online, 3 June 2008, DOI 10.1007/s11082-008-9217-6.
47. Hadley G.R.: Out-of-plane losses of line-defect photonic crystal waveguides, IEEE Photonics Technology. Letters, Vol. 14(5), 642–644, (2002).
48. Lalanne P.: Electromagnetic Analysis of Photonic Crystal Waveguides Operating Above the Light Cone", IEEE Journal of Quantum Electronics, Vol. 38, No. 7, July (2002).
49. Sauvan C., Lalanne P., Rodier J.C., Hugonin J.P., Talneau A.: Accurate modeling of line-defect photonic crystal waveguides, IEEE Photonics Technology Letters, Vol. 15(9), 1243–1245, (2003).
50. Désières Y., Benyattou T., Orobtchouk R., Morand A., Benech P., Grillet C., Seassal C., Letartre X., Rojo-Romeo P., Viktorovitch P.: Propagation losses of the fundamental mode in a single line-defect photonic crystal waveguide on an InP membrane, Journal of Appl. Phys., Vol. 92(5), 2227–2234, (2002).
51. Cryan M.J., Wong D.C.L., Craddock I.J., Yu S., Rorison J., Railton C.J.: Calculation of losses in 2-D photonic crystal membrane waveguides using the 3-D FDTD method, IEEE Photonics. Technology Letters., Vol. 17(1), 58–60, (2005).
52. Jensen J.S., Sigmund O., Frandsen L.H., Borel P.I., Harpoth A., Kristensen M.: Topology design and fabrication of an efficient double 90° photonic crystal waveguide bend, IEEE Photon. Tech. Lett., Vol. 17(6), 1202–1204, (2005).
53. Smajic J., Hafner C., Erni D.: Optimization of photonic crystal structures, J. Opt. Soc. Am. A21(11), 2223–2232, (2004).
54. Sewell P., Benson T.M., Vukovic A., Styan C.: Adaptive simulation of optical ASICs, Proceedings 9th International Conference on Transparent Optical Networks, Rome, July (2007). Vol. 1, 244–248, 2007, IEEE Catalog Number 07EX1796C, ISBN 1-4244-1249-8.
55. Hadley G.R.: High-accuracy Finite-Difference equations for dielectric waveguide analysis II: Dielectric corners, J. of Lightwave Techn., Vol. 20(7), 1219–1231, (2002).
56. Thomas N., Sewell P., Benson T.M.: A new full-vectorial higher order finite-difference scheme for the modal analysis of rectangular dielectric waveguides, Journal of Lightwave Technology, Vol. 25(9), 2563–2570, (2007).
57. Thomas N.: Finite difference methods for the modal analysis of dielectric waveguides with rectangular corners, PhD Thesis, University of Nottingham, (2004).
58. Chiou Y-P., Chiang Y-C., Chang H-C.: Improved three-point formulas considering the interface conditions in the Finite-Difference analysis of step-index optical devices, Journal of Lightwave Technol., Vol. 18, 243–251, (2000).
59. Wykes J.G., Sewell P., Vukovic A., Benson T.M.: Subsampling of fine features in Finite Difference simulations, Microwave and Optical Technology Letters, Vol. 44(1), 95–101, (2005).

13 Plane-Wave Admittance Method and its Applications to Modelling Photonic Crystal Structures

Maciej Dems[1], Tomasz Czyszanowski[1,3], Rafał Kotyński[2], and Krassimir Panajotov[3,4]

[1] Institute of Physics, Technical University of Lodz, ul. Wolczanska 219, 90-924 Łódź, Poland
[2] Warsaw University, Faculty of Physics, ul. Pasteura 7, 02-093 Warsaw, Poland
[3] Department of Applied Physics and Photonics, Vrije Universiteit Brussel, Pleinlaan 2, 1050 Brussels, Belgium
[4] Institute of Solid State Physics, 72 Tzarigradsko Chaussee Blvd., 1784 Sofia, Bulgaria

Abstract. This chapter presents the mathematical basis of the plane-wave admittance method (PWAM), which is a combination of the method of lines and plane-wave expansion. In the first part of the chapter the most important equations are derived and the used admittance transfer procedure is reviewed. In the second part we show the examples of modelling photonic-crystals-based vertical-cavity surface-emitting lasers with PWAM. We analyse the resonant wavelength and modal losses as a function of photonic crystal etching depth. Next we discuss the photonic crystal parameters most suitable for obtaining single-mode regime. Finally, we consider the possibilities of using photonic crystals for stabilisation of the emitted-light polarisation and suggest a new design of birefringent and dichroic VCSEL.

Key words: plane-wave admittance method; Photonic Crystal Vertical Cavity Surface Emitting Lasers (PC-VCSELs); polarisation control

13.1 Introduction

Three-dimensional modelling of photonic crystals, metamaterials or complex subwavelength photonic devices is usually connected with extremely large computational demands. General purpose methods based on finite elements, finite differences, expansion in the Fourier or other function basis are widely known but their use is always strongly limited by the available computational resources. Heavy numerical requirements reflect the need for the full and rigorous treatment of Maxwell equations – including polarisation effects, the presence of field discontinuities at the material boundaries, which in photonic crystals form a sub-wavelength microstructure, huge resonant field enhancements at the boundaries of conducting materials etc. Therefore, the development of novel numerical methods is important, especially such that preserve the generality of their application but at the same time would efficiently make use of any regularity found in the particular structure. While for ideal photonic crystals the regularities could be expressed in terms of simple translational and rotational symmetries, complete photonic devices rather contain a finite number of directly or recursively repeated parts.

The present chapter includes the description of a novel method called the Plane-Wave Admittance Method (PWAM) [1, 2] which tries to answer these challenges. We also provide a number of results related to its application to modelling of Photonic Crystal Vertical Cavity Surface Emitting Lasers (PC-VCSELs) [3]. The PWAM was first introduced by Dems et al. [1], and it shares important elements with the Method of Lines (MoL) [4, 5] and the Plane-Wave Method (PWM), which have been widely used for many years. PWAM is a frequency-domain method, appropriate for the analysis of anisotropic, lossy, and even magnetic or negative refractive index photonic structures surrounded by arbitrarily shaped Uniaxial Perfectly Matched Layers (UPML) [6, 7]. The modelled structure is assumed to consist of parallel layers, and PWAM is especially efficient when certain layers appear multiple times within the stack. Notably, PC-VCSELs state a good example of a photonic device for which PWAM provides significant computational savings as compared to general purpose methods. However, the modelling of other photonic structures is also possible [8], including arbitrarily-shaped ones, although then the layers only correspond to a sufficiently dense discretization of the system along a chosen axis.

PWAM consists of two major steps. First, every layer is analysed with the Plane-Wave Expansion (PWE) formulated with a general material model with a tensor-form, complex-valued permittivity and permeability. This part of the method taken alone [9] allows for efficient modal analysis of waveguide structures such as index-guiding or photonic band-gap guiding anisotropic photonic crystal fibres (PCF) [10, 11] with UPML boundary conditions. In the analysis of fully three-dimensional structures, the PWM-based description of all layers is followed by the matching scheme based on the admittance-matrix transfer method. This part of PWAM is common with the method of lines (MoL).

13.2 PWAM Theory

In this section we show the mathematics laying behind the plane-wave admittance method [1]. As mentioned above, the method is based on the combination of the plane-wave expansion with the method of lines and admittance transfer technique.

13.2.1 General Transmission Line Equations and Plane-Wave Basis

Consider an anisotropic linear dielectric or semiconductor material without any free charges ($\rho = 0$) nor currents ($j = 0$). In such material, the time-independent Maxwell equations read

$$\nabla \times E = -i\omega\mu\mu_0 H , \tag{13.1}$$

$$\nabla \times H = i\omega\varepsilon\varepsilon_0 E , \tag{13.2}$$

where $i^2 = -1$, ω is the angular frequency of light and physical observables are the real parts of E and H.

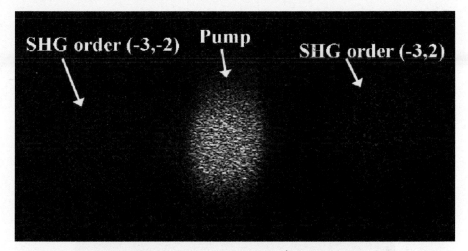

Fig. 3.7 Pump and two simultaneous second harmonic signals from rectangularly poled stoichiometric LiTaO$_3$ crystal. The pump enters nearly parallel to the $(1,0)$ RLV, and two second harmonic beams correspond to the $(-3,2)$ and $(-3,-2)$ QPM orders. This figure is reproduced from the thesis of Nili Habshoosh, Tel-Aviv University (2007)

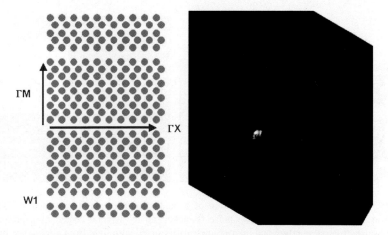

Fig. 4.10 Image of third harmonic signal generated by a W1 waveguide with the line defect along ΓM in the vertical direction and with the pump impinging at normal incidence. The pattern is created on a screen parallel to the sample surface. The *bright spot at the centre* is the aperture for the pump beam. Besides the line defect diffraction along ΓM, different diffraction spots due to the triangular PhC lattice are visible. On the *left* and on the *right* a *curved line* of weak spots between the prominent first order diffraction represents the mixed diffraction orders between triangular pattern and defects

Fig. 6.2 Supercontinuum generation in a photonic-crystal fibre

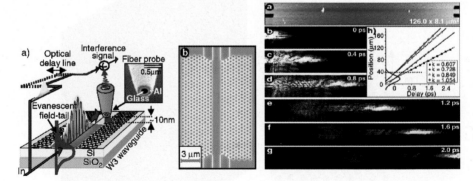

Fig. 9.13 *Left*: (*a*) Schematic representation of a pulse tracking experiment on a W3 PhCW. The evanescent field of a propagating pulse is picked up by a metal coated fibre probe with a subwavelength-sized aperture and interferometrically mixed with light from a reference branch. (*b*) Top view of the PhCW under study, although with a shorter device length. *Right*: A time-resolved NSOM measurement on a W3 PhCW (a 460 nm). (*a*) Topographic image of the structure rotated 90 with respect to Fig. 9.13(b). (*b–g*) The optical amplitude in the W3 PhCW for different reference times (400 ± 1 fs between frames; all frames have the same colour scale). Reprinted from [45]

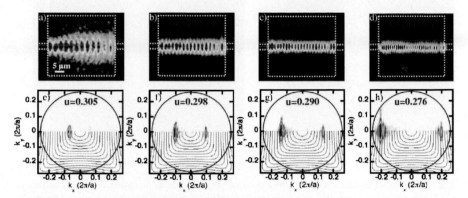

Fig. 9.15 (**a–d**) Near-field image of the Bloch wave mode in a square lattice PhC tile of 80 periods for different reduced energies u. Below each real space image is shown the corresponding far-field image from (**e**) to (**h**). The *thin black curves* in the lower half space of the far-field frame are the 2D PWE theoretical EFSs starting from $u = 0.315$ in the centre and drawn with a step of -0.005. The far-field image is limited by the maximum propagating wave vector $k_{max} = 0.2636 \cdot 2\pi/a$ (the *red circle* in (**e–h**) delimits the output pupil of the objective that defines k_{max}). The corresponding diffraction limited spatial resolution is $0.85\,\mu m$. Reprinted from [49]

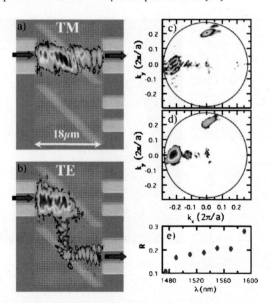

Fig. 9.16 (**a**) and (**b**) near-field image of the TM and TE waves, respectively, propagating in the double slab polarizer beam splitter (*colour coded*: increasing intensity from *blue* to *red*) superimposed with an optical microscope image of the structure (*black* and *white*), (**c**) emission diagram of the TE wave imaged in the Fourier space at $\lambda = 1550\,nm$, (**d**) same as **c**) with the emission of the second splitter filtered in the intermediate image plane, (**e**) reflection coefficient of the first slab deduced from the far-field pattern measured for different wavelengths as in **c**). Reprinted from [49]

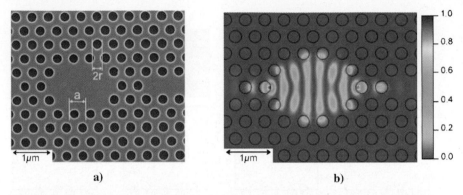

a)   b)

Fig. 10.7 Fabry–Perot-like resonator in a photonic crystal slab. **a)** Scanning electron microscope (SEM) image of the device that was fabricated (at IMEC, Leuven, Belgium) in a silicon on insulator (SOI) technology [5]; silicon ($n = 3.45$) top layer thickness 220 nm, lattice period $a = 440$ nm, air hole radius $r = 270$ nm. **b)** Intensity distribution of resonant mode at $\lambda = 1550$ nm, calculated with a 2D-finite-difference time-domain (FDTD) method, using an effective index method for determining the silicon slab effective index $n_{\mathrm{eff}} = 2.9$ [14]

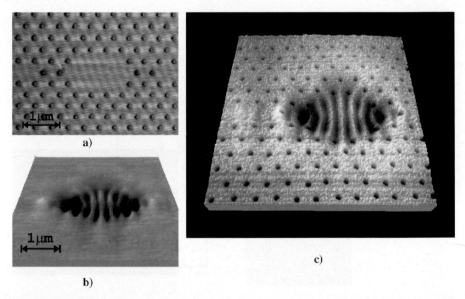

a)

b)   c)

Fig. 10.10 T-SNOM imaging of the photonic crystal resonator shown in Fig. 10.7a. **a)** Topography measured by AFM. **b)** Optical transmission through the resonator at the undisturbed resonance wavelength $\lambda_r = 1539.25$ nm, as a function of AFM tip position; *dark regions* correspond to low light transmission, hence strong interaction of the AFM tip with the optical field. **c)** Composite image, combining data of **a)** and **b)**, showing the location of the optical field distribution with respect to the resonator geometry [14]

Fig. 10.12 T-SNOM image similar to Fig. 10.10, but wavelength, now 1541.5 nm (off resonance). Detuning by the AFM tip could bring it into resonance again, resulting in inverted image contrast [14]

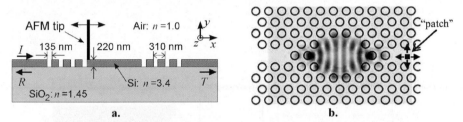

a.   **b.**

Fig. 10.15 Two-dimensional T-SNOM models. **a**) Side view. The resonator is modelled as a slab waveguide cavity enclosed by two short Bragg gratings; the AFM-probe is modelled as a semi-infinite plane perpendicular to and in contact with the slab. **b**) Top view. The slab waveguide outside the photonic crystal holes is modelled by its effective index; the probe is represented by a small "patch" of material having an increased refractive index

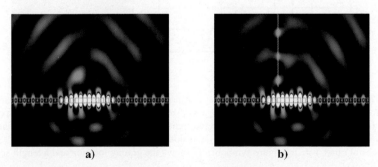

a)   b)

Fig. 10.19 Calculated field distributions at the undisturbed resonance wavelength. **a**) Without probe. **b**) With probe at 0.5 μm height above hot spot [18]

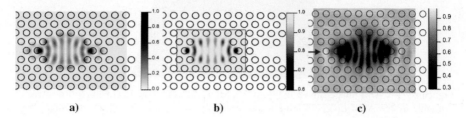

Fig. 10.20 Top-view model, comparison to measurement. **a)** Calculated intensity distribution. **b)** Simulated T-SNOM response by recalculating the model many times for different "patch" locations in order to represent the T-SNOM probe scanning. **c)** Actual T-SNOM measurement data overlaid with structure data [18]

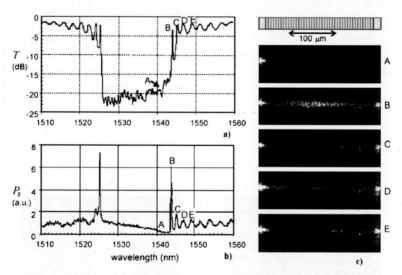

Fig. 10.23 Combined spectral and far-field scattering measurements. **a)** Normalised transmission spectrum of the waveguide grating, showing a stop-band. **b)** Spatially averaged scatter spectrum, derived from camera images. **c)** Typical scattered-light images at 5 different wavelengths, indicated by letters *A–E*, also shown in graphs **a)** and **b)** *A*: inside the stop-band, *B–E* Fabry–Perot like grating resonances of order 1-4. Top figure symbolically shows the grating location and a length scale

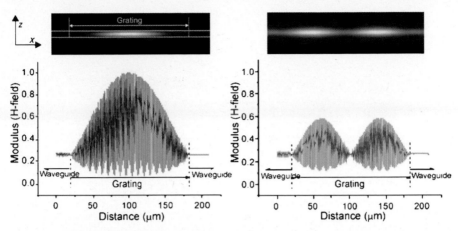

Fig. 10.24 Calculated magnetic field amplitude distribution along the length of the grating. *Top*: field distribution in the grated waveguide cross-section (see coordinate axes in Fig. 10.21). *Bottom*: field distribution along the centre line of the waveguiding layer. Left lowest order mode (corresponding to label *B* in Fig. 10.23); right second order mode (label *C*). It can be seen that the large-scale distribution is actually an envelope of the Floquet–Bloch mode amplitude

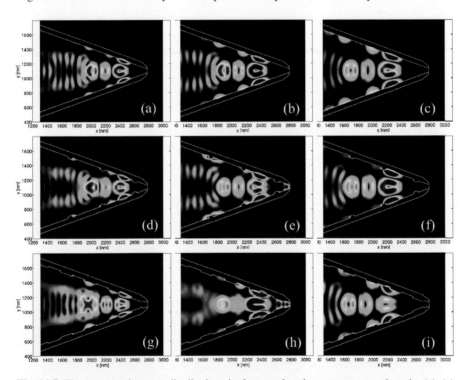

Fig. 11.7 Time-averaged energy distributions in three analysed structures: smooth probe (**a**)–(**c**), and probes with six circular grooves of $\Lambda = 445$ nm and $s = 173$ nm (**d**)–(**f**), and six oval corrugations of $\Lambda = 325$ nm and $s = 87$ nm (**g**)–(**i**). Wavelengths are 450 nm, 510 nm and 600 nm from left to *right columns*, respectively

Now consider an anisotropic material in the Cartesian coordinates. Assuming that both permittivity and permeability of the material can be represented as diagonal tensors, the Eqs. (13.1) and (13.2) can be expanded as

$$-\partial_z E_y + \partial_y E_z = -i\omega\mu_x\mu_0 H_x , \tag{13.3}$$

$$\partial_x E_z - \partial_z E_x = -i\omega\mu_y\mu_0 H_y , \tag{13.4}$$

$$-\partial_y E_x + \partial_x E_y = -i\omega\mu_z\mu_0 H_z , \tag{13.5}$$

$$-\partial_z H_y + \partial_y H_z = i\omega\varepsilon_x\varepsilon_0 E_x , \tag{13.6}$$

$$\partial_x H_z - \partial_z H_x = i\omega\varepsilon_y\varepsilon_0 E_y , \tag{13.7}$$

$$-\partial_y H_x + \partial_x H_y = i\omega\varepsilon_z\varepsilon_0 E_z . \tag{13.8}$$

∂_ξ means partial derivative in the ξ direction. From the Eqs. (13.5) and (13.8) it is possible to express the z components of the electric and magnetic vectors as

$$H_z = \left[-\frac{i}{k_0\eta_0\mu_z}\partial_x \quad -\frac{i}{k_0\eta_0\mu_z}\partial_y \right] \begin{bmatrix} -E_y \\ E_x \end{bmatrix} , \tag{13.9}$$

$$E_z = \left[-\frac{i\eta_0}{k_0\varepsilon_z}\partial_y \quad -\frac{i\eta_0}{k_0\varepsilon_z}\partial_x \right] \begin{bmatrix} H_x \\ H_y \end{bmatrix} , \tag{13.10}$$

where $k_0 = \omega/c = \omega(\mu_0\varepsilon_0)^{1/2}$ is the normalised frequency and $\eta_0 = (\mu_0/\varepsilon_0)^{1/2}$ is the free space impedance. Substituting (13.9) and (13.10) into (13.3), (13.4), (13.6) and (13.7), we can obtain the Generalised Transmission Line (GTL) equations

$$\partial_z \begin{bmatrix} -E_y \\ E_x \end{bmatrix} = -i\frac{\eta_0}{k_0} \begin{bmatrix} \partial_y\varepsilon_z^{-1}\partial_y + \mu_x k_0^2 & -\partial_y\varepsilon_z^{-1}\partial_x \\ -\partial_x\varepsilon_z^{-1}\partial_y & \partial_x\varepsilon_z^{-1}\partial_x + \mu_y k_0^2 \end{bmatrix} \begin{bmatrix} H_x \\ H_y \end{bmatrix} , \tag{13.11}$$

$$\partial_z \begin{bmatrix} H_x \\ H_y \end{bmatrix} = -i\frac{1}{\eta_0 k_0} \begin{bmatrix} \partial_x\mu_z^{-1}\partial_x + \varepsilon_y k_0^2 & \partial_x\mu_z^{-1}\partial_y \\ \partial_y\mu_z^{-1}\partial_x & \partial_y\mu_z^{-1}\partial_y + \varepsilon_x k_0^2 \end{bmatrix} \begin{bmatrix} -E_y \\ E_x \end{bmatrix} . \tag{13.12}$$

In the derivation of above relations we have not used any approximation. Thus they are valid in any point of any structure to be analysed, providing there is no non-linear material. This two equations are the starting point for any further analysis.

Now, consider electromagnetic field in some structure exhibiting two-dimensional periodicity, i.e.

$$\varepsilon(\mathbf{r}) = \varepsilon(\mathbf{r}+\mathbf{a}_i) \quad (i=1,2) , \tag{13.13}$$

$$\mu(\mathbf{r}) = \mu(\mathbf{r}+\mathbf{a}_i) \quad (i=1,2) , \tag{13.14}$$

with $\mathbf{a}_1$ and $\mathbf{a}_2$ being elementary lattice vectors in xy plane. Because of this periodicity the solutions of the Maxwell equations obey Bloch theorem, i.e. both the electric and magnetic fields can be represented as

$$E(\mathbf{r}) = \bar{E}(\mathbf{r})\exp(-i\mathbf{k}_{xy}\cdot\mathbf{r}) , \tag{13.15}$$

$$H(\mathbf{r}) = \bar{H}(\mathbf{r})\exp(-i\mathbf{k}_{xy}\cdot\mathbf{r}) , \tag{13.16}$$

where $\bar{E}(r)$, $\bar{H}(r)$ are periodic functions and k_{xy} is a projection of the mode wavevector into the xy-plane. Furthermore the functions $\bar{E}(r)$ and $\bar{H}(r)$ can be expanded into two-dimensional Fourier (plane-wave) series as

$$\bar{E}(r) = E^g \exp(ig \cdot r) = E^g \, |\varphi_g\rangle \, , \tag{13.17}$$

$$\bar{H}(r) = H^g \exp(ig \cdot r) = H^g \, |\varphi_g\rangle \, , \tag{13.18}$$

where we use the Einstein summation convention[1] and g is the reciprocal lattice vector

$$g = l_1 b_1 + l_2 b_2 \, , \tag{13.19}$$

$$a_i \cdot b_j = 2\pi \delta_{ij} \tag{13.20}$$

with l_1 and l_2 being arbitrary integers and δ_{ij} the Kronecker delta. As we perform Fourier expansion only in xy-plane, the coefficients E^g and H^g are functions of z.

Using the Bloch theorem together with Eqs. (13.17) and (13.18), we can represent both fields as

$$E(r) = E^g \, |\varphi_{g-k_{xy}}\rangle \, , \tag{13.21}$$

$$H(r) = H^g \, |\varphi_{g-k_{xy}}\rangle \, . \tag{13.22}$$

Now consider the GTL equations. As Fourier basis is orthonormal, i.e.

$$\langle \varphi_g | \varphi_{g'} \rangle = \delta_{gg'} \, , \tag{13.23}$$

we can introduce (13.21) and (13.22) into (13.11) and (13.12) and left-multiply them by $\langle \varphi_{g-k_{xy}} |$ to obtain

$$\partial_z \begin{bmatrix} E_y^g \\ E_x^g \end{bmatrix} = -i \frac{\eta_0}{k_0} \left\langle \varphi_{g-k_{xy}} \middle| \begin{matrix} \partial_y \varepsilon_z^{-1} \partial_y + \mu_x k_0^2 & -\partial_y \varepsilon_z^{-1} \partial_x \\ -\partial_x \varepsilon_z^{-1} \partial_y & \partial_x \varepsilon_z^{-1} \partial_x + \mu_y k_0^2 \end{matrix} \middle| \varphi_{g'-k_{xy}} \right\rangle \begin{bmatrix} H_x^{g'} \\ H_y^{g'} \end{bmatrix} \, , \tag{13.24}$$

$$\partial_z \begin{bmatrix} H_x^{g'} \\ H_y^{g'} \end{bmatrix} = -\frac{i}{\eta_0 k_0} \left\langle \varphi_{g-k_{xy}} \middle| \begin{matrix} \partial_x \mu_z^{-1} \partial_x + \varepsilon_y k_0^2 & \partial_x \mu_z^{-1} \partial_y \\ \partial_y \mu_z^{-1} \partial_x & \partial_y \mu_z^{-1} \partial_y + \varepsilon_x k_0^2 \end{matrix} \middle| \varphi_{g'-k_{xy}} \right\rangle \begin{bmatrix} E_y^g \\ E_x^g \end{bmatrix} \, , \tag{13.25}$$

where, in order to simplify notation, we represent $E_y(r)$ as

$$E_y(r) = -E_y^g \, |\varphi_{g-k_{xy}}\rangle \tag{13.26}$$

instead of directly using (13.21).

The matrices in the above equations can be computed easily when both permittivity and permeability are expanded in the Fourier basis. For this purpose, the analysed

---

[1] We use this convention in this whole chapter, unless explicitly stated otherwise.

layer is assumed to be invariant in the z direction. Then we can write

$$\varepsilon_i(\mathbf{r}) = \varepsilon_i^g \, |\varphi_g\rangle \quad (i = x, y) , \tag{13.27}$$

$$\mu_i(\mathbf{r}) = \mu_i^g \, |\varphi_g\rangle \quad (i = x, y) , \tag{13.28}$$

$$\varepsilon_z^{-1}(\mathbf{r}) = \kappa^g \, |\varphi_g\rangle , \tag{13.29}$$

$$\mu_z^{-1}(\mathbf{r}) = \gamma^g \, |\varphi_g\rangle . \tag{13.30}$$

Substituting this into (13.24) and (13.25), we finally obtain the GTL equations in the plane-wave basis

$$
\partial_z \begin{bmatrix} E_y^g \\ E_x^g \end{bmatrix} = -\mathrm{i}\frac{\eta_0}{k_0}
$$
$$
\times \begin{bmatrix} -(g_y - k_y)(g_y' - k_y)\kappa^{g-g'} + k_0^2\mu_x^{g-g'} & (g_y - k_y)(g_x' - k_x)\kappa^{g-g'} \\ (g_x - k_x)(g_y' - k_y)\kappa^{g-g'} & -(g_x - k_x)(g_x' - k_x)\kappa^{g-g'} + k_0^2\mu_y^{g-g'} \end{bmatrix}
$$
$$
\times \begin{bmatrix} H_x^{g'} \\ H_y^{g'} \end{bmatrix} , \tag{13.31}
$$

$$
\partial_z \begin{bmatrix} H_x^g \\ H_y^g \end{bmatrix} = -\frac{\mathrm{i}}{\eta_0 k_0}
$$
$$
\times \begin{bmatrix} -(g_x - k_x)(g_x' - k_x)\gamma^{g-g'} + k_0^2\varepsilon_y^{g-g'} & -(g_x - k_x)(g_y' - k_y)\gamma^{g-g'} \\ -(g_y - k_y)(g_x' - k_x)\gamma^{g-g'} & -(g_y - k_y)(g_y' - k_y)\gamma^{g-g'} + k_0^2\varepsilon_x^{g-g'} \end{bmatrix}
$$
$$
\times \begin{bmatrix} E_y^{g'} \\ E_x^{g'} \end{bmatrix} , \tag{13.32}
$$

where k_x, k_y, g_x, and g_y are the corresponding components of the mode wavevector and the reciprocal lattice vector $\mathbf{g}$, respectively. These equations can be introduced into computer memory for further computations by truncating the Fourier basis $\{\mathbf{g}\}$ at some point. The introduced error will depend on two factors: the distribution of the electromagnetic field and the distribution of the material parameters. The former, as a result of simulation, cannot be directly influenced, although the latter can be convoluted with Gaussian window in order to improve the convergence.

Here, like in Ref. [1] we have assumed a diagonal form of the permittivity and permeability tensors. However, the PWAM may be formulated in the same way for a more general in-plane anisotropy given as

$$
\varepsilon = \begin{vmatrix} \varepsilon_{xx} & \varepsilon_{xy} & 0 \\ \varepsilon_{yx} & \varepsilon_{yy} & 0 \\ 0 & 0 & \varepsilon_z \end{vmatrix} , \quad \mu = \begin{vmatrix} \mu_{xx} & \mu_{xy} & 0 \\ \mu_{yx} & \mu_{yy} & 0 \\ 0 & 0 & \mu_z \end{vmatrix} . \tag{13.33}
$$

For the form of the GTL equations in this case, the details of the corresponding plane-wave expansion, and the exact formulae for $\mathbf{R}_E$ and $\mathbf{R}_H$ operators considered below, please refer to Ref. [9]. We note that the full in-plane anisotropy (13.33) is needed to consider any non-rectangular geometries of UPML.

13.2.2 Eigenmode Determination

Once the GTL equations (13.31) and (13.32) have been derived and represented in their numerical form, we can use them for the determination of the eigenmodes of the analysed structure. To this purpose we first show their solution for a single z-invariant layer and then present a numerically stable technique for connection between separate layers and final computation of eigenmodes. The procedure described below has originally been used with the method of lines [12] and is also applied to the plane-wave admittance method.

In this section we will use the short representation of the vectors and matrices introduced in the previous section. In particular we will name

$$\hat{E} = \begin{bmatrix} E_y^g \\ E_x^g \end{bmatrix},$$

$$\hat{H} = \begin{bmatrix} H_x^g \\ H_y^g \end{bmatrix},$$

$$R_E = \frac{1}{\eta_0 k_0}$$
$$\times \begin{bmatrix} -(g_x - k_x)(g_x' - k_x)\gamma^{g-g'} + k_0^2 \varepsilon_y^{g-g'} & -(g_x - k_x)(g_y' - k_y)\gamma^{g-g'} \\ -(g_y - k_y)(g_x' - k_x)\gamma^{g-g'} & -(g_y - k_y)(g_y' - k_y)\gamma^{g-g'} + k_0^2 \varepsilon_x^{g-g'} \end{bmatrix},$$

$$R_H = \frac{\eta_0}{k_0}$$
$$\times \begin{bmatrix} -(g_y - k_y)(g_y' - k_y)\kappa^{g-g'} + k_0^2 \mu_x^{g-g'} & (g_y - k_y)(g_x' - k_x)\kappa^{g-g'} \\ (g_x - k_x)(g_y' - k_y)\kappa^{g-g'} & -(g_x - k_x)(g_x' - k_x)\kappa^{g-g'} + k_0^2 \mu_y^{g-g'} \end{bmatrix}.$$

It is worth to note, that the whole procedure described below can be applied with any correct representation of the above vectors and matrices. Thus it is applicable not only with the plane-wave expansion, but also with any orthogonal or non-orthogonal basis. In particular the representation of $\hat{E}$, $\hat{H}$, R_E and R_H with the finite-difference approximation is used in the method of lines.

Take (13.31) and (13.32) and represent them in their short form

$$\partial_z \hat{E} = -\mathrm{i} R_H \hat{H} , \tag{13.34}$$
$$\partial_z \hat{H} = -\mathrm{i} R_E \hat{E} . \tag{13.35}$$

By taking the z-derivative of one of these equations and substituting the other one into it, one can write two second-order equations with the fields decoupled

$$\partial_z^2 \hat{E} = -Q_E \hat{E} , \tag{13.36}$$
$$\partial_z^2 \hat{H} = -Q_H \hat{H} , \tag{13.37}$$

where $Q_E = R_H R_E$ and $Q_H = R_E R_H$. In the z-invariant layer, any of the above equations can be solved analytically, provided it is transposed into "diagonalized" coordinates. This is done by determination of all the eigenvalues of either Q_E or Q_H matrices. Because they are similar matrices in the linear algebra sense i.e. $Q_E = R_E^{-1} Q_H R_E$, they share the same set of eigenvalues and it is possible to write

$$Q_E = T_E \Gamma^2 T_E^{-1} , \tag{13.38}$$

$$Q_H = T_H \Gamma^2 T_H^{-1} , \tag{13.39}$$

where Γ^2 is a diagonal matrix of eigenvalues of Q_E and Q_H and T_E and T_H are the matrices constructed by their eigenvectors, respectively. In the following analysis we will use the fact that it is always possible to choose such T_E and T_H that

$$T_E = R_H T_H \Gamma^{-1} . \tag{13.40}$$

Using the above relations we can write (13.36) and (13.37) in their diagonalized forms

$$\partial_z^2 \tilde{E} = -\Gamma^2 \tilde{E} , \tag{13.41}$$

$$\partial_z^2 \tilde{H} = -\Gamma^2 \tilde{H} , \tag{13.42}$$

where $\tilde{E} = T_E^{-1} \hat{E}$ and $\tilde{H} = T_E^{-1} \hat{H}$. The mathematical operations we have just performed can be also explained as a representation of the electric (and similarly the magnetic) field as a linear combination of single-layer eigenmodes. Then the columns of the T_E (T_H) represent the shape of each of such eigenmodes, the diagonal elements of Γ are their propagation constants and the elements of $\tilde{E}$ ($\tilde{H}$) – their amplitudes. Now Eqs. (13.41) and (13.42) have analytical solutions in the form of propagating or standing wave. For the electric field, this solution can be written as

$$\tilde{E}(z) = \cos(\Gamma z) A + \sin(\Gamma z) B , \tag{13.43}$$

where A and B are some constant vectors that have to be determined from the boundary conditions. The application of the admittance transfer technique for this case is shown below.

13.2.3 Admittance Transfer

Having the analytical form of $\tilde{E}$ in a uniform layer of thickness d, it is possible to represent A and B in terms of diagonalized electric field, at $z = 0$ ($\tilde{E}_0$) and at $z = d$ ($\tilde{E}_d$). From (13.43) we have

$$A = \tilde{E}_0 , \tag{13.44}$$

$$B = \sin^{-1}(\Gamma d) \tilde{E}_d - \tan^{-1}(\Gamma d) \tilde{E}_0 . \tag{13.45}$$

Having given $\tilde{E}_0$ and $\tilde{E}_d$ we would like to determine the magnetic field at both boundaries of the layer. For this purpose we will take (13.34) and rewrite it in terms of $\tilde{E}$ and $\tilde{H}$:

$$T_E \partial_z \tilde{E} = -iR_H T_H \tilde{H} ,$$

which, after substituting (13.40), gives

$$i\Gamma^{-1} \partial_z \tilde{E} = \tilde{H} . \tag{13.46}$$

The left-hand-side of the above equation can be computed by taking the derivative of (13.43),

$$\partial_z \tilde{E}(z) = -\Gamma \sin(\Gamma z) A + \Gamma \cos(\Gamma z) B . \tag{13.47}$$

Computing $\tilde{H}_0$ at $z = 0$ and $\tilde{H}_d$ at $z = d$ from (13.46) with substitutions of (13.44), (13.45) and (13.47), we have

$$\tilde{H}_0 = -i \tan^{-1}(\Gamma d) \tilde{E}_0 + i \sin^{-1}(\Gamma d) \tilde{E}_d , \tag{13.48}$$
$$\tilde{H}_d = -i \sin^{-1}(\Gamma d) \tilde{E}_0 + i \tan^{-1}(\Gamma d) \tilde{E}_d , \tag{13.49}$$

which in short can be written as

$$\begin{bmatrix} \tilde{H}_0 \\ \tilde{H}_d \end{bmatrix} = \begin{bmatrix} y_1 & y_2 \\ -y_2 & -y_1 \end{bmatrix} \begin{bmatrix} \tilde{E}_0 \\ \tilde{E}_d \end{bmatrix} , \tag{13.50}$$

where

$$y_1 = -\tan^{-1}(\Gamma d) , \tag{13.51}$$
$$y_2 = \sin^{-1}(\Gamma d) . \tag{13.52}$$

Equation (13.50) gives complete admittance relation for both sides of a single layer. Now consider the whole structure (Fig. 13.1). We divide it into two parts and we derive the admittance relation in the matching plane between these two parts. To this aim we will use the iterative procedure for each part, starting at their outer edges, to determine the relation

$$\tilde{H}_d^{(n)} = Y^{(n)} \tilde{E}_d^{(n)} \tag{13.53}$$

for every layer n. We assume the $\tilde{E}_0^{(1)}$, i.e. the electric field at the beginning of the first layer, to be equal to zero. This is equivalent to putting our analysed structure into a cavity limited by two perfect conductors. In order to correctly estimate the radiating field we need to use perfectly matched layers (PMLs) as the outer-most layers in the structure. In PWAM this is straightforward and requires only a proper choice of material permittivity and permeability. More details on this matter are given in Sect. 13.2.5. So, having the assumption that $\tilde{E}_0^{(1)} = 0$, we can easily see from (13.50) that

$$Y^{(1)} = -y_1^{(1)} . \tag{13.54}$$

To find admittance in other layers we must use the constraint, that both the electric field $\hat{E}$ and the magnetic one $\hat{H}$ are continuous at all the layer boundaries. Thus we

Fig. 13.1 Admittance transfer
in a multilayer structure

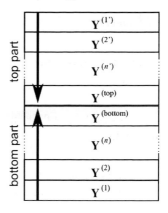

have

$$T_E^{(n)} \tilde{E}_0^{(n)} = T_E^{(n-1)} \tilde{E}_d^{(n-1)} ,$$ (13.55)

$$T_H^{(n)} \tilde{H}_0^{(n)} = T_H^{(n-1)} \tilde{H}_d^{(n-1)} ,$$ (13.56)

or

$$\tilde{E}_0^{(n)} = t_E^{(n)} \tilde{E}_d^{(n-1)} ,$$ (13.57)

$$\tilde{H}_0^{(n)} = t_H^{(n)} \tilde{H}_d^{(n-1)} ,$$ (13.58)

where $t_E^{(n)} = \left(T_E^{(n)} \right)^{-1} T_E^{(n)}$ and $t_H^{(n)} = \left(T_H^{(n)} \right)^{-1} T_H^{(n)}$. Now substituting this into
(13.50) we can state

$$t_H^{(n)} Y^{(n-1)} \tilde{E}_d^{(n-1)} = y_1^{(n)} t_E^{(n)} \tilde{E}_d^{(n-1)} + y_2^{(n)} \tilde{E}_d^{(n)} ,$$ (13.59)

$$\tilde{H}_d^{(n)} = -y_2^{(n)} t_E^{(n)} \tilde{E}_d^{(n-1)} - y_1^{(n)} \tilde{E}_d^{(n)} .$$ (13.60)

Removing $\tilde{E}_d^{(n-1)}$ from these equations and doing some basic transformations,
we can determine the admittance relation for n-th layer

$$\tilde{H}_d^{(n)} = \left\{ y_2^{(n)} t_E^{(n)} \left[y_1^{(n)} t_E^{(n)} - t_H^{(n)} Y^{(n-1)} \right]^{-1} y_2^{(n)} - y_1^{(n)} \right\} \tilde{E}_d^{(n)} ,$$ (13.61)

which finally gives us

$$Y^{(n)} = y_2^{(n)} t_E^{(n)} \left[y_1^{(n)} t_E^{(n)} - t_H^{(n)} Y^{(n-1)} \right]^{-1} y_2^{(n)} - y_1^{(n)} .$$ (13.62)

Using this relation we can determine the admittance matrix on the matching plane,
with the aid of the iterative procedure for both top (we name the obtained matrix
Y^{top}) and bottom (Y^{bottom}) parts of the structure. As both fields $\hat{E}$ and $\hat{H}$ must be

continuous at all planes, then

$$T_H^{\text{top}} \tilde{H}^{\text{top}} = T_H^{\text{bottom}} \tilde{H}^{\text{bottom}}$$

and

$$T_E^{\text{top}} \tilde{E}^{\text{top}} = T_E^{\text{bottom}} \tilde{E}^{\text{bottom}} = \hat{E}^{\text{interface}} ,$$

where $\tilde{E}^{\text{top/bottom}}$ and $\tilde{H}^{\text{top/bottom}}$ are the fields on the interface computed for the top/bottom part of the structure and $T_E^{\text{top/bottom}}$ and $T_H^{\text{top/bottom}}$ are the T-matrices of the layers directly adjacent to the interface. The above relations lead to the following eigenequation that must be satisfied for the whole structure

$$\left\{ T_H^{\text{top}} Y^{\text{top}} \left[T_E^{\text{top}} \right]^{-1} - T_H^{\text{bottom}} Y^{\text{bottom}} \left[T_E^{\text{bottom}} \right]^{-1} \right\} \hat{E}^{\text{interface}} = 0 . \tag{13.63}$$

This equation can be only non-trivially satisfied when the matrix

$$M = T_H^{\text{top}} Y^{\text{top}} \left[T_E^{\text{top}} \right]^{-1} - T_H^{\text{bottom}} Y^{\text{bottom}} \left[T_E^{\text{bottom}} \right]^{-1} \tag{13.64}$$

is singular. In the three-dimensional problem, this matrix is a function of the normalised frequency k_0 and the wave-vector components k_x and k_y. In order to determine the eigenmodes we search such combination of these variables that the function

$$\Phi(k_0, k_x, k_y) = \min \left| \det \left[M(k_0, k_x, k_y) \right] \right| = 0 . \tag{13.65}$$

In our numerical implementation, we fix the two of these variables and search for the third one using the modified Broyden algorithm in a complex plane [13].

An open question is the choice of the matching plane. In general this choice can be arbitrary, however, the numerical error will be the smallest if the electric field $\hat{E}$ is large on the interface. Thus the best position of the matching plane depends on the expected field distribution in the structure.

13.2.4 Determination of the Fields

Once the eigenfrequency of the analysed device is known, the electromagnetic field distribution can be resolved. This task is performed in two steps; first the electric field at the matching interface is computed as eigenvector of (13.64) and then it is determined with an iterative procedure similar to the one presented in Sect. 13.2.3 but performed in a reverse direction.

As already shown in the Sect. 13.2.3, the physical eigenmodes are related to the situation where matrix M (13.64) is singular, i.e.

$$M \hat{E}_{\text{interface}} = 0 . \tag{13.66}$$

This means that the electric field at the interface $\hat{E}_{\text{interface}}$ is an eigenvector of M corresponding to its eigenvalue $\lambda = 0$. Thus the determination of $\hat{E}_{\text{interface}}$ is straightforward with any numerical algorithm for determination of eigenvectors.

To find the electric field distribution inside any uniform layer we must know the field in diagonalized coordinates $\tilde{\boldsymbol{E}}$ at both sides of the layer, namely $\tilde{\boldsymbol{E}}_0^{(n)}$ and $\tilde{\boldsymbol{E}}_d^{(n)}$. For the layer adjacent to the matching interface, the latter one is known and for any other layer it can be easily determined from (13.55). To compute $\tilde{\boldsymbol{E}}_0^{(n)}$ consider (13.50) which can be used to represent $\tilde{\boldsymbol{E}}_0^{(n)}$ as

$$\tilde{\boldsymbol{E}}_0^{(n)} = \left(\boldsymbol{y}_2^{(n)}\right)^{-1}\left[\boldsymbol{y}_1^{(n)}\tilde{\boldsymbol{E}}_d^{(n)} + \tilde{\boldsymbol{H}}_d^{(n)}\right],$$

which using (13.53) gives

$$\tilde{\boldsymbol{E}}_0^{(n)} = \left(\boldsymbol{y}_2^{(n)}\right)^{-1}\left[\boldsymbol{y}_1^{(n)} + \boldsymbol{Y}^{(n)}\right]\tilde{\boldsymbol{E}}_d^{(n)}. \tag{13.67}$$

An application of the above equation requires additional storage space to keep the matrix $\boldsymbol{Y}^{(n)}$ for each layer but provides numerical stability, lacked e.g. by the transfer-matrix method. Having both $\tilde{\boldsymbol{E}}_0^{(n)}$ and $\tilde{\boldsymbol{E}}_d^{(n)}$ it is possible to find the electric field in diagonalised coordinates in any layer using Eqs. (13.43)–(13.45) that together give

$$\tilde{\boldsymbol{E}}^{(n)}(z) = \cos(\boldsymbol{\Gamma}^{(n)}z)\tilde{\boldsymbol{E}}_0^{(n)} + \sin(\boldsymbol{\Gamma}^{(n)}z)\left[\sin^{-1}(\boldsymbol{\Gamma}^{(n)}d^{(n)})\tilde{\boldsymbol{E}}_d - \tan^{-1}(\boldsymbol{\Gamma}^{(n)}d^{(n)})\tilde{\boldsymbol{E}}_0\right],$$
$$\tag{13.68}$$

where z is the relative position in the n-th layer and the vector value in the brackets is constant and can be precomputed and stored in advance to save the calculation time. The magnetic field can be determined directly from the electric one using the admittance relation (13.50) to get $\tilde{\boldsymbol{H}}_0^{(n)}$ and $\tilde{\boldsymbol{H}}_d^{(n)}$ and applying the following relation for finding $\tilde{\boldsymbol{H}}^{(n)}(z)$

$$\tilde{\boldsymbol{H}}^{(n)}(z) = \cos(\boldsymbol{\Gamma}^{(n)}z)\tilde{\boldsymbol{H}}_0^{(n)} + \sin(\boldsymbol{\Gamma}^{(n)}z)\left[\sin^{-1}(\boldsymbol{\Gamma}^{(n)}d^{(n)})\tilde{\boldsymbol{H}}_d - \tan^{-1}(\boldsymbol{\Gamma}^{(n)}d^{(n)})\tilde{\boldsymbol{H}}_0\right]. \tag{13.69}$$

The x and y components of the physical fields can be now easily computed as $\hat{\boldsymbol{E}} = \boldsymbol{T}_E^{(n)}\tilde{\boldsymbol{E}}$ and $\hat{\boldsymbol{H}} = \boldsymbol{T}_H^{(n)}\tilde{\boldsymbol{H}}$ and to get the z-components, it is sufficient to use Eqs. (13.9) and (13.10), which in plane-wave basis can be expanded as

$$E_z^g = \frac{\eta_0}{k_0}\kappa^{g-g'}\left[-(g_y'-k_y)H_x^{g'} + (g_x'-k_x)H_y^{g'}\right], \tag{13.70}$$

$$H_z^g = \frac{1}{k_0\eta_0}\gamma^{g-g'}\left[(g_x'-k_x)E_y^{g'} + (g_y'-k_y)E_x^{g'}\right]. \tag{13.71}$$

13.2.5 Perfectly Matched Layers as Boundary Conditions

An intrinsic property of the plane-wave expansion is the obligation to use periodic boundary conditions. Hence, only the field distribution over a finite area can be represented. This properties can sometimes be very useful, e.g. for the determination of the band gap of bulk photonic crystals. In most cases, however, one needs to model an isolated structure which can radiate into infinity. The only possible way of handling such situations in finite computational domain is the application of absorbing boundary conditions (ABC).

Insofar several methods of introducing ABCs were proposed [14]. The older of them were based on the modification of the Maxwell equations in the boundary region to enforce one-way propagation of light [4, 14]. However, the most of these methods could be applied only with specific computational algorithms, e.g. they required finite-difference representation of the electromagnetic field.

A more modern approach for ABCs is the introduction of absorbing regions at the boundaries. This makes the computational area analogous to the anechoic audio chamber with perfectly absorbing walls as an approximation of an open area. In order to make this approximation precise one needs to make sure that all the field is absorbed and neither reflection nor scattering occurs at the boundaries between the interior and the absorbing layers.

A practical realisation of such layers that obeys the above condition has been proposed by Berenger [7] and named Perfectly Matched Layers (PMLs). The new theoretical medium possesses the property that at its boundary no reflection occurs at any incidence angle. This is achieved with a split-field formulation of the Maxwell equations that threats each electromagnetic field component as a sum of two split parts. These absorbing PMLs have already been applied by many researchers and proved to work very well. However, the Berenger's medium is purely based on mathematical model and thus requires alternation of the Maxwell equations in every practical realisation.

A different approach to realise perfectly matched layers was proposed by Sacks [6]. In his approach, the PMLs are physical medium with uni-axially anisotropic permittivity and permeability. This allows an application of the PMLs (named in this case Uni-axial PMLs – UPMLs) avoiding the non-physical field splitting[2] and the modification of Maxwell equations. For a proof that the boundary between host medium and absorbing layer is reflectionless for any polarisation and angle of incidence, please refer to Refs. [6, 14].

In PWAM we assume a usual definition of UPML – when the vector normal to the UPML boundary is parallel to the $\hat{x}$ unit vector and the permittivity and permeability of the attached simulation area are equal to ε_r and μ_r, UPML consists of the material with permittivity and permeability given as

$$\varepsilon_{\text{UPML}} = \varepsilon_r \begin{vmatrix} s^{-1} & 0 & 0 \\ 0 & s & 0 \\ 0 & 0 & s \end{vmatrix}, \quad \mu_{\text{UPML}} = \mu_r \begin{vmatrix} s^{-1} & 0 & 0 \\ 0 & s & 0 \\ 0 & 0 & s \end{vmatrix}, \quad (13.72)$$

where s can be any arbitrary complex number different than 0. In particular if $s = \kappa - i\alpha/k_0$, then α is the ratio of damping in the PML and κ is responsible for coordinate stretching [16].

Definition (13.72) subject to coordinate transforms through rotations provides the material properties of the UPML regions located at all boundaries of the simulation

---

[2] There have been even attempts to create such medium in laboratory conditions [15].

area. Hence, the UPML rotated counter clockwise by the angle θ with respect to a certain axis is defined as

$$\varepsilon_{\mathrm{UPML}}(\theta) = R(\theta) \cdot \varepsilon_{\mathrm{UPML}}(0) \cdot R(-\theta) \,, \qquad (13.73)$$

$$\mu_{\mathrm{UPML}}(\theta) = R(\theta) \cdot \mu_{\mathrm{UPML}}(0) \cdot R(-\theta) \,, \qquad (13.74)$$

where $R(\theta)$ is the respective rotation matrix in 3D. For instance, rotation around axis $\hat{z}$ has the matrix

$$R(\theta) = \begin{bmatrix} \cos(\theta) & \sin(\theta) & 0 \\ -\sin(\theta) & \cos(\theta) & 0 \\ 0 & 0 & 1 \end{bmatrix} \,. \qquad (13.75)$$

It is easy to verify that the diagonal form of anisotropy allows to surround the simulation area with a rectangular box of UPML, while the in-plane anisotropy (13.33) is sufficient to build complex UPML geometries surrounding subsequent layers.

In PWAM, the PMLs at the boundary perpendicular to the z-axis (vertical PMLs) are treated differently than at the other boundaries. Because of the analytical expansion in the z-direction, these vertical PMLs are matched to the host medium without any numerical dispersion nor discretization-related errors. Therefore it is possible to use a single absorbing layer with any arbitrary value of s and no reflections will occur. On the other hand, on the horizontal boundaries the material parameters are not represented precisely because of the truncated Fourier expansion. This may introduce undesired artificial reflections at the PML interface. To minimise this effect it is beneficial to apply gradual PMLs, i.e. PMLs with s parameter changing gradually between unity and s_{max}.

13.3 Modelling PC-VCSELs with PWAM

In this section we use the plane-wave admittance method for analysis of photonic-crystal vertical-cavity surface-emitting lasers (PC-VCSELs) [17–19]. These are modern designs of VCSELs, in which the light confinement is provided by use of photonic crystal instead of typical oxidation layers. Such an approach can show several advantages as compared to older devices. First of all photonic crystals can be used with materials where oxidation layers are either impossible or very hard to manufacture. Furthermore, they can be used to ensure a single-mode emission of the laser or to stabilise the polarisation of the emitted light.

PC-VCSELs are complex structures, as they may contain over a hundred different layers forming a resonant cavity and two high-reflectivity Bragg mirrors. Furthermore, contrary to the classical VCSEL designs, they cannot be simplified into two-dimensional domain, as they do not posses the axicylindrical symmetry. Hence, their modelling must be performed in three dimensions. If one wishes to use the old well known methods, as plane-wave expansion of FDTD, the necessary numerical effort would be far above the capabilities of personal computer and would require weeks of calculation on super-clusters. On the other hand, the use of PWAM would allow to perform the analysis with good accuracy using a single-processor machine and to achieve the desired results in a reasonable time. Below there are examples of some interesting possibilities of PC-VCSEL designs with this method.

13.3.1 Dependence of the Wavelength and Modal Loss on Etching Depth

The modelled VCSEL structure is an InAlGaAs/InP laser emitting light at wavelength of $\lambda = 1.3\,\mu m$. It consists of three 15 nm-wide $Al_{0.02}Ga_{0.50}In_{0.48}As$ quantum wells inside of an $Al_{0.22}Ga_{0.25}In_{0.53}As$ 3λ cavity. The top and bottom DBRs are 27 and 35 pairs, respectively, of $Al_{0.9}Ga_{0.1}As$/GaAs quarter-wavelength layers. The photonic crystal structure consists of three rings of hexagonal lattice with a single defect inside (Fig. 13.2).

The resonant wavelengths and modal losses, computed for varying etching depths h in cold-cavity device are shown in Fig. 13.3. It can be seen that introduction of deeper holes results in a blue-shift of the resonant wavelength and decrease of the modal losses [20, 21]. Both these effects are subject to saturation that occurs for $h \approx 8\,\mu m$, i.e. with the bottoms of the holes just reaching the bottom DBRs. The

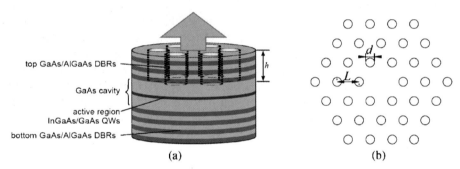

Fig. 13.2 Analysed VCSEL structure: (**a**) Schematic three-dimensional view, (**b**) photonic crystal structure

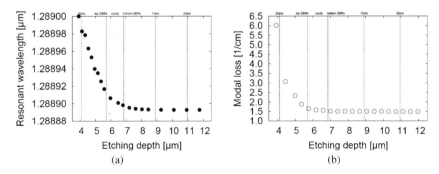

Fig. 13.3 Resonant wavelength (**a**) and modal loss (**b**) as a function of etching depth in PC-VCSEL. The *vertical grids* mark each 10 of top and bottom DBRs and the boundaries of the cavity

Fig. 13.4 A PC-VCSEL mode distribution for $L = 4.6\,\mu m$, $d/L = 0.3$ and $h = 7.2\,\mu m$; (**a**) the vertical distribution of the mode in the whole VCSEL and (**b**) the horizontal intersections in the plane of the active region and the facet

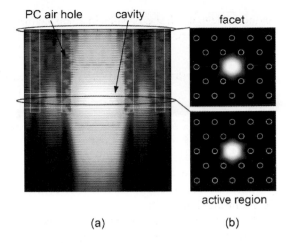

(a)   (b)

explanation of this behaviour is simple: the electromagnetic field has its largest intensity in the cavity; hence, if the holes are etched only in top DBR even a slight increase in their depth corresponds to large relative increase of an overlap between the field and themselves. For deep holes, reaching bottom DBRs, further increase of their depth is not followed by a significant relative change of the overlap. As a consequence the holes etched with a depth of around $8\,\mu m$ can be considered to have the same impact as the ones etched through the whole VCSEL structure.

Figure 13.4 shows vertical and horizontal profiles of the electric field intensity inside the laser. The near-field image at the VCSEL facet is similar to the field distribution in the active region, however, some light leaks out of the cavity through the holes, contributing to undesired interference pattern. This effect can be reduced by the use of bottom-emitting device with mirror layer at the top.

13.3.2 Analysis of Single-Mode Operation of PC-VCSELs

Above we have shown an analysis of the impact of the etching depth on the properties of the fundamental mode in PC-VCSELs. Here, we consider the requirements for single-mode regime of surface-emitting lasers. In classical VCSEL designs such a regime is very hard to achieve without reduction of the lateral dimension of the device and consequently limiting the output power. Application of photonic crystals can help to achieve single-mode operation conditions of the laser through a mechanism similar to the one present in single-mode photonic crystal fibres [10] and, hence, is a subject of intensive study [22–24].

Below we show the results of simulation of InAlGaAs/InP $1.3\,\mu m$ PC-VCSEL, almost identical to the one presented in the previous subsection [25]. The only difference is in the direction of light emission – in this case the laser emits through the substrate in order to reduce the scattering of the output beam on the holes.

In this analysis there are two constraints that have to be fulfilled in order to consider the structure as a single-mode one. The first constraint is non-existence of the

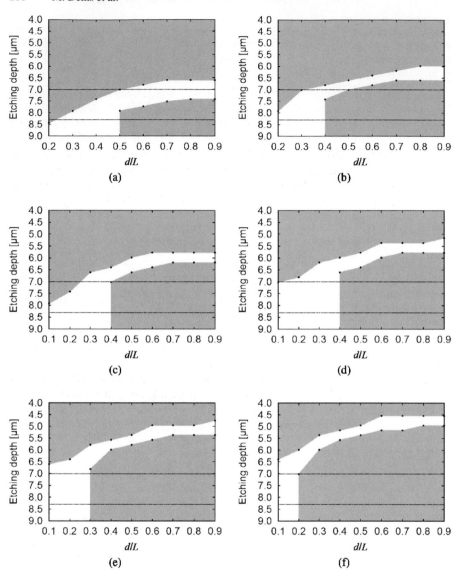

Fig. 13.5 Regions of the combination of hole size and etching depth providing single-mode operation of PC-VCSEL computed for the optical apertures of: (**a**) 2 μm, (**b**) 4 μm, (**c**) 6 μm, (**d**) 8 μm, (**e**) 10 μm and (**f**) 12 μm. *Gray regions* denote either the structure without a well confined fundamental mode or a multi-mode regime. The irregular shape of the regions can be attributed to the crudeness of the numerical modelling. The *thin horizontal lines* denote the boundaries of the cavity

higher-order modes and the second one is the requirement for the fundamental mode to be well confined in the photonic crystal defect. Without the latter condition the efficient work of the laser is virtually impossible, as the fundamental mode suffers from too high losses and is additionally scattered on the holes.

The maps of the parameter combinations providing the single-mode operation are shown in Fig. 13.5. On each of the plots (Fig. 13.5a–f) there are structures with a constant optical aperture, defined as the diameter of the largest circle fitting in the photonic crystal defect. On the horizontal axes there are the relative hole diameters and on the vertical ones, the etching depths. The white regions denote the combination of the parameters providing the single-mode regime. The top grey region corresponds to the fundamental mode not confined to the defect, while the lower grey region marks the multi-mode regime. For each aperture, the same tendency can be observed: for large holes, there is limited etching depth providing the single-mode operation, while for the small ones, it is even possible to create the single-mode VC-SEL with infinitely deep holes. This phenomena can be explained by the fact that the effective index of a photonic crystal waveguide depends on the wavelength and for small holes the effective index contrast is small enough to sustain the single-mode regime.

Figure 13.6 shows the values of the side-mode discrimination, i.e. the difference in the losses between the fundamental and first-order mode near the edge of the multi-mode regime. It confirms the above conclusions: the largest discrimination is for narrow and shallow holes. In addition the loss of the first-order mode sharply increases when approaching the single-mode region. Hence, in the real devices it is possible to obtain the single-mode operation even in the multi-mode region. As a consequence, the boundaries of the single-mode region cannot be considered as sharp and well defined, but they are rather gradual.

In conclusion, we have presented in this section examples of modelling various properties of PC-VCSELs. We have shown that the application of photonic crystals in surface-emitting lasers can be a versatile tool for providing a single-mode regime with high suppression of the side modes. In the following section we consider the differences between fundamental modes with different polarisations in an asymmetric structure and investigate the suitability of using photonic crystals for stabilising not only the lateral distribution, but also the polarisation of the emitted light.

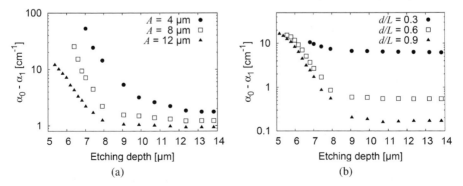

Fig. 13.6 Discrimination of the first-order mode in the PC-VCSEL as a function of the holes depth for different optical apertures (**a**) and hole diameters (**b**)

13.3.3 Polarisation Control of VCSELs with Photonic-Crystals

Traditional, commonly used VCSELs, have geometries with cylindrical symmetry and are usually grown on (100)-oriented GaAs or InP substrate. Thus, they have no a-priori defined polarisation, i.e. the LP-polarised modes can be oriented in any direction perpendicular to the VCSEL symmetry axis. In practise, in real devices, there is always small anisotropy induced by small residual strain introduced during fabrication and the electro-optic effect. Therefore, the electric field of the fundamental mode is almost always linearly polarised (LP mode) along $[011]$ or $[0\bar{1}1]$ crystallographic axis, while the one of the first-order modes is perpendicular to it [26]. During the operation of the laser, a change of temperature, current, or stress can induce polarisation switching, i.e. the VCSEL starts emitting light in the polarisation orthogonal to the previous one. This effect can be a subject to bistability, hysteresis or random dynamics behaviour [27, 28]. Furthermore, optical feedback [29] or optical injection [30], can introduce even more polarisation instabilities in the system.

In general, the polarisation state of the light depends on two main ingredients. The first one is the polarisation characteristics of the gain medium, resulting from different angular momenta of the quantum states of electrons and holes involved in emission or absorption [31, 32]. The second one is the influence of the laser cavity, that can introduce smaller or larger birefringence and dichroism. These two factors can compete or be complementary to each other, providing different types of static or dynamic polarisation behaviour. The change in the polarisation of the emitted light can, thus, be attributed to a change in the number of carriers with different angular momenta [32], but also to the spatial hole burning [33], shifting of gain profiles and cavity resonance wavelengths [27], or thermal lensing [34].

In experiments, the easy to measure quantity, describing the polarisation properties of VCSELs, is the polarisation-mode suppression ratio (PMSR). This is the ratio of the intensity of the emitted light in the two orthogonal polarisations, represented usually in decibels.

In many applications, where the elements of the optical set-up are polarisation-dependent, polarisation switching cannot be tolerated. For this reason, the polarisation stabilisation is an important issue in the design of modern VCSELs. There exist several methods for providing such a control, which aim at introduction of anisotropy to either gain or losses. Anisotropic gain can be achieved through growth on a substrate with crystallographic orientation other than $[001]$ [35] or by in-plane uniaxial strain [36, 37]. On the other hand, anisotropic losses can be introduced by using an asymmetric shape resonator [38, 39] or by making the VCSEL top-DBR polarisation-dependent, e.g. by using metal-semiconductor gratings [40–42] or by directly etching a sub-wavelength grating [43].

Recently, together with the development of photonic-crystal VCSELs, a new method of polarisation stabilisation emerged, namely an application of photonic crystals. For several years, photonic crystals have been successfully used for providing birefringence and dichroism in photonic-crystal fibres, although there are not many works for such photonic crystals application in VCSELs. So far there was one such successful attempt reported [44], where photonic crystal with elliptical holes was applied to provide stable polarisation lasing. In that work, the stabilisation of polarisation with over 20 dB PMSR was achieved with photonic crystal based on hexagonal lattice and air holes elongated along K and M directions, accompanied by an anisotropic current injection.

Below, we perform a detailed analysis of applicability of various photonic crystal configurations for polarisation stabilisation. Contrary to Ref. [44], we investigate not only configuration with elliptical holes, but also some other designs, that have been successfully applied in photonic-crystal fibres for providing large birefringence and single-polarisation waveguiding.

An important issue to reflect on, when considering polarisation control with photonic crystals, is breaking the C_6 point group symmetry. At the first glance, a hexagonal lattice of cylindrical holes does not possesses C_4 symmetry, i.e. after rotating by 90 degrees, the photonic crystal lattice will differ from the original one (Fig. 13.7).

Thus one may assume that this is enough for differentiating the two orthogonal x- and y-polarisations. In fact, this is not true [45]. Consider x-polarised LP mode (Fig. 13.8a) in totally isotropic PC-VCSEL. The electric field vector is oriented in K direction of photonic crystal lattice and, because of the C_6 symmetry, this mode is indistinguishable from other modes marked with white arrows in Fig. 13.8b. The solution of the Maxwell equations allows the existence of any linear combination of these modes, also the one in which the electric field is oriented along y-axis, i.e. in the M direction of the photonic crystal lattice (Fig. 13.8c). Thus the fact, that there exists an x-polarised mode in C_6 lattice, induces an existence of the y-polarised mode degenerated with the original one. Obviously the opposite implication is also true (Fig. 13.8d). This means, that in such a photonic crystal lattice, there is neither frequency separation of the orthogonally polarised modes, nor any difference of their losses.

Hence, the efficient application of photonic crystals for polarisation control requires breaking of the C_6 point group symmetry. This can be achieved by various methods, e.g. by altering the shape of the holes or by changing their arrangement. In this section, we perform numerical analysis of a typical PC-VCSEL. It is a gal-

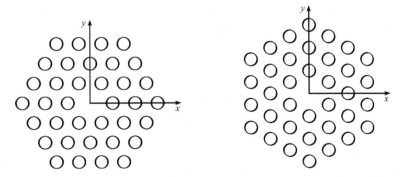

Fig. 13.7 Hexagonal lattice of photonic crystal is not symmetrical when rotation by 90° is considered

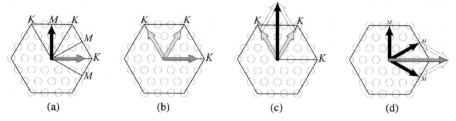

Fig. 13.8 Orthogonal polarisation mode degeneracy for PC-VCSEL lattice with C_6 symmetry. (**a**) x-polarised mode is oriented in Γ-K direction, while y-polarised one in Γ-M, (**b**) C_6 symmetry ensures that all the modes polarised in Γ-K are indistinguishable, (**c**) linear combination of two modes polarised in Γ-K gives a mode polarised in Γ-M, (**d**) linear combination of modes polarised in Γ-M gives a mode polarised in Γ-K

lium arsenide structure designed for operation at 980 nm wavelength, comprising of 24.5 pairs GaAs/AlGaAs top-DBR and 29.5 pairs bottom DBR. The cavity has one wavelength optical length. The optical field confinement is provided by two rings of photonic crystal with hexagonal lattice and single defect inside.

We consider two configurations for providing stable polarisation of the emitted light. In the first step, we analyse a photonic crystal structure with elliptical holes having the ratio of the ellipse radii d_y/d_x varying from 1 to 5. The second photonic crystal structure has cylindrical holes of diameter d_c, however four of the holes have their diameter (d_h) increased (Fig. 13.9) and a small hole of diameter d_s is added inside the photonic crystal defect in order to provide larger birefringence and dichroism [46].

Because the differences in the eigenfrequencies for the two polarisations are very subtle it is necessary to increase the number of plane-waves used in calculations. We have found that satisfactory convergence is achieved with 53^2 plane-waves. The computation time with such a large basis is about two days for a single point on an AMD Opteron processor.

In order to ensure a relatively large polarisation separation, we have chosen small photonic crystal pitch equal to $\Lambda = 1.0\,\mu$m. This provides strong optical confine-

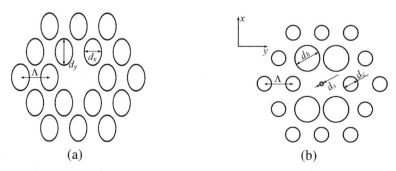

Fig. 13.9 Photonic crystal configurations of analysed structures: (**a**) elliptical holes, (**b**) holes of different diameters

Fig. 13.10 Resonant frequency separation for the
structure shown in Fig. 13.9b
as a function of the photonic
crystal pitch

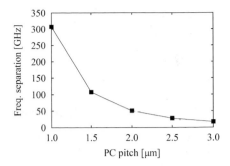

ment of the mode inside the defect. For larger pitches the separation decreases
quickly, as shown in Fig. 13.10. Similar effect can be observed for the modal dichroism.

In the first analysed structure (Fig. 13.9a) the average diameter of the holes has
been kept constant and equal to $d_{avg} = (d_x d_y)^{1/2} = 0.4\,\mu m$, while the ratio d_y/d_x has
been varied. In order to avoid hole overlapping, their longer axes are always oriented
along the ΓM direction. The computed resonant frequency splitting between the two
orthogonally polarised fundamental modes is shown in Fig. 13.11. For ideally cylindrical holes, both x- and y-polarised modes have the same resonant frequencies (in
fact they differ by about 2 GHz, which provides an estimate for the numerical precision at such choice of computational parameters). Increasing the hole ellipticity, the
frequency splitting increases and the x-polarised mode has always lower frequency
than the y-polarised one. Contrary to the resonant frequencies, the modal losses for
the two polarisations remain equal to each other within the numerical precision that
we estimate to be around 3%. The field profiles are also almost identical, however
the field for the x-polarised mode is slightly better confined inside the defect. Therefore, we can conclude that elliptic hole photonic crystal structure has a moderate
impact on the VCSEL polarisation properties, similarly to the experimental results
reported in [44], where a large PMSR could only be achieved with an aid of asymmetric current injection.

As shown above, the elliptical holes are insufficient to provide good stabilisation
of polarisation. However, alternative approaches have been extensively developed

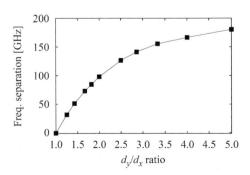

Fig. 13.11 Resonant frequency separation of the x-
and y-polarised modes for the
structure with elliptical holes
as a function of d_y/d_x ratio

for birefringent photonic crystal fibres [47, 48]. The common practise to provide birefringence of the fibre is an asymmetric change in the radius of the photonic crystal holes. An example of such a configuration is shown in Fig. 13.9b. It comprises four holes of increased diameter out of six surrounding the defect. In addition, a small hole in the middle augments the birefringence and increases the polarisation bandwidth [49].

Hence, we analyse a VCSEL with a photonic crystal configuration based on the design of the highly birefringent microstructured optical fibre [46] as shown in Fig. 13.9b. As we will show, such a photonic crystal structure can provide much larger frequency separation between the two orthogonal polarisations than the one with elliptical holes, reaching as high value as 400 GHz. Similarly to the previous configuration, the photonic-crystal pitch is $\Lambda = 1\,\mu\text{m}$. We have found that good results are obtained for photonic crystal hole diameter of $d_c = 0.4\,\mu\text{m}$ and diameter of the large holes $d_h = 0.8\,\mu\text{m}$. The size of the hole in the middle varies in a range $0.01\,\mu\text{m} \leq d_s \leq 0.10\,\mu\text{m}$.

The results of the calculations are presented in Fig. 13.12. Figure 13.12a shows the frequency splitting between the two orthogonally polarised fundamental VCSEL modes. In contrast to the case of elliptic hole PC VCSEL, the x-polarised mode is now the high-frequency one. Also the frequency splitting is much enlarged: it ranges from 310 GHz at $d_s = 0.01\,\mu\text{m}$ up to almost 415 GHz for $d_s = 0.10\,\mu\text{m}$. Figure 13.12b depicts the VCSEL relative loss dichroism, which is computed as $\delta_\alpha = (\alpha_y - \alpha_x)/\alpha_{\text{avg}}$, where α_x and α_y are the modal losses for the x- and y-polarised modes, respectively and α_{avg} is the average value of α_x and α_y. Surprisingly, δ_α is not a monotonic function of d_s, but possesses a minimum at around $d_s \approx 0.06\,\mu\text{m}$.

Similar effect is observed when we consider the relative gain dichroism δ_g, defined analogously to the relative loss dichroism, as $\delta_g = (g_y - g_x)/g_{\text{avg}}$, where g_x, g_y and g_{avg} are the threshold gains of x-, y-polarisation and their average value, respectively. The relative gain dichroism has the smallest value of 5% for $d_s = 0.06\,\mu\text{m}$ and reaches 7% for $d_s = 0.10\,\mu\text{m}$. In addition, the large birefringence can increase the

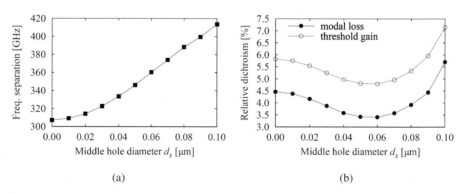

(a)   (b)

Fig. 13.12 Resonant frequency separation of the x- and y-polarised modes (**a**) and relative loss and gain dichroism between x- and y-polarisation (**b**) for the structure with four large holes, as a function of d_s

difference in threshold current. Hence, with the aid of these two effects, a stable-polarisation emission of the laser is possible to achieve. It is worth to note that even for the structure without any hole in the middle ($d_s = 0$) the birefringence and dichroism are quite high.

13.4 Conclusions

In this chapter, we summarise the mathematical foundation of the plane-wave admittance method, which is a combination of the plane-wave expansion and the admittance transfer technique. Plane-wave admittance method is a frequency-domain method, appropriate for the analysis of anisotropic, lossy, and even magnetic or negative refractive index photonic structures. It is especially efficient when the modelled structure consists of parallel layers and some of them appear multiple times within the stack. It allows modelling of three dimensional photonic structures on a single-processor machine and achieving good accuracy results in a reasonable time. In the second part of the chapter we provide examples of modelling of such complex structures comprising hundred different layers. We provide a number of results concerning Photonic Crystal Vertical Cavity Surface Emitting Lasers (PC-VCSELs). We first analyse the impact of the etching depth of photonic crystal holes on the properties of the fundamental mode in PC-VCSELs. Then we show that the application of photonic crystals in surface-emitting lasers can be a versatile tool for providing a single-mode regime with strong suppression of the higher-order modes. Finally, we consider the differences between fundamental modes with different polarisations in an asymmetric photonic crystal structure and investigate its suitability for stabilising the polarisation of the emitted light. We first analyse elliptical hole PC VCSELs and then propose another photonic crystal configuration providing much larger birefringence and dichroism.

References

1. Dems M., Kotynski R., Panajotov K.: Plane-wave admittance method – a novel approach for determining the electromagnetic modes in photonic structures. Opt. Express **13**, 3196–3207 (2005)
2. Dems M.: Plane-wave admittance method and its applications to modeling semiconductor lasers and planar photonic-crystal structures. Ph.D. thesis, Technical University of Lodz (2007)
3. Danner A.J., Raftery Jr. J.J., Yokouchi N., Choquette K.D.: Transverse modes of photonic crystal vertical-cavity lasers. Appl. Phys. Lett. **84**, 1031–1033 (2004)
4. Dreher A., Pregla R.: Analysis of planar waveguides with the method of lines and absorbing boundary conditions. IEEE Microwave Guided Wave Lett. **1**, 239–241 (1992)
5. Helfert S.F., Barcz A., Pregla R.: Three-dimensional vectorial analysis of waveguide structures with the method of lines. Opt. Quantum Electron. **35**, 381–394 (2003)
6. Sacks Z.S., Kingsland D.M., Lee R., Lee J.F.: A perfectly matched anisotropic absorber for use as an absorbing boundary condition. IEEE Trans. Antennas and Propagation **43**, 1460–1463 (1995)

7. Berenger J.P.: A perfectly matched layer for the absorption of electromagnetic waves. J. Comput. Phys. **114**, 185–200 (1994)
8. Dems M., Panajotov K.: Modeling of single- and multimode photonic-crystal planar waveguides with plane-wave admittance method. Appl. Phys. B **89**, 19–23 (2007)
9. Kotynski R., Dems M., Panajotov K.: Waveguiding losses of micro-structured fibres-plane wave method revisited. Opt. Quantum Electron. **39**, 469–479 (2007)
10. Birks T.A., Knight J.C., Russell P.S.J.: Endlessly single-mode photonic crystal fiber. Opt. Lett. **22**, 961–963 (1997)
11. Knight J.C., Russell P.S.J.: Applied optics: New ways to guide light. Science **296**, 276–277 (2002)
12. Conradi O., Helfert S.F., Pregla R.: Comprehensive modeling of vertical-cavity laser-diodes by the method of lines. IEEE J. Quantum Electron. **37**, 928–935 (2001)
13. Press W., Teukolsky S.A., Vetterling W.T., Flannery B.P.: Numerical Recipes in C: The Art of Scientific Computing. Cambridge University Press, New York, second edn. (1992)
14. Taflove A., Hagness S.C.: Computational Electrodynamics: The Finite-Difference Time-Domain Method. Artec House Inc., Boston, second edn. (2000)
15. Ziolkowski R.W.: Time-derivative Lorentz materials and their utilization as electromagnetic absorbers. Phys Rev. E **55**, 1630–1639 (1996)
16. Chew W.C., Jin J.M., Michielssen E.: Complex coordinate stretching as a generalized absorbing boundary condition. Microwave Opt. Technol. Lett. **15**, 363–369 (1997)
17. Dems M., Czyszanowski T., Panajotov K.: Numerical analysis of high Q-factor photonic-crystal VCSELs with plane-wave admittance method. Opt. Quantum Electron. **39**, 419–426 (2007)
18. Czyszanowski T., Dems M., Thienpont H., Panajotov K.: Modal behavior of photonic-crystal vertical-cavity surface-emitting diode laser analyzed with plane wave admittance method. Opt. Quantum Electron. **39**, 469–479 (2007)
19. Czyszanowski T., Dems M., Thienpont H., Panajotov K.: Optimal radii of photonic crystal holes within DBR mirrors in long wavelength VCSEL. Opt. Express **15**, 1301–1306 (2007)
20. Czyszanowski T., Dems M., Panajotov K.: Impact of the hole depth on the modal behaviour of long wavelength photonic crystal VCSELs. J. Phys. D: Appl. Phys. **40**, 2732–2735 (2007)
21. Czyszanowski T., Dems M., Panajotov K.: Optimal parameters of photonic-crystal vertical-cavity surface-emitting diode lasers. IEEE J. Lightwave Techn. **25**, 2331–2336 (2007)
22. Yokouchi N., Danner A.J., Choquette K.D.: Two-dimensional photonic crystal confined vertical-cavity surface-emitting lasers. IEEE J. Sel. Top. Quantum Electron. **9**, 1439–1445 (2003)
23. Danner A.J., Raftery Jr. J.J., Leisher P.O., Choquette K.D.: Single mode photonic crystal vertical vavity lasers. Appl. Phys. Lett. **88**, 091, 114 (2006)
24. Leisher P.O., Danner A.J., Choquette K.D.: Single-mode 1.3 m photonic crystal vertical-cavity surface-emitting laser. IEEE Photon. Technol. Lett. **18**, 2156–2158 (2006)
25. Czyszanowski T., Dems M., Panajotov K.: Single mode condition and modes discrimination in photonic-crystal 1.3 m AlInGaAs/InP VCSEL. Opt. Express **15**, 5604–5609 (2007)
26. Chang-Hasnain C.J.H.J.P., Hasnain G., Von Lehmen A.C., Florez L.T., Stoffel N.G.: Polarization and transverse mode characteristics of vertical-cavity surface-emitting lasers. IEEE J. Quantum Electron. **27**, 1402–1408 (1991)
27. Choquette K.D., Schneider R.P., Lear K.L., Leibenguth R.E.: Gain-dependent polarization properties of vertical-cavity lasers. IEEE J. Sel. Top. Quantum Electron. **1**, 661–666 (1995)
28. Panajotov K., Danckaert J., Verschaffelt G., Peeters M., Nagler B., Albert J., Ryvkin B., Thienpont H., Veretennicoff I.: Polarization behavior of vertical-cavity surface-emitting lasers: Experiments, models and applications. Proc. AIP **560**, 403–417 (2000)
29. Sciamanna M., Panajotov K., Thienpont H., Veretennicoff I., Mégret P., Blondel M.: Optical feedback induces polarization mode hopping in vertical-cavity surface-emitting lasers. Opt. Lett. **28**, 1543–1545 (2003)
30. Gatare I., Sciamanna M., Buessa J., Thienpont H., Panajotov K.: Nonlinear dynamics accompanying polarization switching in vertical-cavity surface-emitting lasers with orthogonal optical injection. Appl. Phys. Lett. **88**, 101, 106 (2006)

31. San Miguel M., Feng O., Moloney J.V.: Light polarization dynamics in surface-emitting semiconductor lasers. Phys Rev. A **52**, 1728–1739 (1996)
32. Martin-Regalado J., Prati F., San Miguel M., Abraham N.B.: Polarization properties of vertical-cavity surface-emitting lasers. IEEE J. Quantum Electron. **33**, 765–783 (1997)
33. Mueller R., Klehr A., Valle A., Sarma J., Shore K.A.: Effects of spatial hole burning on polarization dynamics in edge-emitting and vertical-cavity surface-emitting laser diodes. Semicond. Sci. Technol. **11**, 587–596 (1996)
34. Panajotov K., Ryvkin B., Danckaert J., Peeters M., Thienpont H., Veretennicoff I.: Polarization switching in VCSEL's due to thermal lensing. IEEE Photon. Technol. Lett. **10**, 6–8 (1998)
35. Niskiyama N., Arai M., Shinada S., Azuchi M., Miyamoto T., Koyama F., Iga K.: Highly strained GaInAs-GaAs quantum-well vertical-cavity surface-emitting laser on GaAs (311)B substrate for stable polarization operation. IEEE J. Sel. Top. Quantum Electron. **7**, 242–248 (2001)
36. Jansen van Doom A., van Exter M., Woerdman J.: Strain-induced birefringence in vertical-cavity semiconductor lasers. IEEE J. Quantum Electron. **34**, 700–706 (1998)
37. Panajotov K., Nagler B., Verschaffelt G., Georgievski A., Thienpont H., Danckaert J., Veretennicoff I.: Impact of in-plane anisotropic strain on the polarization behavior of vertical-cavity surface-emitting lasers. Appl. Phys. Lett. **77**, 1590–1592 (2000)
38. Choquette K.D., Leibenguth R.: Control of vertical-cavity laser polarization with anisotropic transverse cavity geometries. IEEE Photon. Technol. Lett. **6**, 40–42 (1994)
39. Ortsiefer M., Shau R., Zigldrum M., Böhm G., Köhler F., Amann M.C.: Submilliamp long-wavelength InP-based vertical-cavity surface-emitting laser with stable linear polarisation. Electron. Lett. **36**, 1124–1126 (2000)
40. Mukaihara T., Ohnoki N., Hayashi Y., Hatori N., Koyama F., Iga K.: Polarization control of vertical-cavity surface emitting lasers using a birefringent metal/dielectric polarizer loaded on top distributed Bragg reflector. IEEE J. Sel. Top. Quantum Electron. **1**, 667–673 (1995)
41. Ser J.H., Ju Y.G., Shin J.H., Lee Y.H.: Polarization stabilization of vertical-cavity top-surface-emitting lasers by inscription of fine metal-interlaced gratings. Appl. Phys. Lett. **21**, 2769–2771 (1995)
42. Berseth C.A., Dwir B., Utke I., Pier H., Rudra A., Iakovlev V.P., Kapon E.: Vertical cavity surface emitting lasers incorporating structured mirrors patterned by electron-beam lithography. J. Vac. Sci. Technol. B **17**, 3222–3225 (1999)
43. Debernardi P., Ostermann J.M., Feneberg M., Jalics C., Michalzik R.: Reliable polarization control of VCSELs through monolithically integrated surface gratings: A comparative theoretical and experimental study. IEEE J. Sel. Top. Quantum Electron. **11**, 107–116 (2005)
44. Song D.S., Lee Y.J., Choi H.W., Lee Y.H.: Polarization-controlled, single-transverse-mode, photonic-crystal, vertical-cavity, surface-emitting lasers. Appl. Phys. Lett. **82**, 3182–3184 (2003)
45. Steel M.J., White T.P., Martijn de Sterke C., McPhedran M.C., Botten L.C.: Symmetry and degeneracy in microstructured optical fibers. Opt. Lett. **26**, 488–450 (2001)
46. Urbanczyk W., Szuplak M., Statkiewicz G., Martynkien T., Olszewski J., Wojcik J., Mergo P., Makara M., Nasilowski T., Berghmans F., Thienpont H.: Polarizing properties of photonic crystal fibers. In: International Conference on Transparent Optical Networks, ICTON 2006, vol. 2, pp. 59–63 (2006)
47. Ortigosa-Blanch A., Knight J.C., Wadsworth W.J., Arriaga J., Mangan B.J., Birks T.A., Russell P.S.J.: Highly birefringent photonic crystal fibers. Opt. Lett. **25**, 1325–1327 (2000)
48. Hansen T.P., Broeng J., Libori S.E.B., Knudsen E., Bjarklev A., Jensen J.R., Simonsen H.: Highly birefringent index-guiding photonic crystal fibers. IEEE Photon. Technol. Lett. **13**, 588–590 (2001)
49. Szpulak M., Olszewski J., Martynkien T., Urbanczyk W., Wojcik J.: Polarizing photonic crystal fibers with wide operation range. Opt. Commun **239**, 91–97 (2004)

Acknowledgements

We would like to express our gratitude to all the contributors to the COST P11 Action that have granted the success of the Action. We want to thank in particular Dr. P. Swiatek, who gave us many suggestions and who encouraged to proceed the effort of coordination in a new Action.

We thank in particular Mariusz Zdanowicz, and Dr. Olga Bolszo for their assistance and priceless technical support to the book realization.

C. Sibilia
M. Marciniak
T. Benson
T. Szoplik

Index

281